S0-CBR-404

VEGETATION DYNAMICS

HANDBOOK

OF

VEGETATION SCIENCE

Editor in Chief

Reinhold Tüxen

1974

Dr. W. JUNK b.v. - PUBLISHERS - THE HAGUE

PART VIII

VEGETATION DYNAMICS

edited by

R. Knapp

1974

Dr. W. JUNK b.v. - PUBLISHERS - THE HAGUE

ISBN 90 6193 189 4

Cover design: Charlotte van Zadelhoff
Printed in The Netherlands by Dijkstra Niemeyer b.v., Groningen

FOREWORD

During the International Botanical Congress in Edinburgh, 1964, Mrs. I. M. WEISBACH-JUNK of The Hague discussed a plan for preparation by her publishing company (Dr. W. Junk b.v.) of an international *Handbook of Vegetation Science*. She proposed a series that should give a comprehensive survey of the varied directions within this science, and their achievements to date as well as their objectives for the future. The challenge of such an enterprise, and its evident value for the further development of vegetation research, induced the undersigned after some consideration to accept the offer of the honorable but also burdensome task of General Editor.

The decision was encouraged by a well formulated and detailed outline for the Handbook worked out by the Dutch phytosociologists J. J. BARKMAN and V. WESTHOFF. A circle of scholars from numerous countries was invited by the Dr. Junk Publishing Company to The Hague in January 1966 to draw up a list of editors and contributors for the parts of the *Handbook*. The outline and list have served since for the organization of the *Handbook*, with no need for major change.

The different burdens of editors and authors have compelled quite different timings for completion of the individual sections. It was consequently decided by Dr. W. Junk b.v. that the parts of the *Handbook* would be published separately as they were ready. Thanks to the tireless and determined service of Prof. ROBERT H. WHITTAKER as editor and to the work of his collaborators as well, the firm can now publish the first part of the *Handbook* to appear, Part V on ordination and classification as means of synthesizing information from a number of communities toward the understanding of vegetation.

The *Handbook* has as its purpose the presentation, through contributions of numerous collaborators from varied fields, of a comprehensive picture of modern vegetation science, including its development, its methods, its discoveries, and its goals, in approaches ranging from plant sociology as a descriptive science based on sharp observation of communities as concrete objects of study in the field (BRAUN-BLANQUET), through the inquiry into casually significant relationships that for many represents its principal goal, to abstract vegetation science that seeks a detached and generalized understanding based on quantitative relationships that may be represented in diagrams, equations, and models.

In this the *Handbook* should reveal the fundamental value of vegetation science for the other disciplines with which it is linked in many ways – not only such other areas of botany as plant geography and systematics, paleobotany and palynology, genetics and evolution, and others, but also soil science and the applied areas of forest, grassland, and wildlife management, water and fishery management and coastal protection, and interpretation of the land's potential for agricultural and other use by man. Not last in this is the decisive importance of vegetation science for nature and environmental conservation.

All these applied fields use vegetation, whether as an object of either harvest or preservation (or both), or as the essential context of the other phenomena of their direct concern, or as an indicator of relationships with which they deal. They all are in some way directed toward the care and maintenance of the green cover of our earth that they seek to use, wisely. For their purposes understanding of plant communities as functional systems and expressions of life-phenomena and the laws that govern them is valuable, or even quite indispensable. The plant cover of the earth is and must remain the basis for all other life.

So we hope to show that for men of many aptitudes and interests concern with the plant cover and its social units is rewarding – not merely in the sense of precedence and preference among individuals, but rather on the basis of the mutual concern of different fields and research purposes that complement and enrich one another. May our *Handbook* be a portrait – a composite portrait as painted by many scientists but harmonious as a whole – or one of evolution's achievements, the plant community as itself a harmoniously functioning, living system.

All our thanks go to the initiators, the editors, and the contributors, as well as Dr. W. Junk, b.v., Publishers, of the Hague. They together are making a lasting contribution to our science; may it also contribute in some measure to solve the grave problems for life on our earth.

Todenmann über Rinteln R. TÜXEN

CONTENTS

Authors and addresses

E. Aichinger
A9020 Klagenfurt-Sandhof, Österreich

V. D. Aleksandrova
Botanical Institute of the Academy of Sciences, Prof. Popov-Str. 2, Leningrad, 197022, U.S.S.R.

J. S. Beard
Department of Geography, University of Western Australia, Nedland, W.A. 6009, Australia

B. A. Bykov
Botanical Institute of the Academy of Sciences of Kazach. S.S.R., Alma-Ata, Kazachstan, U.S.S.R.

R. T. Coupland
Department of Plant Ecology, University of Saskatchewan, Saskatoon, Canada

P. Dansereau
Pavillon Emile-Gérard, Université du Québec, Montréal 101, P.Q., Canada

Z. V. Karamysheva
Botanical Institute of the Academy of Sciences, Prof. Popov-Str. 2, Leningrad, 197022, U.S.S.R.

V. G. Karpov
Botanical Institute of the Academy of Sciences, Prof. Popov-Str. 2, Leningrad, 197022, U.S.S.R.

R. Knapp
Botanisches Institut der Universität, D63 Giessen, Bundesrepublik Deutschland

A. A. Korchagin
Botanical Institute of the Academy of Sciences, Prof. Popov-Str. 2, Leningrad, 197022, U.S.S.R.

H. Lieth
Department of Botany, University of North Carolina, Chapel Hill, N.C., 27514, U.S.A.

J. Major
Department of Botany, University of California, Davis, Calif. 95616, U.S.A.

T. A. Rabotnov
Department of Geobotany, Moscow State University Moscow, U.S.S.R.

F. Reinhold†
formerly: Fürstenbergische Forstdirektion, 771 Donaueschingen, Bundesrepublik Deutschland

F. Stearns
Department of Botany, University of Wisconsin, Milwaukee, Wisc. 53201, U.S.A.

R. Tüxen
Arbeitsstelle für Theoretische und Angewandte Pflanzensoziologie, D 3261 Todenmann, Bundesrepublik Deutschland.

R. H. Whittaker
Section of Ecology and Systematics, Div. Biological Sciences Cornell University, Ithaca, N.Y. 14850, U.S.A.

A PREFACE AND INTRODUCTION

R. Knapp

A PREFACE AND INTRODUCTION

Syndynamical aspects were important in vegetation science since its early times until present days. During some periods and in certain countries, it actually dominated vegetation research, e.g. in North America and in England during the first decades of this century. In recent years, studies on vegetation dynamics seem to be increasing in general importance and in amplitude of scope, namely by three groups of reasons:

(1) Now many years have passed since the first accurate vegetation surveys (réleves, Vegetationsaufnahmen, detailed mapping etc.). By repeated analysis of areas studied in earlier times with appropriate methods, it is possible to ascertain vegetational changes on an exact and inductive basis. Beyond various important new results, theories and hypothetical constructions of early periods of this century can be replaced now by evaluations based on reliably analyzed facts.

(2) The possibilities in experimental studies of plants under controlled conditions become amplified more and more. Well equipped experimental fields, climatically controlled greenhouses, phytotrons and similar arrangements may be mentioned only. Experimental studies become increasingly relevant for vegetation science. Connexions between results of field studies on vegetation and of experiments under controlled conditions on the basis of simultaneous work with both groups of methods show highly promising views (e.g. WENT 1957, ELLENBERG 1954, KNAPP 1954, 1960, 1967, HARPER 1961, GIGON 1971). New aspects of causal explanations of vegetation dynamics are emerging by the results of such comprehensive work, beyond the scope of emphasis on restricted numbers of environmental factors (e.g. light, water etc.).

(3) Vegetational changes affect more and more the life of mankind directly. Deterioriation of plant cover in consequence of unfavorable influences of industry, traffic, inadequate land use and other disappointing effects of certain economical developments becomes extraordinarily problematic. The task to avoid ultimate damages is increasing in urgency. On the other hand, the numbers of cases are amplifying, in which a stop of certain infavorable influences appeared not to be sufficient to regain a vegetation of a desirable condition. In such cases, additional actions are necessary to release a development of a vegetation feasible for environmental

protection, for recreational aspects and for other purposes. Thus, the necessity to control vegetational changes becomes highly obvious. Also wildlife management in extension to plant communities is rising in importance. The theoretical and practical basis of a great part of the actions necessary in the fields mentioned is extensively connexed with scientific vegetation dynamics.

The present volume has the aim to collect reviews on the intentions, on the foundations, and on the methods of studies on vegetation dynamics. Also the special features of syndynamical work in certain parts of the world are explained in a number of reports. Additionally, it is intended to discuss a choice of the most important results and terms in vegetation dynamics. Therefore, the presentation attempts an intermediate position between a review volume and a textbook. For this reason and moreover in consequence of the separate contributions by different authors, the material cannot be arranged purely in a systematic textbook way. The definitions of the technical terms in vegetation dynamics are interspersed in the various articles. But an elaborate register will be helpful to find the definitions of the technical terms easily without much loss of time.

It was originally planned to include on the volume a subchapter with articles characterizing the main features of syndynamics in the great vegetational zones of the world. But by reasons of limited size for the volume, such articles had to be omitted for the most part. This omittance seemed to be justified, at least partly, for most vegetational zones, since many examples of successions and their causes from different parts of the world are described in the various articles. Only the special treatment of certain important features of syndynamics in the humid tropics, in the tundra, and in some arid or semi-arid regions seemed to be indispensable in a more accentuated form and is consequently accomplished by separate articles. Again, the index may help to find the examples given in the various articles of the volume on syndynamical attributes of particular vegetational zones and of certain groups of plant communities.

Some decades ago, certain publications emphasizing vegetation dynamics were crowded with a special technical terminology. These publications are liable for a reputation, surviving still in present days, of a notorious connexion of successional considerations with an unusual high number of such technical terms. It was an aim of the editor to reduce the use of technical terms to a proportion adequate and necessary for the clarifications of the theoretical and practical aspects of vegetation dynamics. However, it seemed necessary to mention and to define all syndynamical technical terms of any importance and frequence in present and earlier literature.

The editor planned originally to contribute only one article. But to avoid incompleteness and inadequate delay of the issue of the volume, he had necessarily to report on some fields not covered by the other articles, whose treatment seemed to be indispensable in a handbook volume on syndynamics.

In this connexion, it may be mentioned that some considerations relating to vegetation dynamics have to be included in other parts of the "*Handbook of Vegetation Science*", in consequence of its general plan. For instance, the historical development of the research work and of the theories of vegetation dynamics will be treated in the part "History of Vegetation Science".

The figures are mostly published in this volume for the first time. These figures are otherwise unpublished graphs or photographs of the authors of the articles concerned. They can be recognized by the abbreviations "Or." (= original figure) at the end of the figure's explanations or by no special reference notice.

B. KINDS OF CHANGES IN VEGETATION

1 KINDS AND RATES OF CHANGES IN VEGETATION AND CHRONOFUNCTIONS

J. Major

Contents

1 KINDS AND RATES OF CHANGES IN VEGETATION AND CHRONOFUNCTIONS

1.1 Kinds of Changes in Vegetation

Instead, in a hierarchical classification of vegetation changes, we shall be primarily concerned with only one specific level: the duration in years (Table I). In fact, long-term plant successions

TABLE I. Kinds of changes in vegetation in connexion with duration (years). Modified from YAROSHENKO (1946, 1953: 123, 1961: 195ff.), MAJOR (1951: 398) ALEKSANDROVA (1964: 300, 1962: 11—12), SUKACHEV & DYLIS (1964: 475 ff., 482 ff.), DAUBENMIRE (1968: 228).

DURATION (years)	KIND OF CHANGE
Daily:	Daily rhythms of the component species: transpiration, photosynthesis, e.g.
1	Seasonal. Related to the ontogeny of particular species of the community, changes in tolerances in different growth stages, aspection.
Ca. 4.	Small rodent cycles possibly related to cycling of necessary minerals from the animal to the plant population. (Cf. PITELKA & SCHULTZ 1964.)
Few 10's.	Weather cycles producing changes in productivity, seed supply, cover, abundance of annuals.
Several 10's.	Changes in species' vitality with age, ontogenetic changes in perennials with age (RABOTNOV 1950).
Several to many 10's.	Replacement cycles within the community, WATT's (1947) pattern and process cycles, related to limited life spans of species. (Cf. DAUBENMIRE 1968: 224—8.)
100's.	Changes leading to development of a new community from the former one(s). Directional and therefore predictable as the above shorter than changes are not. A t_0 can be defined. The independent factors of the environment are constant; therefore differs from the allogenic changes below.
Few 100's.	Microevolution producing ecotypes. (Cf. BRADSHAW 1965, 1969, ALEKSANDROVA 1964: 200.) The flora changes on a micro-scale.
Several 100's.	Change caused by man's activities in some factor of the environment producing changes in the vegetation: burning or fire control, mowing, forestry (as distinguished from "timber harvesting").
1000's.	Natural changes in factors of the environment. Historical changes related to changes in climate, physiography, flora or fauna, slow modifications of soil parent material by geological processes.
Many 1000's.	Evolution of coenoses related to changes in flora by immigration, extinction, or evolution.

must overlap in time the shorter term changes in vegetation. For instance, the range manager's "trend" is a loose usage of the idea of plant succession, and trend in range condition has always been difficult to separate from fluctuation in condition due to short-term climatic cycles.

On a longer time scale, the results on vegetation of a change in climate form the basis for palynological interpretations of climatic changes during glacial and interglacial periods. Such historical changes (MAJOR 1958) have been confused with other vegetation changes in some palynological reports. FIRBAS (1949: 275—277) distinguishes clearly between the two.

Taking up the other independent factors determining ecosystems in addition to climate, changes in vegetation related to sudden additions of a loessial layer in grassland during a widespread drought (dust bowl) or of fertilizer, giving a new t_0, have been well-documented. Similarly, changes in vegetation follow a rise or a drop in the water table, even a sudden shift in snow accumulation patterns (BILLINGS 1969). Changes in flora or fauna have produced changes in vegetation dramatized by the chestnut blight in the eastern U.S. and the Tessin, by the introduction of exotic mammals into New Zealand and Australian ecosystems, and by man at various levels of civilization in Neolithic Denmark or England or modern California. Fire can conceivably change in frequency and intensity without man's help or simply be episodic with long intervals between events, in which case the effects on vegetation would be similar to the well-known effects of fire set by man or the effects of fewer fires caused by man's efficient fire control.

It follows immediately from the general equation and our knowledge of history that changes of vegetation with time, when all other factors are constant, is only one kind of chronological change in vegetation. Climofunctions ($v = f(cl)_{p,r,o,py}$), lithofunctions ($v = f(p)_{cl,r,o,py}$), topofunctions ($v = f(r)_{cl,p,o,py}$), biofunctions ($v = f(o)_{cl,p,r,py}$), and pyrofunctions ($v = f(py)_{cl,p,r,o}$) are all possible over time, in non-seral vegetation, when the effects of successional time are zero, where the vegetation is zonal or climax and $dv/dt \rightarrow 0$, when successional t is many hundreds of years; but in the individual, named functional relationships the factors cl, p, r, o and py themselves change historically. In a broad sense then $v = f$ (time) where $cl = f$ (time), $p = f$ (time), etc. but the time scale in this case would be of a larger order of magnitude as compared to the lengths of time in what we differentiate as successional changes (Table I).

Thus, the changes in post-glacial vegetation of the last 10,000 years were not predictable. They would have become so if and when the nature and extent of the climatic changes were known because plant ecologists have some generalizations about relationships of vegetation to climate.

When the factors of the environment, upon which vegetation depends, change, then obviously the properties of the vegetation must change also. However, changes in the independent factors of the environment are imposed on the vegetation from outside the plant community. The changes are allogenic. They have no known, predictable regularity. They are functionally exogenic to the biogeocoenose in SUKACHEV's terminology (SUKACHEV & DYLIS 1964: 468, 483—4). Secular (LÜDI 1930: 515; MAJOR 1951: 398), long-term changes of this kind are separated from short-term vegetational changes, but rather discussed under the history of vegetation since the data do not allow reconstruction of individual plant communities, by BRAUN-BLANQUET (1964: 705) and ALEKSANDROVA (1964: 301—2). They are not amenable to "field geobotanical" studies. ALEKSANDROVA would split short-term changes of individual plant communities ("particular change", 1964: 302) into catastrophic, anthropogenic, and plant successional (see below). The two former are allogenic. "Sudden" is more appropriate than "catastrophic" according to YAROSHENKO and GRABAR (1969: 3).

The kinds of allogenic, phylogenetic changes in vegetation discussed above contrast with those producing the ontogeny of vegetation which we can isolate for study by finding situations where all the independent factors of the environment but time are constant (MAJOR 1951: 398). We call these changes successional. Seral vegetation in a habitat with the independent factors constant develops from one plant community into another (ALEXSANDROVA 1962, 1964: 306). In TANSLEY's terms (1920: 136, 1929: 680) these are autogenic changes in vegetation (also SUKACHEV & DYLIS 1964: 483). They are characterized in Table 1. ALEKSANDROVA's discussion (1962, 1964) of this concept is full, logical, complete, and balanced. ELLENBERG's basic paper on the dynamics of vegetation, which was read at the Rinteln Symposium organized by TÜXEN in 1967, evidently agrees with most of the positions taken above (KORCHAGIN, GORCHAKOVSKY & MATVEEVA 1968).

1.2 Vegetation Properties

But what vegetation properties (v_1, v_2, etc.) can be measured in such plant successions? What do we really know about autogenic plant succession? It progresses from open vegetation to closed, from simple, one-layered communities to complicated, many-layered ones. Species diversity increases with time (or does it decrease?). Biomass increases, at least up to a point. Productivity may not increase at all. Life forms change from annuals or short-lived herbs to long-lived herbs, shrubs, or trees depending on the levels of climate, soil parent material, relief, flora, etc. obtaining in that ecosystem. In some ecosystems conifers replace broad-leaved trees, or evergreens replace deciduous. In leaching (humid) environments soils become more acid so calcifuges increase in proportional representation in the species list, weighted by abundance or not. In arid environments acid parent materials become less acid. Xeric environments become more mesic and wet environments also often become more mesic, so the proportions of both xerophytes and hydrophytes decrease in favor of mesophytes. The soil part of the ecosystem is enriched with organic matter — with C, N, in many cases with Ca, K, Na, Mg, P and S in organic form. Leaching may be less as increased plant cover evapotranspires more water, or leaching may increase very locally from stem flow (GERSPER 1968, GERSPER & HOLLOWAYCHUCK 1970, 1971, MINA 1967).

Note that the changes mentioned above have very seldom been quantified. Unless both v and t are described in quantitative terms, it is impossible to derive rates of change. Many excellent studies of plant succession have more or less quantitative data on vegetation, but they lack an absolute time scale. CLEMENTS in his excellent study of the life history of lodgepole pine (*Pinus contorta* ssp *latifola* forests following forest fires in Colorado could only order his stands sequentially in time (1910). ELLISON's study (1954) of primary succession leading to herbaceous meadows in the subalpine zone of the Wasatch Plateau in Utah is undoubtedly correct in its ordering of investigated stands into a chronosequence but there are no measurements of time whatsoever. In addition, rocky and rock-free substrates were thrown into the same sere so perhaps it is just as well that the measures of vegetation were qualitative. On the other hand, IVES' (1941) photographic record of the recovery of spruce fir forest (*Picea engelmanni—Abies lasiocarpa*) in the Colorado Rockies has a splendid time scale with accurate dates but no quantitative measures of the vegetation. The dilemma is well-illustrated by BRAUN-BLANQUET & JENNY's (1926, BRAUN-BLANQUET 1964: 666—8) classic study of vegetation succession and soil development in the perhumid Swiss Alps. Because the parent material was limestone, the oldest, most developed soils with the most organic matter in their profiles must be the most leached, the most acid. Both soils and their characteristic vegetation could therefore be arranged in a chronosequence. But the time scale is not absolute and the detailed records on the vegetation are difficult to put quantitative values on. In addition, ELLENBERG has suggested (1953) that the soils were really not all parts of one homogeneous sere, but that the most developed soils could develop only on limestone moraines with an admixture of siliceous rocks.

Photographs have been effectively used, or provide some necessary documentation, in many successional studies (IVES 1941, ELLISON 1954, SHANTZ & TURNER 1958, PETTERSSON 1958, FREY 1959, PHILLIPS 1963, HALL, SPECHT & EARDLEY 1964, BRAUN-BLANQUET 1964: 670—1, HASTINGS & TURNER 1965, HELLER 1968) as have maps (FUKAREK 1961, J. BRAUN-BLANQUET, WIKUS, SUTER & G. BRAUN-BLANQUET 1958—cf. BRAUN-BLANQUET 1964: 630—633, ALEKSANDROVA 1964: 343—4) or permanent chart quadrats (BRAUN-BLANQUET 1964: 620, 690—3, ELLISON 1954, LEONTIEV 1952, BECHER 1963, WOIKE 1958, COOPER 1923, 1931, 1939 cf. ALEKSANDROVA 1964: 349—356 and BRAUN-BLANQUET 1964: 614). However, these pictorial records of vegetation are not quantitative, and we cannot use them in determining rates of plant succession even though their time axis leaves nothing to be desired.

On the other hand, many successional studies have a very poor time scale which again makes it impossible to calculate rates of change. Most indirect methods have this defect.

Finally, even if a time scale is available, the investigator may have had to put together different sites to make a chronosequence. There is usually no assurance that, even though the various factors of the environment have been constant over time in the region, all the members of this chronosequence have in fact developed under the same climate, from the same parent material, on the same relief, from the same flora, and have had the same fire history — or lacking sameness that differences in these factors have had no effect on the course and amount of succession, however measured.

1.3 Chronofunctions

Given that a chronofunction, $v = f(t)_{cl, p, r, o, dy}$, has been used to describe vegetation, we can specify that the time involved since t_0 is less than a few thousand years since the hypsithermal period probably, and the last glacial certainly, put an absolute upper limit on the length of time during which one plant succession under one

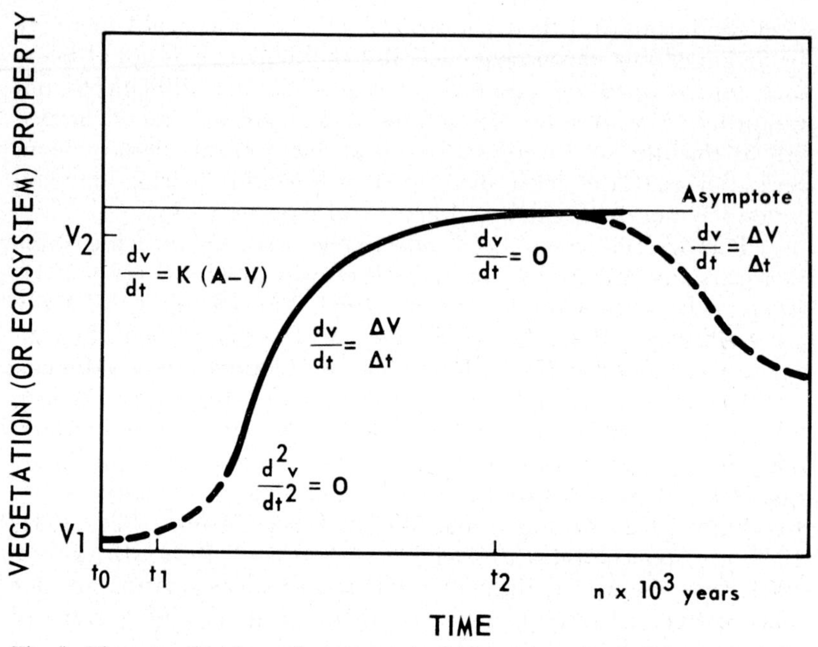

Fig. 1. Diagram of a chronofunction, $v = f(t)_{cl,p,r,o,pv}$. Dashed sections of the curve are most unreliable. For explanation see text.

climate can have proceeded. A generalized curve is shown in Fig. 1. Here the solid part of the curve has $dv/dt = k(A—v)$. The rate of change is greatest at the beginning and decreases as the asymptote, A, is approached. The result of a sufficiently long period of plant succession is stable vegetation with $dv/dt \rightarrow 0$.

Unfortunately for simplicity the real data we have bear out FAEGRI's conclusion (1933: 241) that the rate of succession is slow just after t_0 and not a maximum. It is universally agreed that $dv/dt \rightarrow 0$ as succession proceeds. Therefore the curve must have an inflection where dv/dt is a maximum or $d^2v/dt^2 = 0$. But also it has been pointed out that dv/dt is < 0 after some time. If v represents biomass, then not only must productivity become less with time, but it must become negative.

This decrease in many measures of vegetation after a period of time, a real climax, was long ago suggested by LÜDI (1923, 1930: 532) and also by PACHOSKI (1921) and SUKACHEV (1928: 127). LÜDI's ideas were based on his own studies of succession in the perhumid Swiss Alps (1921) and on the soil-vegetation studies of BRAUN-BLANQUET & JENNY (1926) in a nearby area. BRAUN-BLANQUET amplifies this idea, with other examples (1964: 651). Additionally, the curves for soil nitrogen as a function of time at

Glacier Bay in perhumid, southeastern coastal Alaska show a maximum N content somewhat before 100 years from t_0. At this time the alder (*Alnus crispa* ssp. *sinuata* has been overtopped by poplar (*Populus balsamifera* ssp. *trichocarpa* which in turn is overtopped and replaced by the sitka spruce (*Picea sitchensis* and then by hemlocks (*Tsuga heterophylla* and *Ts. mertensiana* with the latter being more abundant northward) on podsol soils (Crocker & Major 1955: 433, Ugolini 1966: 72, 1968: 132 & 134, Stevens 1963). Zach (1950) and Lawrence (1958) described the further course of this plant succession and soil development as leading to muskeg with bog pools. Sjörs (1963) broadens the idea to paludification in the boreal forest in general. Even on dry land Plochmann (1956) describes the breaking up of a forest canopy as the even-aged, old trees die. Probably productivity and even biomass decrease when regeneration of trees is beginning in the openings. Heusser (1956: 276) describes a similar mature forest on centuries-old terrain beyond the Little Ice Age position of the Mt. Robson glacier in the Canadian Rockies. Finally Cooper (1960, 1961) has described a similar breakup of old *Pinus ponderosa* stands in Arizona at some 300 years of age. In another kind of ecosystem of cold, continental climates Benninghoff (1952) and Drury (1956) suggested a downward trend of biomass in plant succession on Alaskan terrain underlain by permafrost. Viereck (1970) documents this course of autogenic succession leading to permafrost formation within the solum. The succession moves from xeric to mesic to hydric plant communities — from willow (*Salix alaxensis* stands on freshly deposited river gravels to balsam poplar (*Populus balsamifera*), to white spruce (*Picea glauca*) with feather mosses, to black spruce (*Picea mariana*) with *Sphagnum*. Heilman (1966) gives further details on the succession and on quantitative soil nitrogen changes during its course. In tropical Brazil Freise (1938) suggests a succession on lateritic soils leading to less biomass when the laterite is exposed to drying which irreversibly dehydrates its component iron and aluminum oxides.

Loucks (1970) for deciduous forests of Wisconsin, USA, adduces data on primary productivity which fit part of Fig. 1. The maximum productivity occurs at just under 200 years. Natural catastrophes can break this regular pattern at any time, and, since species diversity also shows a similar maximum, such perturbations are important in an evolutionary/as well as a purely biomass productivity sense. Krajina (1969: 77) came to a similar conclusion for boreal *Picea glauca* forests of British Columbia which are seral to *P. mariana*. Forest fire therefore returns such sites over the long run to a more productive condition. There is a great deal of evidence, from many kinds of ecosystems, that if plant succession goes on long enough it results in less productivity and even in lower biomass in some cases.

Studying wetlands zonation in oxbow lakes in Alberta, Canada and equating it to the classical hydrarch succession VAN DER VALK and BLISS (1971) found that both biomass and productivity increased from submerged through emergent plant communities but then decreased in the adjacent (subsequent) *Carex* meadow community.

Unfortunately plant physiological data and ideas are not much help in formulating or even checking a theoretical production curve as in Fig. 1. However, using a compartment transfer model, GORE & OLSON (1967) have produced on a computer curves of biomass accumulation similar to Fig. 1. In a thoughtful paper OLSON (1964) has discussed the basic assumptions and problems involved in computer simulation of the way an ecosystem accumulates organic matter. We should note that the logistic form of the curves in Fig. 1. is very poorly reproduced by current assumptions in the equation describing the slope of the logistic, and the equation's constants cannot be interpreted biologically when applied to plant succession (PIELOU 1969: 19—21).

1.4 Rates of Plant Succession

When we attempt to pull rates of plant succession out of the literature, we find few useable data. Data for individual species are not very helpful since most species in a succession are replaced. A species invades, increases in cover, density, biomass, reproduction, etc. and then decreases over time. For an individual species to measure succession alone it must be a dominant, an edificator, or highly constructive, but the meaning of these terms is operationally unclear.

We should use vegetational data. Biomass is an example, but there are few data [RODIN & BAZILEVICH 1965 (1967), BUDYKO & YEFIMOVA 1968]. We can use data on those soil properties intimately related to vegetation, namely amount of organic matter (C, N, organic P), pH, depth or weight of litter horizon, soil temperature. Most of the standard measurements of vegetation such as cover, leaf area index, density, frequency may quickly reach a maximum and remain invariant through further succession, be meaningless when species of different life form replace one another, or be meaningless to begin with (frequency). "Importance values" cannot be transferred between life forms and at best simply obscure the basic data. Phytosociologists do have a way out of this dilemma. They know, or think they know, which species respond similarly, as changes in dependent properties of the ecosystem, and they can in fact use increases and decreases of such species as measures of dependent ecosystem properties. Thus, calcicoles decrease and calcifuges increase in abundance as succession proceeds on a calcareous substrate in a leaching environment. Xerophytes decrease in abundance in many temperate zone successions which produce progressively more mesic habitat. There is a strong hint of this procedure in some of BRAUN-BLANQUET's diagrams (1964: 666, 700).

Finally, there is a qualitative difference in rates of primary plant succession on substrates which are clastic versus those which are more or less solid, on regosols versus lithosols. On the latter plant colonization is in cracks, and the vegetation spreads over the hard rock surface without affecting it or being affected by it. On clastic materials, whether primary or secondary, succession is of at least an order of magnitude faster than on the solid rock.

We will talk about rates of succession not only in terms of dv/dt but much more loosely. If we know the state of the vegetation at the beginning and at some later time when the rate of change has slowed down considerably, this difference in vegetation divided by the time interval is a rate of overall succession. In Fig. 1, $(v_2-v_1)/(t_2-t_1) \sim \Delta v/\Delta t \sim dv/dt$.

The following rates of plant succession (dv/dt) I have picked off various curves of $v = f(t)$ in the literature. If the authors did not plot their data, I did. In general the rates are given for the steepest section of the curve. Where only two points are given the rates must be $\Delta v/\Delta t$.

If forests can be arranged in a chronosequence, i.e. with independent site factors more or less effectively constant, we can use some biomass, N content, etc. data which were gathered primarily for forest measurational purposes. The technical literature on forest growth contains much additional data which could also be used if the economic measure of m^3 of wood were converted to biologically meaningful biomasses. The re-working of foresters' yield tables would be very profitable. Below-ground parts can be included when we have descriptions of such elementary structural features of plants as their variation of shoot/root ratios with age, site, stocking, etc. The contributions of understory vegetation should not be underestimated. In particular, the understory makes a contribution to mineral nutrient content of the biomass which is far greater in proportion than its contribution to the biomass itself [SCOTT 1955, PYAVCHENKO 1960, REMEZOV, SAMOYLOVA, SUIridova & Bogashova 1964, Rodin & Bazilevich 1965 (1967)].

An attempt has been made to roughly characterize the physical (mainly climatic) factors of the ecosystems for which data are presented. More details can, hopefully, be found in the references. In most American ecological work the flora has been neglected as an independent ecological factor. In much extra-American work a good first approximation to the flora is given in the species list accumulated from many stand surveys (Aufnahmen, relevés.)

The rates of plant succession are presented in this order: For biomass including organic matter on or in the soil, for other vegetational properties such as species composition — useful only in determining times to equilibrium, for measures intimately related to biomass such as N and C, then pH, then other mineral elements such as Ca, K, Mg, P. Within these categories vegetational data are considered first, then ecosystem, then soil, including litter. Data are given in dry weights as $g/m^2 \cdot yr$ or the chemical elements in $g/m^2 \cdot yr$ (Major 1970).

To determine rates of plant succession we need to solve the equation $v = f$ (cl, p, r, o, py, t) (Jenny 1941, 1946, 1958, 1961, 1965, Jenny, Salem & Wallis 1968, Major 1951, Crocker 1952, Olson 1958) in the form $v = f(t)_{cl, p, r, o, py}$ where v is a quantitatively measured vegetation property, f indicates a functional relationship exists, t represents time elapsed from a chosen t_0, cl represents regional climatic parameters as a group, p the soil parent material, r the effects of relief, o the biota which in the case of its plant parts is the summation of kinds of disseminules provided by the regional flora, and py a fire factor independent of man's pyric propensities. The properties represented by v are dependent on the independent environmental factors or groups of factors on the right side of the equation. The case we are interested in has cl, p, r, o, and py constant at particular levels.

Rates of plant succession are values of dv/dt in the equation $v = f(t)_{cl, p, r, o, py}$. If vegetation becomes stable, if $dv/dt \rightarrow 0$, then time to, e.g., 95 % of equilibrium would be a valuable meaning to give to "rate of plant succession".

2

DIFFERENCES BETWEEN FLUCTUATIONS AND SUCCESSIONS

Examples in Grassland Phytocoenoses of the U.S.S.R.

T. A. Rabotnov

DIFFERENCES BETWEEN FLUCTUATIONS AND SUCCESSIONS

Examples in Grassland Phytocoenoses of the U.S.S.R.

Every year is peculiar in meteorological, hydrological and other conditions, important for plant growth, and as a consequence of this the changes in phytocoenoses occur from year to year or from one period to another one. This form of dynamics of phytocoenoses is named fluctuation. Fluctuations are especially pronounced in grassland, where they were studied by many plant ecologists (PACHOSKY 1917, RAMENSKY 1925, SHENNIKOV 1930, 1941, RABOTNOV 1955 and others). Fluctuations, as differentiated from successions, are characterized by (a) differently orientated changes in phytocoenoses from year to year or from one period to another one, (b) reversibility of changes, (c) in typical cases stability of floristic composition, absence of invasion of new species.

In the fluctuations, the more or less considerable changes occur in productivity, in quantitative ratio of the components, in composition and importance value of the dominants (up to their replacement), in morphological structure of sward, in the balance of substances and energy.

In some fluctuations the changes are so great, that they can be taken for successions, if their final result — "return to initial position" — is unknown. Fluctuations occur simultaneously with successions, therefore reversibility of fluctuational changes is usually imperfect, phytocoenoses return not to their original state, but to the state near to original position. In the same cases demarcation between fluctuations and successions is relative. Dynamics of phytocoenoses connected with short-term or cyclic changes in climate returning to state near to the original one are necessary to be considered as fluctuations or to be distinguished as special form of dynamics of phytocoenoses.

Sometimes the changes during "climatic cycles" are not accompanied by return to "original state", as a result of complete or almost complete dying off of species dominating at the beginning of the "cycle" or as a consequence of invasion of new more competitive species. Such changes should be considered as successions.

Successions caused by cutting down of forests or ploughing up of steppes etc. include, usually, return of vegetation to original position. Therefore it is advisable at the demarcation successions from fluctuations to take into consideration the duration of period essential for reestablishment of original vegetation. Short-term reversible changes can be considered as fluctuations, long-term ones as successions. As approximated limiting duration for reversible changes related to fluctuations can be taken duration of "eleven year cycles" of sun activity.

Fluctuations, appearing coincidentally with successions, influence the successional changes accelerating or retarding replacement of one phytocoenosis by another one. It is not unusual that invasion of the species in the successions occurs in the periods of fluctuational disturbances of phytocoenoses. However invasions taking place in the fluctuations not always lead to successions. Therefore effective and ineffective invasions can be distinguished. "False invasions" sometimes observed is spreading of species occurring previously in dormant (including viable seeds in soil) or in suppressed state.

The primary causes of fluctuations are changes from year to year or from one period to another one (a) of meteorological conditions, (b) of hydrological conditions, (c) of effects of human activity. All these causes influence directly or indirectly (by means of changing of vegetation) zoo- and microbial-components of coenoses and are responsible for considerable changes from year to year in populations and activity of these components. Changes in zoo-components can also induce fluctuations of phytocoenoses. Fluctuations can be also caused by peculiarities of life cycle of some plants. However such fluctuations are connected closely with changes in environment from year to year. The following types of fluctuations can be distinguished depending on the causes of their origin: (a) ecotopical, connected with changes in ecotope (climate, soil, hydrology), (b) anthropogenic, caused by differences in man's activity from year to year, (c) zoogenic, as a result of mass reproduction of some animals, chiefly insects and voles, (d) phytocyclical, connected with peculiarities of life cycle of some plants and (or) with their nonuniformity from year to year in reproduction by seeds, (e) phytoparasitical, caused by mass reproduction of parasitic plants, chiefly fungi and flowering parasites.

Distinctness of ecotopical fluctuations is determined by changes from year to year in meteorological and hydrological conditions, depending on climate of district and on hydrology of site where phytocoenosis is situated. Fluctuations are more pronounced in the districts with more continental climate; they are less marked in the sites with the stable water conditions. In flooded parts of vallies fluctuations depend not only on changes in meteorological conditions but also on differences in duration of inundation and in sedimentation.

Distinctness of fluctuations depend also on plants forming

phytocoenoses. In the forests more stable phytoclimate is created than in grasslands. Woody plants are more weather-resistent than herbaceous ones. Respectively, fluctuations in forests are less marked than in grasslands.

Changes in quantitative ratio between components of phytocoenoses from year to year are connected with alteration in number and vitality of their individuals. Changes in seed production, in number of arised seedlings and in their mortality are observed at the fluctuations everywhere. In some phytocoenoses fluctuations are followed basically by changes in vitality of components while in other ones episodical or periodical changes in number of mature individuals are observed as a result of their mass dying off.

Differences in vitality of herbaceous plants appear in alteration of number and vigour of tillers, in ratio between generative and vegetative shoots, in productivity of individuals, and in some species in ratio between individuals in active and dormant state.

The fluctuational changes in underground parts of grassland phytocoenoses were not studied. Considerable fluctuations in number of viable seeds in grassland soil from year to year were observed.

In accordance to the form of changes in phytocoenoses, the following types of fluctuations can be distinguished:

a) Obscure fluctuations, characterized by little changes in quantitative ratio of components. They have no significant importance for life of phytocoenoses and for their utilization by man.

b) Oscillatory fluctuations (oscillations) shorttermed (duration 1—3 years) changes in productivity and in the ratio of the components, followed by repeated from year to year, reversible replacement of dominants, connected with changes in their vitality. For example: dominance of *Bromus inermis*+*Agropyron repens* (dry year) ⇄ dominance of *Alopecurus pratensis* (wet year).

c) Phytocyclic fluctuations. Characterized by periodical dominance of one or two mono-oligocarpic species, as a rule along with dominance of other plants. The luxuriant development of individuals of such species are followed by their mass dying off as a result of completion of their life cycle. For example: periodical dominance of *Trifolium pratense* and *T. hybridum* in meadows ("clover years").

d) Digressive — demutation fluctuations, connected with mass dying of one or several dominants, as a result of large and sometimes long-term deviations from average meteorological and (or) hydrological conditions or as a consequence of mass reproduction of phytophages. Such disturbances are followed by spreading of species with little competitive ability, usually reproducing themselves effectively vegetatively (experiments: L. G. RAMENSKY).

The dominance of explerents is short-term and phytocoenosis returns to its "original state". Duration of such fluctuations can be rather long (5—10 years). For example: dominance of *Alopecurus pratensis* → dominance of *Ranunculus repens* (result of stagnation of water) → dominance of *Alopecurus pratensis*.

Every type of phytocoenoses is characterized by one or several forms of fluctuations. For complete understanding of nature of grassland phytocoenoses it is necessary to study their fluctuational states.

The investigation of fluctuational changes in grassland phytocoenoses is of great importance for grassland management. One should consider the fluctuations in grasslands for the development of methods of improvement and rational utilization of meadows and pastures.

C. METHODS OF SYNDYNAMICAL ANALYSIS AND OF CONCLUSIONS IN VEGETATION DYNAMICS

3 GROSSRESTE VON PFLANZEN, POLLEN, SPOREN UND BODENPROFILE IN IHRER BEDEUTUNG FÜR SYNDYNAMIK UND SYNCHRONOLOGIE

R. Tüxen

Contents

3 GROSSRESTE VON PFLANZEN, POLLEN, SPOREN UND BODENPROFILE IN IHRER BEDEUTUNG FÜR SYNDYNAMIK UND SYNCHRONOLOGIE [1])

Auskünfte für den Nachweis von Pflanzengesellschaften und ihrer Entwicklung in früheren Abschnitten der Erdgeschichte liefern pflanzliche Großreste (Blätter, Holz, Früchte, Samen usw.), sowie Pollen und Sporen und in Sonderfällen auch gut erhaltene Reste von tierischen Organismen, die eng an bestimmte Biozönosen gebunden sind, wie z.B. Thecamöben u.a. Auch auf Grund der Koinzidenz von Pflanzengesellschaften mit makromorphologisch auffallenden pedologischen Erscheinungen kann aus fossilen Boden-Profilen auf das ehemalige Vorhandensein dieser Gesellschaften geschlossen werden.

Die Möglichkeit, mit Hilfe pflanzengeographisch-florengeschichtlicher Erwägungen die Herkunft und Entstehung von Pflanzengesellschaften zu erklären (vgl. z.B. BRAUN-BLANQUET 1923: 202) soll hier nur angedeutet werden, bleibt aber außerhalb unserer Betrachtungen. Ebenso haben wir rein pflanzensoziologische Arbeiten, die anthropogen bedingte und ihre potentiell natürlichen Ausgangsgesellschaften behandeln, (wie z.B. Niederwälder aus syntaxonomisch abweichenden Hochwäldern), hier kaum berücksichtigt, sondern unsere kurze Betrachtung aus der Sicht des Pflanzensoziologen auf palynologische Befunde beschränkt, ohne unbedingte Vollständigkeit zu erstreben.

Die pflanzensoziologische Auswertung aller dieser Ergebnisse ist allerdings von mancherlei Voraussetzungen abhängig und verlangt vielseitige palynologische, pflanzensoziologische, synökologische und pedologische Erfahrungen. Denn es handelt sich ja hier nicht um den Nachweis einzelner Gehölz- oder anderer Arten und ihrer Mengenanteile an der Pflanzendecke, sondern um Pflanzengesellschaften, die nur in Ausnahmefällen als solche vollständig erhalten sind, oder deren Bindung an bestimmte Bodenprofil-Ausbildungen nur selten so eindeutig ist, daß sie mit Sicherheit in enger Fassung nachgewiesen werden könnten.

3.1 Grossreste von Pflanzen

Großreste finden sich in See-Ablagerungen (z.B. Mudden,

[1]) Den Herren Prof. Dr. BEUG, Göttingen und Dr. MÜLLER, Hannover danke ich herzlich für ihre wertvolle Beratung.

Kieselgur) Torfen und Kohlen. Sie können zum Teil makroskopisch erkannt oder bestimmt werden. Oft sind aber auch Aufbereitungsarbeiten und sorgfältige mikroskopische Vergleiche mit rezenten Pflanzen nötig. Mit zunehmendem Alter der Ablagerungen wird die Bestimmung und soziologische Deutung der Reste in der Regel schwieriger, weil manche von ausgestorbenen Arten herrühren oder heute nicht mehr bestehenden Pflanzengesellschaften entstammen können.

Vielfach lagern diese Reste nicht mehr am Ort ihrer Entstehung, sondern sie wurden durch fließendes Wasser oder durch Wellenbewegung in stehenden Gewässern verschwemmt und gesondert, durch Wind befördert oder auch vom Menschen zusammengetragen (Abfall-, Misthaufen und dergl.). Arten aus verschiedenen Pflanzengesellschaften wurden dabei vermengt und manche Teile beschädigt. Auch wurden bei weitem nicht immer alle Arten einer Gesellschaft in gleicher Weise konserviert (vgl. Braun-Blanquet 1964: 712), denn nicht alle haben gleiche Erhaltungsfähigkeit. Manche Arten erscheinen daher stark gehäuft, andere fehlen ganz.

Dennoch sind diese Quellen soziologisch aufschlußreich wenn sie mit der gebotenen Umsicht ausgewertet werden (Beijerinck 1929, Knörzer 1964—1971, Körber-Grohne 1967, Lang 1962, 1967, R. Tüxen 1956, 1957a, Willerding 1960). Auch Holzkohlen-Funde können das Waldbild zur Zeit ihrer Entstehung beleuchten (z.B. Neuweiler 1925, Popovici 1932—1934, Müller-Stoll 1936, Fukarek 1955), sind aber in jüngerer Zeit offenbar weniger untersucht und nicht erschöpfend pflanzensoziologisch ausgewertet worden.

Manche Stapel-Funde dürften sogar bestimmte Gesellschaften erkennen lassen, wie solche aus Heu- oder Streulagen der Wurten, die von Salzwiesen (Juncetum gerardi) oder aus Abfall-Resten von Acker-Wildkrautgesellschaften der Chenopodietalia albi oder der Aperetalia spica-venti stammen. Denn selbst bei Berücksichtigung der Tatsache, daß eine Verschiebung in der Artenverbindung der Gesellschaften im Laufe der Zeit stattgefunden hat, lassen sich syntaxonomische Einheiten nicht zu hohen Alters doch mit Hilfe ihrer Kennarten auch dann erkennen, wenn nicht alle Begleitarten erhalten sind. Denn es war gewiß immer ausgeschlossen, daß z.B. *Spergula arvensis* oder *Papaver* spec. in einer Wiese oder gar in einem Wald oder *Juncus gerardi* mit anderen Halophyten außerhalb des Armerion maritimae oder sehr nahe verwandter Gesellschaften wuchsen. Mit Recht weist Knörzer (1971) darauf hin, daß man zur Deutung früherer Pflanzengesellschaften von den heutigen ausgehen müsse, und Janssen (1970: 187) macht deutlich, daß ein Vergleich fossiler und rezenter Vegetation am besten auf Grund der Artenkombination durchgeführt werden kann.

Genauer ist die Entwicklung bestimmter Pflanzengesellschaften in Torflagern zu verfolgen, in denen Reste ihres gesamten Artenbestandes konserviert sein können. Aus solchen Lagerstätten lassen sich denn auch aufschlußreiche Erkenntnisse über die Syndynamik und die Synchronologie enger gefaßter torfbildender Gesellschaften gewinnen. Hier kann die Berücksichtigung der Rhizopoden und anderer tierischer Reste wertvolle Erkenntnisse vermitteln (GROSSPIETSCH 1967. Vgl. dazu jedoch GROSSE-BRAUCKMANN 1962a: 108, WEST 1964, JANSSEN 1970: 187).

Einzelne Großrest-Lagerstätten geben zunächst nur örtlich und zeitlich begrenzte Einblicke in den allgemeinen Verlauf der Gesellschafts-Bildung und -Entwicklung, der sich erst durch Verknüpfung mit anderen Funden aufklären läßt.

3.2 Pollen und Sporen. Pflanzensoziologie und Pollenanalyse

Weit verbreiteter, aber schwieriger soziologisch deutbar als Großreste von Einzelpflanzen oder gar von Pflanzengesellschaften in situ (vgl. JANSSEN 1970: 187) sind die fast allgegenwärtigen Sporen und Pollen ihrer einzelnen Arten. Die Pollenkörner sind sehr klein (10—250 μ), die meisten erreichen nur Durchmesser von 20—35 μ. Sie werden in großer Zahl erzeugt, größenordnungsweise 100 Millionen je Quadratmeter. Die Außenwand (Exine) der meisten Pollenkörner ist chemisch sehr widerstandsfähig. Mit einem guten Licht-Mikroskop lassen sich in der Regel Familien, oft auch Triben, Gattungen oder Untergattungen, die Arten dagegen fast nur bei monotypischen Gattungen sicher erkennen. Ihre Untersuchung, die durch die Palynologie seit langem intensiv betrieben wird, hat tiefe Einblicke in den allgemeinen Floren- und Vegetationswandel vor allem nach der Eiszeit, aber auch älterer Zeit-Abschnitte eröffnet.

3.2.1 MENGENVERHÄLTNISSE, ERHALTUNG, UMLAGERUNGEN

Sporen und Pollen werden nicht von allen Arten in gleicher Menge hervorgebracht und noch weiniger in gleicher Weise verbreitet. Windblüter lassen ihre Pollenkörner weit durch die Luft verfrachten, soweit sie nicht in der Krautschicht eines Waldes blühen, (wie manche *Cyperaceen* und *Gramineen*). Die Pollenkörner von Insektenblütern gelangen dagegen nur zum kleinen Teil in höhere Luftschichten und damit zu weiterer Verdriftung.

Ein Pollen-Spektrum aus einem bestimmten Horizont gibt keineswegs ein reines oder vollständiges Bild einer Pflanzengesellschaft, denn die Pollenkörner einer Probe stammen aus vielen Gesellschaften, deren Arten aber nie in ihrem wirklichen Mengenverhältnis vertreten sind. Um die Aussagekraft der gefundenen Pollen-Anteile über die Mengen der sie liefernden Arten in der umgebenden Vegetation beurteilen zu können, wurden schon frühzeitig Studien vor allem für die Baumarten begonnen und bis in die jüngste Zeit mit mehr und mehr gesicherten Ergebnissen für viele Arten fortgesetzt. (Vgl. z.B. Erdtman 1921, Bertsch 1933, 1935, Hesmer 1933, Pfaffenberg u. Hassenkamp 1934, Pohl 1937, Rempe 1937, Mothes et al. 1937a, b, Florschütz 1941, Iversen 1947, Jonassen 1950, Fukarek 1961, Proctor & Lambert 1961, Heim 1962, Mullenders 1962, Bastin 1964, Pop, Boşcaiu, Raţiu, Diaconeasa 1965, Janssen 1966, 1967 Andersen 1967, 1969, 1970, Wright 1967, Janssen 1970, Schwaar 1970 u.a.).

Bei Godthaab in West-Grönland verglich Iversen (1947) den Pollen-Niederschlag in rezenter Gyttja mit der prozentualen Arten-Beteiligung in der heutigen Pflanzendecke. Die Abweichungen sind z.T. erheblich.

Die wiederholten pollenanalytischen Untersuchungen von Rohhumus oder von Moosdecken in Wäldern haben nach van Zeist (1970: 139) ebenfalls noch keinen voll befriedigenden Aufschluß über des Verhältnis von Baumpollen zu der wirklicher Verteilung der sie liefernden Holzarten erbracht Mehrere Autoren (zuletzt z.B. Janssen 1967b, 1970, Andersen 1970 und van Zeist 1970) erörterten unter kritischer Berücksichtigung älterer Versuche einen Korrektur-Faktor (R-Wert), mit dem die gefundenen Pollen-Mengen multipliziert die wahren Mengen-Anteile der Pollenspender an der Pflanzendecke wiedergeben sollen.

Pollenkörner werden auch nicht von allen Arten und in den verschiedenen natürlichen Einbettungsmitteln gleich gut erhalten, und nicht zuletzt müssen auch hier Bestimmungs- und Erkennungsschwierigkeiten berücksichtigt werden. See-Ablagerungen, *Sphagnum*-Moore und Rohhumus-Lagen gelten als die besten Pollen-Archive. Sand- und Lößböden, sowie Eis bewahren ebenso die in sie hineingelangten Pollenkörner, deren Erhaltung und damit Erkennbarkeit hier allerdings oft viel weniger gut ist als in Torfen oder Mudden. Zudem werden in Mineralböden nicht selten sekundäre, d.h. umgelagerte ältere Pollenkörner gefunden (Iversen 1936) oder jüngerer Pollen in Wurzel-Tüllen oder durch Regenwürmer, Insekten und andere wühlende Tiere in tiefere Lagen befördert, so daß sie sich dort mit solchen älterer Ablagerung ver-mengen können (vgl. DEWERS 1926, 1933, 1934/5, MOTHES et al. 1937a, b, WELTEN 1958, 1962: 331). Daß selbst rezente Samen in Regenwurm-Gängen oder Wurzelbahnen in eine gewisse Tiefe und damit in prähistorische Fundhorizonte gelangen können, haben MEYER und WILLERDING (1961: 26) gezeigt (vgl. WILLERDING 1966a: 25, 1970: 297).

Hier wäre auch an die älteren und in ihrer Auswertbarkeit und Beweisführung umstrittenen Schriften von BEIJERINCK (1933, 1935) und JONAS (1935, 1954) zu erinnern, deren zunächst nahezu algemeine Ablehnung (vgl. z.B. SELLE 1940) nicht die Wiederaufnahme von Pollenanalysen in Mineralböden durch niederländische, belgische und andere Palynologen verhindert hat, durch deren Arbeiten bemerkenswerte Ergebnisse erzielt werden konnten (SCHRÖDER 1934, WATERBOLK 1954, WELTEN 1958, 1962, MUNAUT 1959, MULLENDERS & MUNAUT 1960, GÉHU & PLANCHAIS 1965, HEIM 1966, COÛTEAUX 1967, MUNAUT, DURIN & EVRARD 1969 u.a.). Dennoch bleiben manche Palynologen in der Auswertung von Pollenanalysen aus Mineralböden vorsichtig und zurückhaltend.

Ein normales Pollen-Spektrum gibt zunächst — etwa wie eine Flora — einen Überblick über die in einer bestimmten Zeit, (die mit den scharfsinnigen palynologischen Datierungsmethoden oft erstaunlich genau zu ermitteln ist), in der Umgebung vorhandenen Pflanzenarten, soweit ihre Pollen (und Sporen) sich erhalten haben und bestimmt werden konnten, und, wenn man die Pollen-Erzeugung und die Transport-Möglichkeiten berücksichtigt, bis zu einem gewissen Grade auch über ihre Mengen-Verteilung.

3.2.2 BEDEUTUNG KLEINSTER MOORE

Je ausgedehnter ein waldfreies Moor oder ein See sind, in denen Pollenkörner erhalten blieben, desto höher ist der Anteil der aus weiteren Entfernungen herangeführten Pollen und desto weniger scharf sind infolge der Vermengung der Pollenkörner während ihrer Verfrachtung bestimmte Pflanzengesellschaften, denen sie entstammen, als solche erkennbar. Kleine und kleinste Moore aber, die in Söllen, verlandeten Altwässern oder ähnlichen Bildungen entstanden, empfangen vorwiegend Pollen aus den sie bedeckenden und unmittelbar umgebenden Gesellschaften, während die weiter entfernt wachsenden mit zunehmendem Abstand einen immer geringeren Anteil des Pollen-Niederschlags liefern können (vgl. TAUBER 1965, VAN ZEIST 1970: 133). IVERSEN (1941:

27, 1960: 10) hebt den Trenn-Wert kleinster, steilwandiger vermoorter Teiche als Pollen-Archive hervor, die im Naturzustand ganz vom umgebenden Regional-Wald überdacht sind und damit die Pollenspende örtlicher Randwälder oder -bäume ausschließen (vgl. dazu Firbas 1949: 21).

Nachdem Lüdi (1930: 704), Hesmer (1931: 558), R. Tüxen (1931: 86, 1935: 103, 1956, 1957a), Losert (1953) u.a. auf diese Auswertungsmöglichkeit kleinster Moore für die Erkennung bestimmter fossiler, in ihrer nächsten Nähe wachsenden Pflanzen-Gesellschaften hingewiesen hatten, wurde sie öfter zum Nachweis eng gefaßter Waldgesellschaften von örtlicher Verbreitung ausgenutzt (Hesmer 1933: 518, Mothes, Arnold u. Redmann 1937a, b, Budde 1938, Selle 1939, Iversen 1941: 27, Firbas 1949: 21, Pfaffenberg 1952: 27, Janssen 1960: 62, Scamoni 1966: 262 u.a.).

3.2.3 Pflanzensoziologische Eichung von Pollendiagrammen

Solche glücklichen Sonderfälle für die Erkennung einzelner Vegetationstypen vergangener Zeiten sind aber selten. In der Regel steht für die Pflanzensoziologische Auswertung nur ein schwer zu entwirrendes Gemisch von Pollenkörnern vieler, wenn nicht aller in der Umgebung des Pollen-Lagers vorkommender Pflanzengesellschaften zur Verfügung. Um die soziologische Aussagekraft von Pollendiagrammen zu steigern, versuchte Menke (1968, 1969) eine Eichung von Pollen-Diagrammen an rezenten Pflanzengesellschaften durch Oberflächenproben [vgl. dazu Firbas (1949: 21), Proctor & Lambert (1961), West (1964: 51)].

Trautmann (1952), Faegri (1954) und Iversen (1958, 1960, 1964) zeigten an Hand von Pollenanalysen im Rohhumus lokale Wald-Sukzessionen.

3.2.4 Einfluss des Menschen auf die Vegetation

Im Laufe der Zeit tritt der Anteil an natürlichen oder naturnahen Gesellschaften um so mehr zurück, als der Mensch Ersatzgesellschaften an ihrer Stelle zu erzeugen begann. Die natürliche Vegetation großer Erdräume ist der Wald. Der Mensch lichtete seine geschlossene Decke je nach der Widerstands- und Erneuerungskraft der verschiedenen Wald-Gesellschaften und der Eignung ihrer Standorte für seine Zwecke zunächst punktförmig, dann auf immer größeren Flächen auf, die sich mit Ersatz-Gesellschaften (Rasen, Heiden, Äckern u.a.) bedeckten. Aber auch die Rest-Wälder blieben nicht frei vom Einfluß des Menschen, der sie degradierte oder ihre Regeneration über neue Initial-Phasen zu Sekundär-Wäldern zuließ.

Die geomorphologisch gegebenen Standorte bedingen ein Mosaik natürlicher Vegetationstypen. In Gebieten, in denen Eiszeiten die Vegetation vernichteten, entwickelte sich die natürliche Pflanzendecke nach dem Verschwinden der Eises mit wachsender Artenzahl der Flora je nach den gegebenen Standorten zu immer feiner differenzierten Gesellschaften. In diesen florengeschichtlich bedingten Vorgang griff der Mensch mit steigender aber für die einzelnen Gesellschaften verschiedener Wirkung abwandelnd ein (vgl. Tüxen, R. 1931, 1967: 39, v. Zeist 1970). Die verschiedenen standörtlich bedingten Waldgesellschaften entwickelten und verhielten sich verschieden gegenüber dem Menschen und umgekehrt.

"Bei der Auswertung des Pollendiagramms muß das jeweilige lokale Standortsmosaik in Rechnung gestellt werden; denn es bildet eine wesentliche Voraussetzung für den Pollenniederschlag und die verschiedenartigen Einwirkungen des Menschen auf die Vegetation" (Burrichter 1970: 207. Vgl. dazu R. Tüxen 1931: 88, Janssen 1970: 191).

"Je mehr die Bereicherung der natürlichen Pflanzendecke Mitteleuropas seit der Eiszeit fortschreitet, desto spezieller müssen auch die Möglichkeiten menschlicher Ansiedlung und der damit verbundenen Ausnutzung bestimmter Pflanzengesellschaften werden. Wir wollen uns zwar nicht verhehlen, daß schon in früheren Zeiten, d.h. etwa im Mesolithikum, in dem erst ein Teil der heute vorhandenen natürlichen Pflanzengesellschaften entwickelt und diese vielleicht erst schwach ausgeprägt waren, schon deutlich erkennbar bestimmte Standorte mit ihrer Vegetation von der menschlichen Siedlung bevorzugt waren (in Nordwestdeutschland z.B. lichte Kiefernwälder auf Fluß-Dünen).

Aber mit dem Eindringen der wärmeliebenden Holzarten in Mitteleuropa und dem ganzen Gefolge ihrer Begleiter und dem darauf folgenden Einwandern der Schattenhölzer, Buche, Hainbuche, Fichte und Tanne, muß sich die Pflanzendecke — nicht nur in ihrem physiognomischen Aspekt, sondern auch in ihrer Zusammensetzung, und ganz besonders auch in ihren Nutzungsmöglichkeiten für den Menschen — mehrfach gewandelt haben. Die eintönigen Birkenwälder der frühen Nacheiszeit oder die darauf folgenden Kiefernwälder des frühen Boreals dürften je nach Klima, Relief und Bodenverhältnissen verschiedenen Unterwuchs beherbergt haben. Aber mit der Zunahme der Artenzahl, sowohl der Holzgewächse als auch der Gräser und Unkräuter, mußten sich feinste Standortsunterschiede durch besondere neue Artenkombinationen, d.h. Pflanzengesellschaften, mehr und mehr voneinander abheben.

Das Eindringen von Buche, Hainbuche, Fichte und Tanne aber wandelte erneut durch die hohe dynamisch-genetische Kraft, die diesen Holzarten als biotischer Faktor innewohnt, Gesicht und Zusammensetzung der schon vorhandenen Pflanzengesellschaften ab. Dabei haben diese Holzarten, ebenso wie die früher schon vorhandenen, stets ganz bestimmte, ihnen zur Entfaltung ihrer vollen Konkurrenz-Kräfte besonders zusagende Standorte besiedelt und andere gemieden, das schon angedeutete Mosaik der Vegetationseinheiten nun noch schärfer hervortreten lassend.

Je später also die Besiedlung eines Gebietes durch eine seßhafte Bevölkerung erfolgte, desto sorgfältiger mußte die Wahl bestimmter Standorte, d.h. bestimmter Pflanzengesellschaften, für alle die verschiedenen notwendigen Bedürfnisse, die mit der Siedlung verbunden sind (Wohnplatz, Jagd, Weide, Ackerbau usw.), getroffen werden.

So zeigt sich die Bedeutung der natürlichen Pflanzendecke und ihrer Einheiten, der Pflanzengesellschaften, für alle diese Fragen: Sei es, daß sie direkt in ihrer floristischen Zusammensetzung den Ausschlag gaben, oder aber, daß sie unbewußt als Indikatoren für bestimmte wirtschaftliche Verwendungsmöglichkeiten des Bodens nach der Zerstörung der natürlichen Vegetation oder aber auch für die Eigenart des Klimas betrachtet und gewertet wurden". (R. Tüxen 1939: 21—24, vgl. auch 1966: 260/1).

3.2.5 Entwaldung und die Rolle der Restwälder als Pollenspender

Der Einfluß des siedelnden Menschen auf die verschiedenen natürlichen Waldgesellschaften, durch den bestimmte Einheiten stärker vernichtet wurden als andere, die lange unberührt blieben (vgl. z.B. Buchwald u. Losert (1953: 136), Fukarek (1955: 56), Iversen (1964: 140), Burrichter (1970)), kann nach Burrichter (1970: 206) im Pollen-Bild eine scheinbare Sukzession vortäuschen.

Bestimmte Waldgesellschaften wurden in moor-reichen Gebieten auch das Opfer der Moor-Transgression, die eine Verschiebung im Flächen-Verhältnis der noch vorhandenen Waldgesellschaften bewirkte [R. Tüxen (1931: 90); vgl. dazu aber Schubert (1933: 89, 136), Overbeck (1950: 86), Janssen (1960: 94)]. Iversen (1960: 11) weist auf die entgegengesetzte Möglichkeit hin, daß nach Austrocknung oder Verlandung der Wald ehemalige Seeböden besiedelt.

Möglich ist die Erkennung der zunehmenden Entwaldung eines Gebietes aus dem Verhältnis von Baum- zu Nichtbaum-Pollen. Der Beginn des Ackerbaues und die Begründung von menschlichen Siedlungen zeigen sich deutlich neben dem Auftreten von Getreidepollen auch an Stellarietea- und Artemisietea-Arten, die zumindest dann viel häufiger werden.

Ein eindrucksvolles Beispiel für die Umwandlung des Querco-Carpinetum in Kulturland durch die mittelalterliche Rodung ist der pollenanalytischen Untersuchung des Otter-Sees durch H. Müller (1970: 40) zu entnehmen (vgl. a. Steinberg 1944). Trautmann (1966: 261) schränkte diese Befunde auf Alt-Siedlungslandschaften ein (vgl. dazu H. Mayer 1966: 262).

Die soziologische Beurteilung der Restwälder sowohl in ihrer Natürlichkeit als auch nach ihrer syntaxonomischen Stellung ist nur in Ausnahmefällen, z.B. in Kleinst-Mooren, durch Pollenanalysen allein möglich weil sonst die Gesellschaften nicht getrennt und die menschlich bedingten Abwandlungen der natürlichen Waldgesellschaften nur mittelbar aus den Pollen-Diagrammen erschlossen werden können. Beug (brieflich) weist darauf hin, daß man während einer Siedlungslücke — bei sorgfältiger Methodik und gutem Material — die Vorgänge pollenanalytisch studieren kann, welche die reale Vegetation nach Schluß der Siedlungszeit in Richtung auf die damalige potentiell natürliche Vegetation führten. (Meistens dürfte allerdings der Endzustand nicht erreicht worden sein.)

3.2.6 Potentiell natürliche Vegetation

Noch weniger aber kann die heutige potentiell natürliche Wald-Vegetation entwaldeter Heide-, Grasland- oder Acker-Flächen aus den Misch-Spektren der Restwälder abgeleitet werden, zumal diese auch keineswegs alle natürlich sind.

Aber auch, wenn man die Pollen-Spektren aus einer Zeit geringen menschlichen Einflusses auf die Waldzusammensetzung in Mitteleuropa, also am Ende des Neolithikum, für die Konstruktion der heutigen potentiell natürlichen Vegetation zugrunde legen will, so müssen dabei neben der in Altsiedlungs-Landschaften schon erfolgten Zerstörung mancher Bestände bestimmter siedlungsfreundlicher Waldgesellschaften, die seit dieser Zeit bis heute eingetretenen irreversiblen Standortsveränderungen berücksichtigt werden, durch welche die heutige von der damaligen potentiell natürlichen Vegetation nicht unerheblich abweichen kann. Die potentiell natürliche Vegetation als das gedachte Mosaik der jeweils möglichen Initial-, Folge- und Schlußgesellschaften wird sich ja im Laufe der Zeit um so mehr gewandelt haben müssen, als solche irreversible Standortsveränderungen durch natürliche oder anthropogene Ursachen stattgefunden haben.

Klimaänderung, Erosion, Sedimentation (z.B. Auelehm, Flugsand u.a.), Bodenreifung (Podsolierung), Eutrophierung (J. Tüxen 1958), Verlagerung der Erosionsbasis, Änderungen im Wasserhaushalt (Küstensenkung-Meeresanstieg, Versumpfung, Austrocknung), Vulkanismus, Bergstürze u. andere Einflüsse lassen die heutige potentiell natürliche Vegetation (ohne jemals real vorhanden gewesen zu sein) wesentlich von früheren potentiell natürlichen Zuständen abweichen (R. Tüxen 1961), was sich pollenanalytisch nur schwer oder gar nicht nachweisen läßt.

Zudem sagt der Nachweis von bestimmten Baum-Pollen allein für die Konstruktion der heutigen potentiell natürlichen Vegetation nichts über die soziologische Gruppierung dieser Arten aus. Denn aus dem Vorherrschen z.B. der Buchen-Pollen allein läßt sich nicht entscheiden, ob sie dem Melico-Fagetum, dem Luzulo-Fagetum, dem Querco-Carpinetum oder dem Fago-Quercetum enstammen, in welchen die Buche überall herrschen kann (vgl. Firbas 1954: 195). Hier hilft aber die Kenntnis von der Standortsbindung der heutigen realen naturnahen Waldgesellschaften und ihrer Verteilung in der Umgebung des Pollen-Lagers weiter.

In der Pflanzensoziologie hat der Begriff der potentiell natürlichen Vegetation (R. Tüxen 1956, 1957a) die ehemalige Klimax-Vorstellung weitgehend ersetzt. Die heutige potentiell natürliche Vegetation ist die Gesamtheit der Pflanzengesellschaften, die sich einstellen würden, wenn der menschliche Einfluß ausgeschaltet sein würde. Entsprechendes gilt auch für diejenige früherer Zeitabschnitte. Die potentiell natürliche Vegetation enthält also die Endstufen der möglichen Sukzessionsreihen der Pflanzengesellschaften eines Gebietes und diese selbst. Sie ist darum das Richtmaß für die Beurteilung der syndynamischen Vorgänge in den Pflanzengesellschaften.

Für den Vergleich der pollenanalytischen Befunde sowohl mit der heutigen als auch mit einer ehemaligen potentiell natürlichen Vegetation darf man nicht vergessen, daß dies gedachte Begriffe sind, während die Pollenkörner von der realen Vegetation stammen, die um so weniger mit der potentiell natürlichen übereinstimmt, als der Mensch jene verändert hat. Der Pollenanalyse sind aber nur die Pollenspektren der real vorhandenen Gesellschaften zugänglich, die nach ihrer "Entzerrung" und Entflechtung das Bild des Mosaiks der jeweiligen realen Vegetation wiedergeben können.

Reale und natürliche Vegetation sind identisch bis zum Beginn des menschlichen Einflusses auf die Vegetation, der an Stelle der natürlichen die Ersatz-(Substitutions-)Gesellschaften erzeugt.

Die ursprüngliche reale Vegetation ist nach dem Einsetzen menschlicher Einflüsse in bestimmter Stärke, d.h. seit der "Urlandschaft" im Sinne von R. Tüxen (1931), nicht mehr der potentiell natürlichen Vegetation gleichzusetzen.

Nachdem R. Tüxen (1931) den Weg zur Rekonstruktion von Waldgesellschaften früherer Zeiten gezeigt hatte, (vgl. auch Janssen 1962, 1970) und in NW-Deutschland Pfaffenberg (1952: 41 u.a.) wiederholt die pflanzensoziologischen Vorstellungen von den potentiell natürlichen Waldgesellschaften bestätigt hatte, haben Buchwald & Losert (1953), Trautmann (1957, 1966—1969), Schwaar (1969) und Burrichter (1970) Karten der potentiell natürlichen Vegetation unter Berücksichtigung aller dieser Gesichtspunkte mit den Ergebnissen der Pollenanalyse verglichen und fast immer die überhaupt mögliche Übereinstimmung gefunden.

Trautmann (1957: 294) schrieb nach dem Vergleich seiner Karten der potentiell natürlichen Vegetation des Egge-Gebirges mit zwei dort von ihm hergestellten Pollendiagrammen: "So sind die pflanzensoziologischen Vorstellungen von der natürlichen Vegetation der Egge und ihrer angrenzenden Landschaften durch die pollenanalytischen Befunde voll bestätigt worden".

Auch in Nord-Amerika konnten van Zeist & Wright (1967) die Übereinstimmung pollenanalytischer Befunde mit der natürlichen Vegetation vor der Besiedlung durch die Weißen dartun.

Janssen (1960) gab eine der gründlichsten pflanzensoziologischen Auswertungen von Pollendiagrammen zahlreicher Moore aus dem südniederländischen Löß-Gebiet (Limburg) und konnte dabei manche der heutigen Gesellschaften (Assoziationen und Verbände) schon frühzeitig nachweisen.

Seitdem hat die Zahl der palynologischen Arbeiten, die sich mit dem Nachweis bestimmter Pflanzengesellschaften in früheren Zeiten beschäftigen, erfreulich zugenommen. (Zahlreiche Beispiele dafür sind unten angeführt. S. 267 ff.).

Palynologische Untersuchungen haben die frühere pflanzensoziologische Auffassung von der Bedeutung der Buche in der heutigen potentiell natürlichen Vegetation und von der Erzeugung von Eichenwald-Gesellschaften durch den Menschen, die zunächst von der Pflanzensoziologie verkannt worden waren (R. Tüxen 1930, 1967c: 116, vgl. dazu Firbas 1954: 195) berichtigt. Vielleicht ist aber neuerdings das Pendel zuweilen etwas zu weit zugunsten der Buche ausgeschlagen. Die sorgfältige Berücksichtigung der Bodenprofile und ihrer Koinzidenz mit den Waldgesellschaften wird hier das rechte Maß finden lassen.

3.2.7 Nachweise säkularer Sukzessionen

Wenn mit der Buchenfrage auch die Vorstellung von einer säkularen Sukzession der Waldgesellschaften im atlantischen Klima-Bereich W-Europas (R. Tüxen 1967c) berichtigt werden mußte und von der Pflanzensoziologie vorerst nicht weiter verfolgt wurde, so tauchen in der palynologischen Literatur immer wieder Hinweise auf Wandlungen basiphiler oder neutrophiler Vegetationstypen zu azidophilen oder eutropher zu meso- bis oligotrophen Pflanzengesellschaften auf, die eine weitere Verfolgung dieser Vorgänge und ihrer Tendenz doch fruchtbar erscheinen lassen. Gesellschaften der Potametea wurden durch solche der Litorelletea ersetzt (z.B. Andersen 1967: 110, Behre 1966: 82, vgl. p. 269, Perring 1967: 258) Caricetalia davallianae-Gesellschaften gingen in solche der Caricetalia fuscae über (Rybniček & Rybničkova 1968), eine Sukzession, die durch viele weitere Beispiele aus den Mooren bis zur Ausbildung der Oxycocco-Sphagnetea belegt worden ist.

Höchst bemerkenswert ist aber die von Iversen (1958, 1960: 12, 1964) im jütischen Postglazial nachgewiesene und auch von Andersen (1967) für ein Interglazial gefundene und mit geistreichen ökologischen Argumenten pflanzensoziologisch gedeutete säkulare Sukzession von Querco-Fagetea- zu Quercetea robori-petraeae- und Vaccinio-Piceetea-Gesellschaften. Unter dem Einfluß des atlantischen Klimas bewirkte die Ansammlung des Waldhumus eine Auslaugung der oberen Horizonte (auch ohne Mitwirkung der Buche), die eine entsprechende Veränderung der Waldgesellschaft nach sich zog. "Die Bodenbildung war zweifellos ein entscheidender Faktor für die großen quartären Entwicklungszyklen, und wir können erwarten, große Unterschiede in den Vegetations-Sukzessionen und in der Ausbildung der physiokratischen Stadien zu finden gemäß der Variationen der lokalen Klima-, Boden- und Geländeverhältnisse" (Andersen (1967: 113). [Vgl. dazu Firbas (1949: 160).]

Wenn auch nach Janssen (1970) die pflanzensoziologische Auswertung von Pollenanalysen schwieriger ist, als sie zunächst zu sein schien, so bleibt doch die Palynologie bei umsichtiger Anwendung einer der wichtigsten Forschungszweige für die Aufklärung der Säkular-Sukzession unserer heutigen Wald-Gesellschaften, die allerdings nur von erfahrenen Fachleuten beider Seiten zugleich ohne Überschätzung der einen oder der anderen Seite erfolgreich betrieben werden kann. "Die richtige Entzerrung und Deutung der Pollendiagramme gelingt um so besser, je gründlicher der Bearbeiter auch pflanzensoziologisch und ökologisch geschult ist".

"Wir können keine völlige Gleichheit der Vegetation der älteren Nachwärmezeit mit einer zukünftigen, auf den heutigen Zustand folgenden natürlichen Vegetation voraussetzen. Wohl aber müssen wir von den Vorstellungen über eine solche natürliche Vegetation fordern, daß sie sich an die historisch nachweisbare, von der menschlichen Nutzung zunehmend bestimmte Vegetationsentwicklung in einer ökologisch völlig verständlichen Weise anschließen lassen. Hierin liegt also u.a. der Wert pollenanalytischer Untersuchungen für die pflanzensoziologische Sukzessionslehre". (FIRBAS 1954: 194/5).

Am fruchtbarsten wird die Verknüpfung dieser Disziplinen, wenn sie von einer Person beherrscht werden.

3.3 Bodenprofile und Pflanzengesellschaften (Koinzidenz-Methode)

Aus der Tatsache, daß Pflanzengesellschaften und bestimmte makromorphologisch wahrnehmbare Merkmale der Bodenprofile miteinander koinzidiert sind, lassen sich auch aus fossilen Profilen Rückschlüsse auf die Pflanzengesellschaften ziehen, die das Profil zur Zeit seiner Ausbildung besiedelten. Diese Folgerungen werden um so sicherer sein, je mehr die biogenen Vorgänge bei der Profilausbildung gegenüber den geogenen von Bedeutung waren.

Morphologisch stark differenzierte Profile, wie sie in altpleistozänen Quarzsanden im subatlantischen Klima-Gebiet NW-Europas vorkommen (Altmoränengebiet der Niederlande, Nordwestdeutschlands und Schleswig-Holsteins) geben eine Fülle von Beispielen für die Anwendung dieser Methode (vgl. ZEIDLER 1956: 238).

So gelang es diese "Schrift des Bodens" für den Nachweis ehemaliger Bestände des Querco robori-Betuletum, des Fago-Quercetum, der Geschichte des Calluno-Genistetum, des Ericetum tetralicis, von Festuca ovina-Trockenrasen und Ruderal-Gesellschaften zu entziffern (R. TÜXEN 1957b, 1960). Auch begrabene fossile Profile lassen sich bis zu einem gewissen Grade in dieser Weise deuten und erlauben die auf ihnen einst wachsende Pflanzengesellschaft zu rekonstruieren (R. TÜXEN 1967a). Anstehende fossile Profile des letzten Interglazials, die heutige Ersatz-Gesellschaften tragen, können für die Deutung der heutigen potentiell natürlichen Vegetation problematisch werden, weil noch nicht erwiesen ist, daß sie trotz morphologischer Übereinstimmung mit rezenten Profilen heute auch die zu diesen gehörende Waldgesellschaft tragen würden (z.B. Fago-Quercetum).

Die Ergebnisse dieser Untersuchungen müssen mit denen der Pollenanalyse und des Großrest-Studiums ebenso in Einklang gebracht werden, wie mit den pflanzensoziologisch-syndynamischen Erfahrungen.

So zeigen sich die Palynologie wie auch die morphologische Pedologie in Verbindung mit der Synökologie in der Hand des Erfahrenen als wertvolle Kriterien für die Beurteilung synchronologischer Fragen. Dieses Zusammenwirken mehrerer Disziplinen — auch Klimatologie, sowie Urgeschichte und Siedlungskunde sind nicht zu entbehren — ist bezeichnend für die synthetische Arbeitsweise der Pflanzensoziologie oder doch einer besonders fruchtbaren Richtung unserer Wissenschaft.

3.4.1 Summary

Plant Macrofossils, Pollen, Spores and Soil Profiles in Relation to Syndynamics and Synchronology

Macrofossils of plants relevant for synchronological and syndynamical conclusions can be found in lake sediments, in peat and in coal deposits. It has to be considered that the plant material is often transported by water, wind or by man before fossilisation and that the various species are differently conservable.

In staple founds (e.g. originating from hay and stall litter etc.) defined plant communities can be recognized, namely by the occurrence of remnants of characteristic species.

More conclusive results on the successional and synchronological development of hygrophilous associations are attainable in peat deposits.

Analysis of pollen and spores can be applied in more places. But the phytosociological evaluation of the results is generally more difficult than in macrofossil findings. A pollen spectrum of a defined horizon is usually a combination of pollen and spores shed in several plant communities. The propagation, production, and conservation of pollen of the particular species is highly different; consequently, the percentages of pollen of the various species deviate considerably from the quantitative relations (e.g. coverage) of species in living stands of the vegetation concerned.

Lake deposits and *Sphagnum* bogs are the best pollen archives. Pollen analysis in mineral soils yielded also remarkable results in certain cases; but its much more critical evaluation is emphasized by some authors due to possibilities of inadequate conservation and overturnings by soil organisms or by other effects. Bogs of limited size or small ponds afford the best conditions for phytosociological conclusions on pollen analysis. Such small bogs and ponds receive pollen mainly by plant communities growing in situ or in nearest neighbourhood.

Considerations on influences of man are highly important for the phytosociological evaluation of the pollen diagrams. The plant communities of different sites were influenced by man in different ways and in different intensities. This diversification was apparently growing in the course of postglacial vegetational changes and in connexion with raising numbers of immigrated species in consequence of improving climatic conditions.

By destruction of forests since neolithic periods in consequence of expanding agriculture, certain woody plant communities decreased highly in area. Instead of it, percentages of other associations were growing. Marked changes in percentages of species represented in pollen spectra can be caused completely by such human influences. Agricultural activities are well recognizable by pollen of cereal crop species and of certain weeds.

Pollen analysis in connexion with other work (namely studies on present vegetation and its site conditions) is important for the evaluation of the potential naturalvegetation. The potential natural vegetation contains the terminal and seral stages of successions possible in an area. The potential natural vegetation of present times is often not identical with that one of the last period with relatively unimportant human influence. The reasons for this fact are climatic changes, soil erosion, sedimentations, eutrophication, changed water economy etc. The results of pollen analysis and work on plant communities were in best agreement in studies considering all these complications and special views (examples in the German text).

By means of pollen analysis, some secular successions could be well confirmed. In marsh and bog vegetation in a number of cases, replacement of basiphilous vegetation by acidophilous plant communities has been found in studies applying these methods. Some cases of secular successions from eutrophic forest associations to dystrophic acidophilous woody vegetation by influences of humid oceanic climate seem to be well proved in the same way.

Since plant communities and certain macro-morphological characters of soil profiles are coinciding, it is possible to recognize the local vegetation living at the time of the formation of a fossile soil profile. This method is highly useful in regions with marked morphological differences in soil profiles, e. g. in areas of the ancient glacial morains in The Netherlands and in Northwestern Germany. Former stands of deciduous forests, of heath vegetation and of other associations are detectable by means of these *coincidence methods* in these areas.

4 SYNDYNAMICAL ANALYSIS AND CONCLUSIONS BY MEANS OF THE PRESENT VEGETATION STATUS, OF EARLIER RECORDS AND OF REPEATED STUDIES ON PERMANENT PLOTS

R. KNAPP

Contents

4 SYNDYNAMICAL ANALYSIS AND CONCLUSIONS BY MEANS OF THE PRESENT VEGETATION STATUS, OF EARLIER RECORDS AND OF REPEATED STUDIES ON PERMANENT PLOTS

4.1 Comparisons of Areas Settled by Plants since Periods of Different Length

Most important and most obvious syndynamical methods are comparative studies of the vegetation on areas settled by plants since periods of different length. The condition for such inductive studies is the knowledge of the number of years since the start of settlement of the successional initial stage. However, these numbers are not reliably detectable in most cases. Therefore, the following lists collect examples for situations and references, which offer possibilities of exact determinations of the numbers of years since the beginning of first settlement by plants (initial stages). But additional restrictions in many cases of known numbers of years concerned are disturbances of successional development since its start in a more or less uncontrollable way which exclude such areas for successfull syndynamical studies.

(1) On morains and other sediments of recessing glaciers (Coaz 1887, Cooper 1923, 1931, 1939, Faegri 1933, Friedel 1938a, b, Lüdi 1945, 1958, Crocker & Major 1955, Crocker & Dickson 1957, Baxter & Middleton 1961, Jochimsen 1963, 1970, Stork 1963, Persson 1964, Decker 1966, Zollitsch 1969).

(2) On land formerly flooded permanently by water, e.g. on newly emerged islands (Birger 1906, Hayrén 1914, 1931, Dansereau 1954b, Voderberg 1955, Luther 1961, Einarsson 1967, Vartiainen 1967), on polders (Feekes 1936, Bakker 1951, 1960, Feekes & Bakker 1954, Iwata & Ishikuza 1967), or on the ground of desiccated natural and artificial lakes (Ant 1967a, b, Schwickerath 1952, Burrichter 1960, Salisbury 1970).

(3) On lava fields, on tuff and on other volcanic deposits (Treub 1888, Penzig 1902, Ernst 1907, Takenouchi 1923, Yoshii 1932, 1939, 1940, Miyai 1936, Yamakawa & Nakamura 1940, Skottsberg 1941, Rivals 1952, Asa 1952, Keay 1959, Klausing 1959, Léonard 1959, Lebrun 1960, Tezuka 1961, Tagawa 1964, 1967, Einarsson 1967, Eggler 1969).

(4) On abandoned cultivated areas (old fields) studied extensively in the United States (Costello 1944, Keever 1950, Evans & Cain 1952, Bormann 1953, Bliss & Linn 1955, Launchbaugh 1955, Rice, Penfound & Rohrbauch 1960, Knapp 1965,

Bazzaz 1968, McCormick 1968, Davis & Cantlon 1969, Parenti & Rice 1969, Parks 1969, Ashby & Weaver 1970, Olmstead & Rice 1970, Buell et al. 1971, Shure 1971, Wee 1971).

Historical and related records are mostly the source of the years and days of volcanic eruptions, island emergence and other events. Also early topographical surveys can be useful. In some cases, features of sedimentation (e.g. periodic clay deposits, "Bändertone", "Varven") are means of the determination of the passed period. Methods with radiocarbon and other radioactive isotopes become increasingly important in recent years. (Textbooks on methods of exploration of the time function: Bowen 1958, Zeuner 1958, Franke 1969 et al.).

Fig. 1. Bottom of a valley north of Valdivia, Chile, flooded since the geologic-tectonic events in connexion with the catastrophic submarine earthquakes in the year 1960 (May). The dead trees of flooded former forests are partially visible (namely in whitish tint in front of the darkish slopes in the background). In the foreground, open stands of early stages (with *Scirpus riparius*, *Juncus div. spec.* et al.) of new aquatic successions (hydroseres), whose start is well documented by the time of the earthquakes (5 years ago at the time of the exposure of the photograph). Or.

The species composition, the vegetation structure, the inter-relations with particular factors of soil and microclimate, and the duration of the different successional stages can be explored by comparison of the vegetation and its environment on places which are differentiated only by the length of the periods mentioned above and which are otherwise primarily uniform in site conditions. Material on this basis has been published more and more in the course of the last decades (references mentioned above).

In the majority of syndynamical work based on supposed similar principles, particularly in papers of earlier times, the length of period since the start of successional series is mostly estimated rather superficially. Conclusions from properties of vegetation and of soil profiles are extended and transferred from often more or less remote areas on which the length of period since first settlement by plants is established by methods mentioned. Also such studies applied with the necessary criticism can give desirable results and ideas; but their more or less hypothetical character has to be realized always. On the other hand applied in an uncritical, unexperienced and preoccupied way, they can be the reason of untenable hypothetical ideas and of ample misconclusions.

A rather great difficulty and a source of error in these methods, regardless the exactity of the determination of the time function, is the avoidance of comparisons of places which are not really comparable in their primary site conditions. High experience in the knowledge of local interrelations between vegetation versus soil and microclimate is indispensable for correct conclusions in this way. Additionally, it is necessary to ascertain that the macroclimate has not changed essentially since the initial stages of the successional series concerned.

4.2 Conclusions from Relics of Earlier Vegetation

Isolated old trees or small woody patches in an otherwise totally open area can demonstrate the possibility of successions

Fig. 2. Scattered trees, a small grove, and a stump (in the foreground) of *Eucalyptus* as relics of former woodlands in the hills east of Adelaide, South Australia; now mainly tree-less range land. (*Eucalyptus* tree species of the region e.g. *Eu. leucoxylon, Eu. paniculata, Eu. viminalis*, in valleys *Eu. rostrata.*) Or.

ending with forest vegetation. The trees can be relics of former forests destroyed by influences mostly anthropogenic in such cases. But these conclusions are only applicable when the vegetation with the trees is not restricted to local special favorable sites.

Other relics of various vegetation units can be used in the same way to get an idea of possibilities of successions and several vegetation changes (CLEMENTS 1934). E.g. patches of certain perennial grassland indicate potential growth possibilities of bunch grass prairie in some interior areas of California covered in present times nearly exclusively by annual (therophyte) vegetation (CLEMENTS 1934, KNAPP 1965b).

But also all these conclusions are to be applied very critically. E.g. meticulous studies are necessary on the identity of the soil and water conditions on the site of the relic vegetation and of the surrounding areas less covered by plant communities. Ample measurements and considerations have to affirm the definite absence of any site factor combination in the place concerned which could be the reason for the local occurrence of a vegetation more dense and higher than in the adjacent areas.

Soils are deteriorated very often irreversibly by erosion or other events, or even the local climate has changed since the forests flourished of which the isolated trees are relics. Furthermore, it is

Fig. 3. Abandoned farm land in the dry plains near the lower Murray, Australia. On the left and in the background: remnants of the natural vegetation (Mallee Scrub with *Eucalyptus*). The site conditions of the abandoned fields are changing by heavy wind erosion of the most valuable parts of the soil in consequence of the destruction of the natural vegetation cover. By this reason, a complete regeneration of the original vegetation cannot take place on the abandoned fields. (Mallee *Eucalyptus* species of the region: *Eu. bicolor*, *Eu. dumosa*, *Eu. incrassata*, *Eu. oleosa*, *Eu. uncinata* et al.) Or.

well possible that an isolated tree with an extensive root system can exist under conditions unfeasible for a more or less closed forest of the same species or of a species combination of the same ecological group. A forest needs generally much more water than an isolated tree.

It seems to be nearly unnecessary to mention that trees are of no indicator value mentioned growing near farm houses or gardens on places watered occasionally by intention or by chance. Also trees growing by the side of broad paved roads have only highly limited indicator value; they are supplied by a surplus of water which cannot penetrate in the tarred or otherwise condensed surface of these roads.

4.3 Conclusions from the Age Classes of a Stand

The age class distribution of the tree species can be instructive for successional trends in forests. This method depends mainly on climate causing vessels of periodically alternating smaller and greater lumina (annual rings, Jahresringe), which are a basis of exact and uncomplicated age determination. It is applicable for this reason only to most forests outside the tropical cancers and to tropical

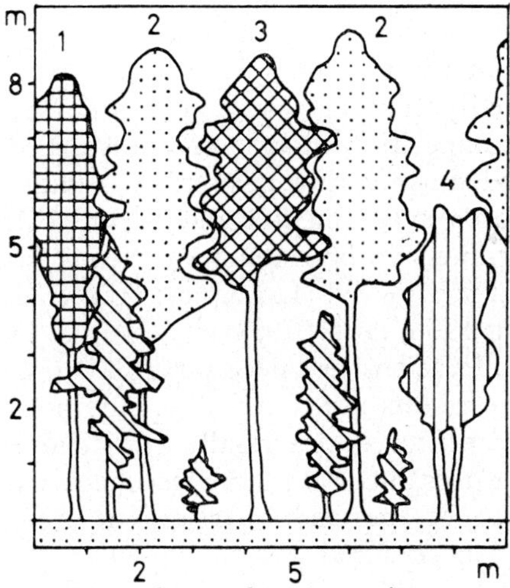

Fig. 4. The upper tree layer of a woody regeneration stage of a Central Alaskan forest is composed by deciduous species (1 = *Populus balsamifera* = *tacamahaca*, 2 = *Betula alaskana* 3 = *Populus tremuloides*). These species are deficient among the lower and younger individuals. Dominance in the upper tree layer will be attained by Picea glauca (diagonally hatched) in the course of successional development, now already prevalent among the low tree individuals. (Profile W of Fairbanks, breadth of transsect 0,75 m.) O1.

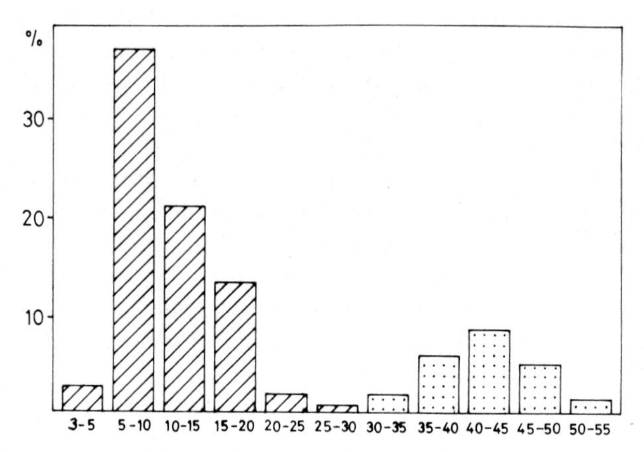

Fig. 5. Percentages (%) of various stem diameter (in cm) classes among the trees of a deciduous forest in the northern Taunus, Germany. Beech (*Fagus sylvatica*, (hatched columns), prevalent among the younger trees, will replace the oaks (*Quercus petraea*, dotted columns), now dominant in the upper tree layer, respectively among the oldest individuals with large stem diameters. Or.

woodlands in areas with periodically long dry seasons. In tropical rain forests with a continuous development of timber, the age of trees can be estimated in most cases only more or less indirectly and with limited exactity.

Species represented only by old individuals will be replaced by species with many vigorous young specimens. Stands with species present evenly in all age classes will presumably not change considerably in composition in case of undisturbed development; they can be regarded as more or less terminal stages of successions. This method affords exact measurements and calculations. It is also useful for beginners without much hesitation in contrast to the methods described previously. (Examples: HEYWARD 1939, SCHULZ 1960, DAUBENMIRE 1968, HORN 1971.)

Critical considerations are necessary on the extension of the area on which the results are applicable. Again a finding is only representative for places of essentially identical site conditions. Also the previous treatment and development can be a reason of special attributes of age class distributions, especially in intermediate and early successional stages.

4.4 Comparisons of Earlier and Recent Maps, Photographs and Other Documentations

Comparisons of earlier and recent topographical maps and

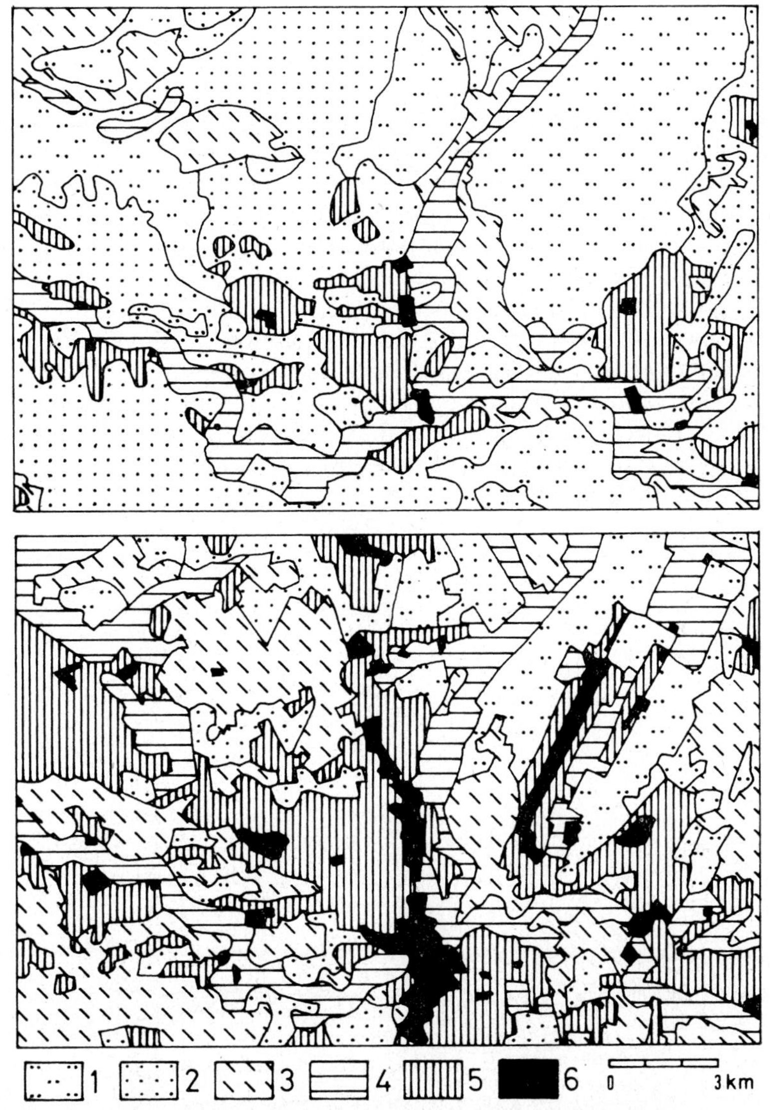

Fig.s. 6, 7. Land surveys at 1800 (above) and at 1955 (below) demonstrate fundamental vegetational changes in the environments of Gifhorn in the Northwest German plains. Bogs (1, mainly Oxycocco-Sphagnetea) and dwarf shrub heath vegetation (2, mostly with dominant *Calluna vulgaris*, Calluno-Genistion) were disappearing for the greatest part. Forest vegetation covers at 1955 an area of more than the double size as compared with 1800 (as a result of afforestations, but also of spontaneous successions on former bog and heath sites). Synchronously, the acreage of anthropogenic meadows and pastures (4, Molinio-Arrhenatheretea), of the tilled fields (5, with weed vegetation of Secalinetea and Polygono-Chenopodietalia), and of the areas with buildings and gardens near the houses (6) was enormously increasing.

photographs are used increasingly in the last years for syndynamic research work. Series of detailed topographic maps surveyed at different times ago and also useful photographs from earlier years are much more available today than some decades ago. These comparisons show exactly certain vegetation changes.

Unfortunately, only few details of species composition are recognizable mostly on the topographical maps and on the photographs. Much more details of vegetational changes afford vegetation maps surveyed at different numbers of years ago on the basis of vegetation units (plant communities) differentiated by differential and characteristic species or similar methods (BRAUN-BLANQUET et al. 1958, SEIBERT 1958, KNAPP n.p.).

The limits of usefulness of photographs for syndynamical studies, especially compared with the methods described in the next paragraphs were studied in a series of work in a detailed way (BECHER 1963, KNAPP n.p.).

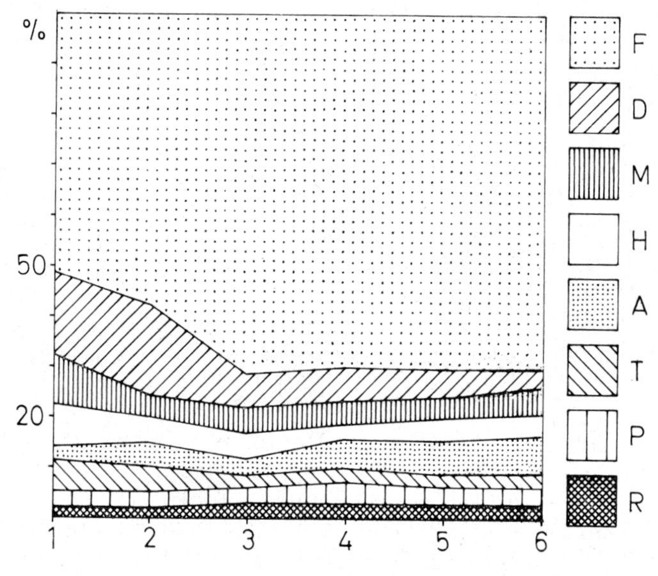

Fig. 8. Coverage percentages (%) of the 10 most important species in a permanent plot (100 m²) of a hay meadow (Arrhenatherion) near Marburg, Germany, at different times (on abscissa: 1 = August 1966, 2 = June 1967, 3 = October 1967, 4 = June 1968, 5 = October 1968, 6 = October 1972). With exception of higher percentages of orchard grass (*Dactylis glomerata*) at 1966 and June 1967, only minor changes of the coverages during the six-years period. F = *Festuca rubra*, M = *Achillea millefolium*, H = *Holcus lanatus*, A = *Agrostis tenuis*, T = *Trisetum flavescens*, P = *Poa pratensis*, R = *Rumex acetosa*. Or.

4.5 Permanent Plots and Other Permanent Observation Areas

The most objective basis of studies on vegetation changes are measurements and protocols by means of permanent observation areas. Permanent plots are areas of limited size with exactly marked boundaries, often quadratic in form (permanent squares, permanent quadrats, Dauerquadrate). The composition, structure and other attributes of the vegetation within such plots are observed and measured in certain time intervals. The results become increasingly informative with the extension of the observation period and with the number of observation terms.

The plot size mostly applied is limited to a few square meters

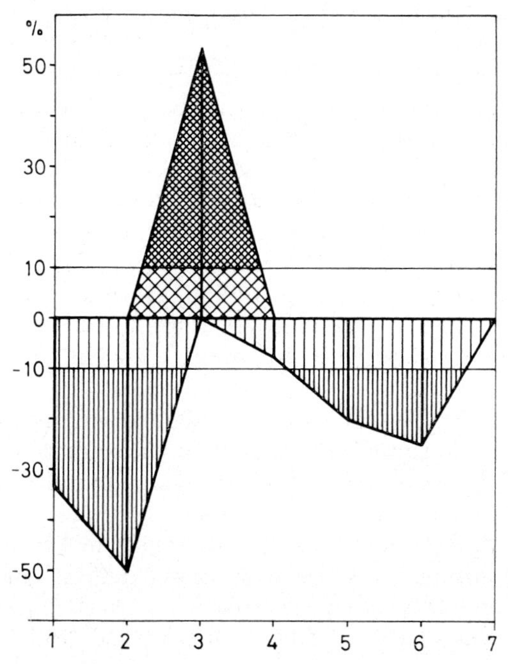

Fig. 9. Species changes in a permanent plot (1 m^2) in a railway area in the Wetterau, Germany. Ordinate-axis = additional species (with living shoots) or disappeared species (—) in % of species numbers of the precedent survey; abscissa = survey times: 1 = October '70, 2 = May '71, 3 = July '71, 4 = Oct. '71, 5 = May '72, 6 = July '72, 7 = Oct. '72. In connexion with rapid successional changes, high instability of the species composition during most of the survey period: disappearances of species during the winter 1970/71; more than 50 % additional species at July 1971 as compared with beginning of May 1971; later on, disappearances of species only. At the end of 1972 the succession has approached a steady state (Dauergesellschaft); no species changes from July to October 1972. (Early stages Polygonion avicularis, last stages of the survey Lolio-Plantaginion majoris.) Or.

(m²). In moss and lichen vegetation, but also in plant communities composed mainly by small therophytes, plot sizes between 25 and 2500 cm² can be recommendable. On the other hand, permanent plots for studies of tree canopies have to include areas between 100 and 1000 m² or even more.

The species composition within the permanent plots can be checked with the various methods of vegetation analysis (relevés, Vegetationsaufnahmen) (Braun-Blanquet 1928, 1964, Christiansen 1937, Doing-Kraft 1954, Jochimsen 1963, Runge 1969a,b, c, Lüdi 1940, Betz & Cole 1969, Heller 1969, Pegau 1970, Stüssi 1970, Londo 1971, Westhoff n.p.). The determination of coverage percentage (%) and of the productivity (by measurement of dry weight and certain other methods) of the particular species is highly useful (Knapp 1955b, 1968a, 1969b, Woike 1958, Becher 1963, Runge 1963a, b).

Exact mapping of the species distribution within the permanent plots (micro-mapping) is most informative, but needs much time applied in a number of plots adequate for comparative conclusions (Weaver 1918, Braun-Blanquet 1931, Cooper 1923, 1931, 1938, Alechin 1934, Weaver & Albertson 1936, Lüdi 1940, Leontiev 1952, Ellison 1954, Voderberg u. Fröde 1957, Braun-Blanquet, Wikus & Sutter 1958, Frey 1959, Sochava, Lipatova & Gorshkova 1962, Becher 1963, Knapp 1965b, Orshan & Diskin 1970, Londo 1971, Watt 1971, Bornkamm 1961).

Experimental work with an analysis on the reasons of vegetation changes and on their dependence on various environmental and anthropogenic factors can be extended in permanent plot studies. For this purposes besides permanent squares without any experimental operations (control squares), parallel permanent plots can be treated with various nutrient solutions, with artificial shading, with differential clipping (differences in height and time intervals of clipping, verschiedene Schnitthöhen und Schnitthäufigkeiten) and with other controlled influences (Knapp 1967a, Becher 1963, Beran 1964, Stillger 1971, Golf 1973).

4.6 Conclusions from Immigrating and Soil Borne Seed Populations

Condition for settlement of a plant species by germination is that its seeds are or become present on the place concerned. In primary successions the seeds have to be mainly transported by different means to the places. This seed immigration can be studied by different methods. Certain boxes feasable for catchment of seeds can be applied (WENDELBERGER & HARTL 1969). They will be most useful for seeds predominantly transported by wind.

Living seeds can be also in the soil in considerable amounts. They will germinate as soon as conditions get favorable for this process, e.g. after ploughing or digging of a grassland sward, or after forest fires. Seeds of many species remain dormant for partially long periods until breaking of the dormancy by extreme temperatures and other special influences or until completion of after-ripening processes of various nature. All these living seeds in the soil can be of paramount importance in successions starting on areas already previously covered by vegetation destroyed by any disturbance or catastrophe. The species composition of early successional stages can be dominantly influenced by plants originating from such soil borne seeds. Longevity of seeds can be very high (more than 60 years, CROCKER 1938, BARTON 1961). Therefore, light demanding species dominant after clear cuts or after death of old trees on previously forested places can develop in many cases from seeds available in situ in the soil.

The living seed populations in soil can be analyzed by induction of germination and consequent identification of the emerging seedlings. It is necessary for this purpose to apply various light and temperature conditions, vernalization (low temperature chocks), certain chemical and mechanical treatments (e.g. for breaking of hard seed coats) and all means promoting germination under the special situation. (Ref.: CHAMPNESS & MORRIS 1948, GUYOT & MASSENOT 1950, KNAPP 1954, 1967a, BUDD et al. 1954, NUMATA et al. 1964, LIVINGSTON & ALLESSIO 1968, PYATIN 1970.)

4.7 Considerations on Long Distance Immigration of Plant Species

The immigration of plant species into a country, province or great reclamation area (e.g. the Dutch polders, VAN DER TOORN et al., 1969) is highly important for vegetation changes, mainly for early successional stages. Such immigrations are always reasons for any vegetation change and for new successions. Therefore studies on such processes should be within the scope of syndynamic work.

Recent long distance immigrations can be of overwhelming influence on syndynamics of areas with original vegetation not or insufficiently resistant against influences connexed with modern agriculture and pasture management. This is the case in a great part of the world. The relatively highest resistance mentioned has the vegetation of most parts of Europe, Asia and Northern Africa.

Therefore, the tendency of underestimation of the importance of such immigrations on successions by scientists of these regions is quite understandable.

The resistance of vegetation against newly invading species (neophytes) coincides mostly with the age of introduction of agriculture, namely plough tillage, and pasture economy in the area concerned. In regions with ancient agriculture in the concept just mentioned, plant communities well adapted to these anthropogenic influences could be developing in the course of millenia. Thus, no or only few ecological niches are remaining on the agricultural sites of these areas open to species eventually immigrating.

Additionally, certain members of natural ecosystems can influence the vegetation structure in a way similar to some effects of modern economical land use. For instance, grazing indigenous animals, namely ungulates (e.g. wild *Equidae*, *Capridae*, *Ovidae*, *Bovidae* s.l. with the various sub-families of antelopes etc.) exert influences similar to those of domestic pasture livestock. Therefore, the original vegetation of regions with indigenous faunas without such animals and with agriculture introduced in relatively recent times is often totally replaced in wide areas by plant communities completely or mainly composed of species immigrated or introduced in recent periods (e.g. New Zealand, figure 10, Hawaiian Islands, St. Helena).

Fig. 10. An area near Palmerston North, New Zealand, with prevalent pasture vegetation, composed nearly completely of species introduced or immigrated to New Zealand from other continents, mainly from Europe, since c. 1770. In the forest patches in the background, namely in the left, plant communities with mainly indigenous species are better preserved. Or.

In the Hawaiian Islands waves of successions were connexed with the immigration of highly competitive species from other parts of the world during this and the last century (EGLER 1942, KNAPP 1965b). New immigrations and their vegetational importance were recorded most impressively from Northern Europe (e.g. Finland: NIEMI 1969, HEIKKINEN 1969, SUOMINEN 1970; Northern Great Britain: RICHARDSON 1970). They are supposed to be more important in this part of the continent than in other regions of Europe, because the flora is rather poor in species and "unsaturated" under certain aspects. In the German hill country (Deutsche Mittelgebirge), many ruderal species invaded areas of higher elevations during the last decades (RUNGE 1959, GEISLER 1967, KNAPP 1971d). Early stages of successions on roadsides, in quarries and in industrial areas are highly influenced by this process.

Such immigrations have to be investigated by repeated detailed distribution mappings over periods as long as possible. Determination of the age of the oldest individuals is a method to accomplish an immigration mapping in one operation in woody species.

5 SPECIAL METHODS OF SUCCESSION ANALYSIS IN EURASIAN TUNDRA VEGETATION

V. D. Aleksandrova

Contents

5 SPECIAL METHODS OF SUCCESSION ANALYSIS IN EURASIAN TUNDRA VEGETATION

The specific features of the methods of studying successions in tundra vegetation are associated with the specifity of the factors causing vegetational changes in the tundra and with the peculiarity of the course of these changes. Two main groups of factors are distinguished: the natural and anthropogenic factors.

5.1 The Study of Natural Successions of the Tundra Vegetation

Distinct dynamic nature of vegetation, particularly characteristic of the tundra, is associated with cryogenic phenomena of different kind, such as the appearance of frost cracks, the movement of the ground and of the soil water at the time of repeated freezing and thawing. It results in the presence of a persistent layer of permafrost in the sorting of soil, in frost-induced dislocations of different kind, in the phenomena of bulging, solifluction, formation of ice insertions in the form of lenses and streaks and also in the deep penetration of ice in the form of lodes into the ground. These processes lead to the extensive development of structural (stone nets, stone rings, medallions etc.), hummocky and polygonal formations, as well as to the thermokarst phenomena in places of thawing of the ice lodes earlier accumulated.

The cryogenic factors are most important in the dynamic of the microforms of the tundra surface, which are one of the most principal causes of the unceasing dynamics of the vegetation, proceeding even in the communities of the climatic climax of the tundras (CHURCHILL & HANSON 1958 and other authors). The action of cryogenic factors is to a certain extent combined with that of wind and water erosion and of biotic factors: the vital activity of burrowing animals and of the vegetation itself. The uneven growth of the organic matter in the polydominant communities of the tundra, containing many tussock forms (*Eriophorum vaginatum*, *Carex lugens*, *Deschampsia brevifolia* etc.) and cushion plants (*Dryas octopetala*, *D. punctata* etc.), is favorable for the development of hummocks of different size or for the unevenness of the turf surface of the tundra. The accumulation of plant remains becoming mineralized extremely slowly at low soil temperatures is also favourable for these processes. Burrowing animals, mainly lemmings and other small mammals, also play a great part in the development of micro- and nanorelief by means of making burrow-holes and treading paths between hummocks etc. Such an instability of the habitats and the consequent dynamic character of the vegetation is the main cause of the natural successions inherent in the vegetation of tundras.

The most reliable data on the course of successions can be obtained by means of direct observations of the same sample area during several years. The studies of this kind were carried out by NORIN at the Research Station of the Komarov Botanical Institute of the Academy of Sciences of the U.S.S.R. in the East-European forest-tundra of the Komi A.S.S.R. (IGNATENKO & NORIN 1969). In 1960 NORIN made a detailed plan of a 1 m^2 sample plot of tundra with bare loamy patch and the surrounding vegetation. In 1965 he repeated the drawing of the plan. It was observed, that during the five years' interval the outline of the loamy patch has changed: on one side it was owergrown with *Empetrum hermaphroditum* and on the other side with *Polytrichum piliferum*, *Cladonia mitis*, *Vaccinium vitis-idaea*, *Calamagrostis lapponica*; in its other part, however, the area of the bare earth patch has extended in consequence of the death of some mosses and lichens that had adjoined it on that side. These observations corroborated the author's conclusion about the pulsation of the borders of the bare earth patches and about the cyclic character of the changes of the vegetation.

However, most investigations of natural successions of the tundra vegetation are based only on comparative observations. This method also affords some valuable evidence elucidating the course of natural successions provided the number of observations is sufficiently large and all the natural environmental conditions are taken into consideration. The investigators were most interested in the dynamics of the vegetation of tundras with spot-medallions (SUKACHEV 1911, GRIGORYEV 1925, 1946, SOCHAVA 1930, ANDREEV 1933, GORODKOV 1935, 1950, 1958, GOVORUKHIN 1936, 1950, TIKHOMIROV 1957, ALEKSANDROVA 1962, et al), in the dynamic of the vegetation in relation to the thermokarst phenomena in the tundra (GORODKOV 1956, TIKHOMIROV 1958, ALEKSANDROVA 1963, et al.), in the development of hillocks and of tetragonal formations in mires (SAMBUK 1933, ZUBKOV 1932, BOGDANOVSKAYA-GIENEF 1938, ANDREEV 1938, PYAVCHENKO 1955, ALEKSANDROVA 1956, 1963, et al.), in the paludification of tundras and of tundra water bodies (ZUBKOV 1932, SAMBUK 1933, GORODKOV 1956, 1958, et al.), in the colonization of river and marine alluvia (KOR-

CHAGIN 1935, GORODKOV 1935, 1938, 1956, 1958, ALEKSANDROVA 1956, et al.).

The concept of natural successions of vegetation includes also those successions that are caused by animals, such as the formation of tundra short-grass meadows in places enriching with nitrogen by animals, e.g. near the bird colonies on sea coasts, around the "watch-posts" of predatory birds and in places of concentration of burrow-holes of lemmings, of arctic foxes and of other burrowing animals. Successions of this type were described by a number of authors (ANDREEV 1930, 1931, GORODKOV 1935, ALEKSANDROVA 1956, TIKHOMIROV 1959, 1960, et al.). Successions taking place in the course of overgrowing of rocks and screes in the tundra zone also attracted the attention of a number of investigators beginning with BAER (1838).

5.2 The Study of Anthropogenic Successions of the Tundra Vegetation

Anthropogenic successions of vegetation in the tundra are most widely represented by successions of vegetation resulting from the pasturing of reindeer, also by succession on the burns (in cases of destruction by fire of the lichen cover etc.), the appearance of ruderal plants and weeds near towns, villages and other human settlements, the transformation of tundras into meadows in places of camping areas of reindeer breeders and the artificial transformation of tundras into meadows by means of land-improvement, extermination of shrubs with herbicides etc.

The methods of studying the effect of the reindeer pasturing on vegetation, particularly on that of lichen tundras, are elaborated at the Institute of Agriculture of the Far North (at present situated in Norilsk) beginning from the thirties of this century. Besides the comparative observations of pastures of different degrees of exploitation, methods of direct observation of the changes of vegetation caused by grazing have been elaborated. For this purpose sample plots are arranged along the "pathway" (a wide strip of the land serving as the route for a grazing herd of reindeer). For each of these sample plots a control plot is selected as similar to it as possible, on which all the plants are cut down for measure their vegetative mass. After the passage of the herd all the sample plots are thoroughly examined, all the speciments bitten by reindeer are recorded and then all the plants in the sample plot are cut down for the purpose of comparison with the cuts on control plots (IGOSHINA 1934, 1937, GLINKA 1938, 1939, SALASKIN 1937,

Igoshina & Florovskaya 1939, Kovakina 1958, et al.). Successions of the vegetation taking place in the course of regeneration of the lichen cover destroyed by grazing or damaged by fire were studied by a number of authors (Gorodkov 1926, Sochava 1933, Andreev 1933, Rabotnov 1936, Igoshina 1939, et al.). It has been established that the regeneration of the lichen cover in places where it was completely destroyed requires from 20 to 50 years. The changes of the tundra vegetation in the sites of human settlements are described by Gorodkov (1939, 1958), Dydina (1954), Dorogostaiskaya (1968) et al. The artificial transformation of the tundra into meadows for purpose of improvement of pastures and creation of hay meadows in the tundra zone was promoted by the scientific workers of the Institute of Agriculture of the Far North (Dydina 1954, 1957, Savkina 1951, 1953, 1960, et al.), of the Komi Branch of the Academy of Science of the U.S.S.R. (Khantimer 1951, 1964, et al.) and also by some other investigators.

6 METHODS OF SUCCESSION ANALYSIS IN EURASIAN STEPPES AND SEMIDESERTS

Z. V. Karamysheva

6 METHODS OF SUCCESSION ANALYSIS IN EURASIAN STEPPES AND SEMIDESERTS

The forms of successions of vegetation in steppes and semideserts (= subdeserts) are extremely diverse; the most important ones are local successions of spontaneous and anthropogenic genesis (consecutive local successions according to the classification of ALEKSANDROVA (1964).

Spontaneous successions caused by climatic factors include the processes resulting from the progressing salinization of soil and subsoils, the processes of erosion, such as the spreading of the network of ravines etc. Certain successions inherent in different natural zones, such as the successions on the stony sites, screes, eluvial deposits, sands etc., proceed peculiarly under arid conditions and are characterized by peculiar initial stages, by participation of representatives of certain life forms of plants etc. Thus, in no other Eurasian natural zone the burrowing activity of animals is so conspicuous as in the steppe zone (VORONOV 1954) and in no other zone the influence of these animals on the vegetation is so strong (LAVRENKO 1952). Very widespread in the steppes are the successions caused by man's agricultural activity, such as the re-establishment of vegetation on fallow lands, the pasture digression and demutation, the changes of vegetation by mowing and irrigation, the successions on the burned areas etc.

There are a few general and special text-books on the methods of studying successions in steppes and semideserts (LAVRENKO 1959, PROZOROVSKYI 1938, ALEKSANDROVA 1964, LARIN 1952, KARAMYSHEVA 1960 and also some papers) and a great number of monographs and papers containing the descriptions of different concrete forms of successions. As it can be seen from the analysis of all this literature, there are no special (specific) methods of studying successions in arid regions. As in other zones, the ordinary geobotanical methods are used here; these methods are based on the records of qualitative characteristics of the vegetation, its floristic composition, layer- and synusial structure, on the horizontal distribution of plants, on the biologo-morphological characteristics of prevailing, initial, relict species (characteristic features of their propagation and vegetative reproduction, specific features of their root systems etc.). Although these methods afford no possibility for any quantitative evaluation of the stability of serial communities, for absolutely accurate determination of the duration of different stages, nevertheless, being applied as a single complex and combined with the analysis the physico-geographical factors and with the observation of the dynamics of the landscape as a whole, they usually permit to establish objectively both the causes and the direction of successions, to determine approximately the stages of successions, to find out the role of separate ecological factors in the dynamics of the vegetation etc. Such a complex approach to the investigation of different aspects of vegetation, the use of several methods, the efforts to establish the relationship between the processes taking place in the vegetation and the genesis of soils, the recent geomorphological processes etc., all this is particularly characteristic of the geobotanical investigations in Russia.

It is worthwhile to mention here only the methods most widespread in the U.S.S.R., those that are used for studying the successions in steppes and semideserts and those that develop most intensely during the last few years.

1) The most ample information on successions can be obtained *by the method of geobotanical profiles.* These profiles can be marked either by sight or by means of instruments, used for the topographical surveying (the level, the eclimeter). The locality for such a profile as well as its length and direction depend on the specific features of the object studied and on the purpose in view. Thus, if the object of the investigation is the effect of desalinization on the vegetation, the profile begins from the water (lake, sea) level and extends up to the placor in sense of a watershed plain, built of loamy deposits, moistened by the atmospheric precipitation and covered by zonal type of soil and vegetation.

If the object is the pasture digression and demutation of the vegetation, the profiles should be started from human settlements and extend to the area where the vegetational cover is not changed by grazing. The successions on the stony sites and the effect of erosion on the vegetation are studied by means of profiles traversing areas with a uniform geological structure (lithologically homogenous soil-forming rock) with the same type of soil-forming process, i.e. in the *definite type of landscape.* Repeated laying of profiles under similar natural conditions, the comparison of the descriptions of the vegetation on these profiles, as well as their analysis concurrently with the data, characterizing the physico-geographical conditions along these profiles, all this permits to outline erosional-genetic or successional series (Lavrenko 1938, Karamysheva 1960a). The principles studied by means of profiles should be verified in the simultaneous phytogeographical investigations.

This can be illustrated by the example of the works of Ka-

TYSHEVTZEVA (1955) in the northern part of the Caspian Area, BEIDEMANN (1957), as well as BEIDEMANN & PREOBRAZHENSKY (1957) in the Kura-Araxian Plain, containing the analysis of successions proceeding in the course of desalinization of soils and subsoils as the result of the recession of Caspian Sea and of melioration. In the first of these works (KATYSHEVTZEVA 1955) the description of the vegetation along 17 profiles (from 5 to 50 km long), traced from the water level up to placors with the semidesert vegetation, are compared with the data on the soil structure and composition. The result of this work was the establishment of the stages of successions and of their duration. The latter was determined precisely on the basis of the geological evidence concerning the age of the deposits. In the works of BEIDEMANN and PREOBRAZHENSKY the regime of longterm inundation in the past and the stages of the development of vegetation were traced in the course of the simultaneous studying of vegetation and soils in profiles, based mainly on the organic remains in the mineral part of soils (fresh-water and marine shells, carboniferous plant remains, reed rhizomes etc.). The precise age of the stages was also established on the basis of the geological data.

In the investigations of primary and secondary successions on stony sites the following procedures are carried out concurrently with the description of the vegetation in profiles: the measurement of the depth of the location of the bed-rock, the study of the degree of development of the soil cover, the study of the recent geomorphological processes and of the biology of plants (the characteristic features of the root systems of initial and prevailing species, the modes of vegetative propagation etc. (SEMENOVA-TJAN-SHANSKAYA 1954, YAROSHENKO 1954, DOCHMAN 1954, LAVRENKO 1957, KARAMYSHEVA 1960, 1961, 1963 and other works). The relationship between the successions on stony sites on the one hand, and the lithology of the bedrocks, their age, the characteristic features of the deposition, the recent geomorphological processes etc. on the other hand are established in the course of these investigations.

The same method (i.e. the method of profiles) with the drawing of horizontal projections of fragments of plant communities and of separate plants, excavation and drawing of burrow-holes etc. can be used for obtaining the data on the succession in marmots' and gophers' burrow-holes (FORMOZOV et al. 1954, LAVRENKO 1952, LAVRENKO & YUNATOV 1952).

For studying the successions taking place under the influence of grazing, the profiles are stretched from small human settlements over the area with a homogenous soil cover and the same type of relief up to the area, where the vegetation is only insignificantly affected by grazing. For the establishment of the stages of pasture digression the intensity of exploitation, the season and the duration of the most intense grazing and other conditions should be taken into consideration (LARIN 1952); the same methods are used for studying the stages of successions in ravines (KALASHNIKOV 1936, VYSHIVKIN 1953 and other papers).

2) The method of investigating the vegetation *in the sample plots* is as widespread as the method of profiles. The sample plots are usually situated along profiles; but sometimes they are not attached to any profiles. Thus, in studying the effect of fires on the steppe vegetation, the sample plots are arranged in two parallel rows: one on the burned area, another on the area not damaged by fire. When studying the successions on dumps and waste heaps, the sample plots are situated from the old part to the new part of the dump etc. Stages of post-fire successions under different natural conditions are described in many papers; this literature is reviewed in the works of LAVRENKO (1940, 1950), RODIN (1946), and IVANOV (1958). In the studies of the re-establishment of the virgin steppe vegetation on fallow lands except the descriptions of the vegetation on the sample plots situated on fallow lands of different age, the following facts should be taken into consideration: the time during which the area studied was occupied by the last crop; the character of the soil cultivation and the subsequent exploitation of the fallow lands etc. (LAVRENKO 1940). The literature on this problem is reviewed in the papers of LAVRENKO (1940), SEMENOVA-TJAN-SHANSKAYA (1953) and IVANOV (1958).

Certain other methods of studying successions recently (1950—1960) accepted are also to be mentioned.

3) *The method of mapping.* Plans and drawings are used quite frequently in the studies of the dynamics of the vegetation (VORONOV 1954, KARAMYSHEVA 1960b, BEIDEMANN 1962 and many other papers). One of the variants of this method also used frequently is *the repeated mapping* of the same area. For example, the map-schemes of the distributions of vegetation along the Caspian Sea coast composed in 1929 and 1947 show distinctly the successions of the vegetation in this area (BEIDEMANN 1962). During the recent years the scientists composing large-scale geobotanical maps try to use the dynamic principle, i.e. not only to reflect in these maps the geographical distribution of vegetation and its relationship with the environment, but also to demonstrate the dynamic categories of the vegetation (serial, of varying stability, climax communities), to reveal their successional interrelations and the dependence of dynamics of the vegetation on the environment. The mapping method of investigating the dynamics

of the vegetation, like all the methods mentioned above, is an indirect one; however, it has a number of advantages. One of them is the necessity of a complete survey of the entire area, in the course of which all the plant communities, both serial and climax ones, are studied in detail and plotted on the map.

The main stages of successions determined for a certain part of the area and erosional-genetic series elucidated in definite types of landscape are subsequently tested in the course of the mapping of the other part of area with similar natural conditions. The comparison of geobotanical maps with certain other special maps, such as geological, geomorphological and pedological, contributes to a more objective knowledge of the causes of successions and of the dependence of the dynamic status of the vegetation on the dynamics of the environment. Thus, the mapping method is the most objective and obvious one (on account of its graphic technique) for investigating the dynamics of the vegetation.

There are a number of papers in which the mapping method is used in the studies on the dynamics of the vegetation: in Central Kazakhstan (Karamysheva & Rachkovskaya 1962), in Transbaikalia (Isachenko 1965), in Transuralia (Iljina 1968). The evaluation of the mapping methods is made and the possibilities of its application to the investigation of the dynamics of the vegetation are considered in detail in the paper of Gribova & Isachenko (1972).

4) Similar to the mapping method in the technique of field studies and in the simultaneous investigation of all the components of the landscape is the *landscape method.* The classical work of Popov (1914) is an excellent example of the use of this method. In the course of the studies of the evolution of the microrelief and soils and of the comparison of this process with the changes of the vegetational cover in different depressions the author has established the genesis and the evolution of the "kolki" (groves) in the steppe-region. The deciphering of aerial photographs and the establishment of the dynamic trends based on certain indirect characteristics (landscape features) can be regarded as a particular instance of the use of the landscape method.

5) During the recent years in the studies of the successions of vegetation the methods of *bio-morphological investigations of plants* in plant communities are used more extensively. Such investigations are undertaken usually in several communities representing the consecutive stages of a succession. Including a great variety of bio-morphological characters, such as the number of assimilating and the size of generative shoots, the number of flowers, the structure of the biomass etc. They afford some objective data on the

changes taking place in the vegetational cover under the influence of grazing, fire etc. Of particular interest is the study of the age composition of the population of the dominant species in the dry-steppe pastures of the Central Tien-Shan Mountains (TRULEVICH 1962, 1966). The object of these investigations was the effect of grazing on the vegetation. The studies were carried out on 15—20 1 m^2 plots in each stage of the successional series, investigated with the method of RABOTNOV (1950a, 1950b).

7 THE USE OF THE AMERICAN GENERAL LAND OFFICE SURVEY IN SYNDYNAMICAL VEGETATION ANALYSIS

F. Stearns

Contents

7 THE USE OF THE AMERICAN GENERAL LAND OFFICE SURVEY IN SYNDYNAMICAL VEGETATION ANALYSIS

7.1 Introduction

The North American continent was settled in a relatively short time by active and energetic people who in the process drastically altered the original native vegetation. Over most of the continent few fragments of the pre-settlement forests remain. These few are widely scattered and of limited value for ecological interpretation. Even before European settlement the American Indian was responsible for changes in vegetation, particularly through the use of fire.

The records of the General Land Office survey have proven an invaluable source of information on the original vegetative cover before European settlement. The surveys provide qualitative information in the form of notes made by each surveyor on the vegetation and agricultural suitability of the land. Quantitative information consists of data on the recorded bearing or witness trees.

The surveys began in Ohio in 1785 and progressed slowly westward in response to the demand for land and need for a framework on which to record land ownership. The history and procedures of the survey have been summarized in an excellent review by BOURDO (1956) and are described in detail by STEWART (1935).

The surveyors' records did not describe merely undisturbed vegetation. Disturbance and catastrophe were common before Western settlement and in fact even before advent of the American Indian. Vegetation recorded in the surveyors' notes included old growth forest, grassland, windfallen timber, swamps and a great variety of successional communities.

7.2 The Land Survey

The Ordinance of 1785 established the land survey to divide the public lands of the United States into uniform townships six miles on each side for sale to settlers. Each township was further divided into 36 sections, each one mile square (2.59 km²). As the survey progressed, baselines and principle meridians were established in each state and the survey proceeded from these coordinates. Meridians (north-south lines) and baselines (east-west lines) were supplemented by parallels or correction lines (Fig. 1). The ex-

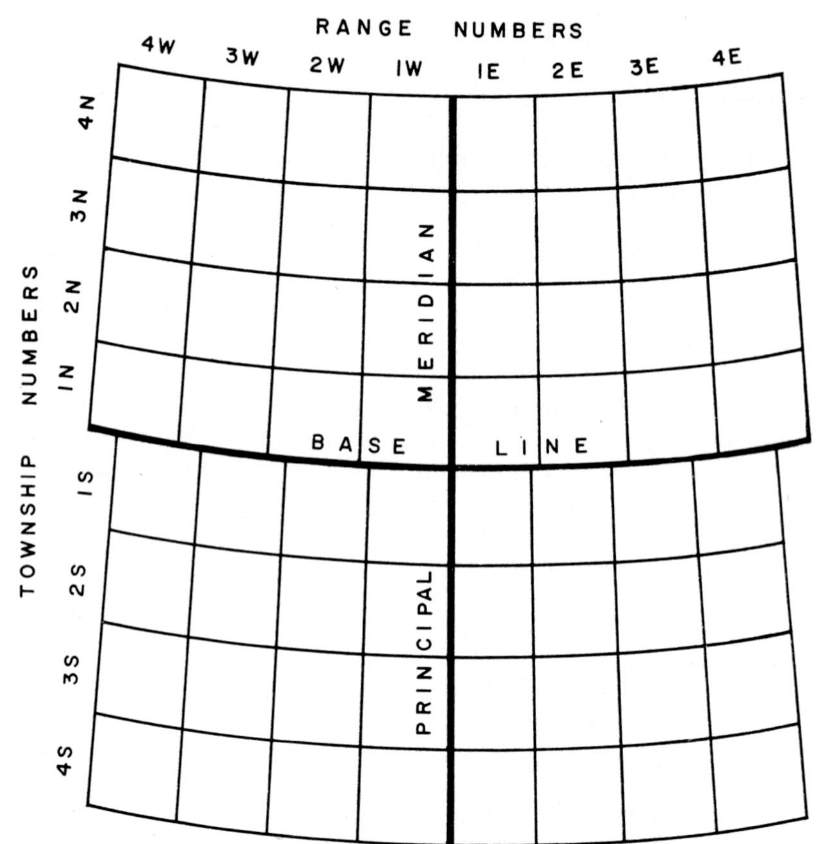

Fig. 1. Arrangement of base line, principal meridian and standard parallels showing numbering of townships.

terior or township lines were surveyed first and corners were marked at one mile (1.609 km.) intervals. Later the interior lines, marking out individual sections were surveyed. The surveyor located, by compass or transit and tape or chain, the placement of the corner or intersection of four section lines. Each corner was marked with a post or monument and further located by blazing nearby trees which were designated as witness trees. The location of the corner was scribed or cut on each witness tree. At least two, but more often four, witness trees were selected in different quadrants at each section corner. Two trees were used as witness trees at each quarter corner, i.e. the halfway point between section

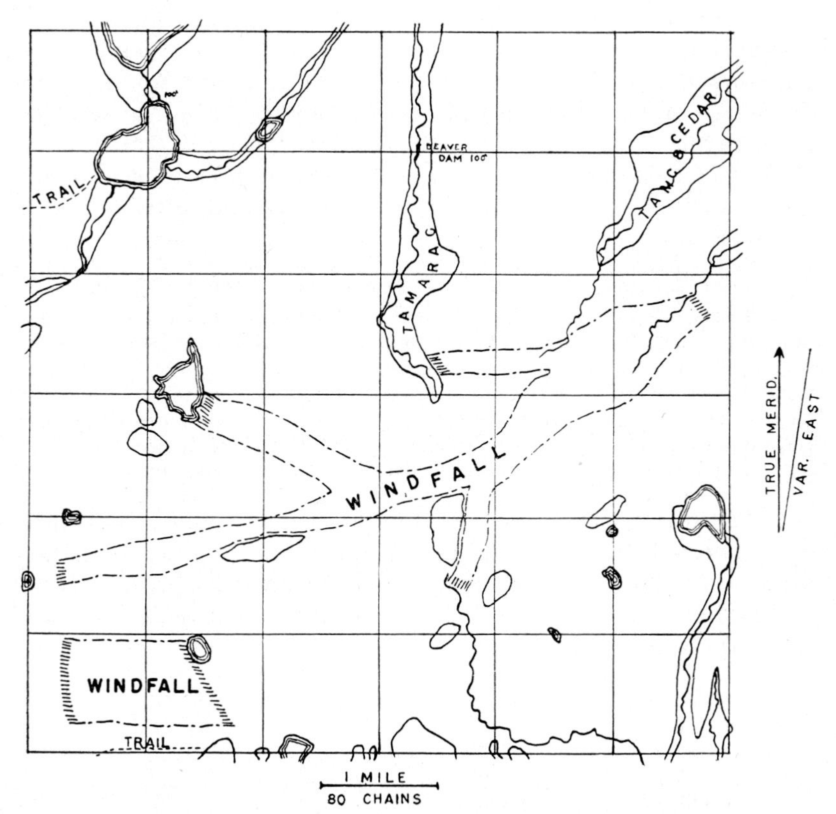

Fig. 2. Township 35 N., Range 13 E., as mapped in 1859 by Deputy Surveyor James McBride. Taken from the original on file in the Office of the Commissioners of Public Lands, Madison, Wisconsin. Reprinted by permission from Ecology 30: 355, (STEARNS 1949).

corners. A sketch map of each township was included (Fig. 2). The survey notes include the species, diameter and distance and direction from the corner for each witness tree and comment on soils, vegetation and other features as illustrated in the following excerpts:

> A line in the jack pine barrens of northwestern Wisconsin in Township 46 North, Range 10 West "... going north between sections 22 and 23, in August, 1854"
>
> "40:00 Set quarter section post
> Black Pine 6 N 34 E 22 links
> Black Pine 6 S 43 W 6 links
>
> 80:00 Set post corner to Sections 14, 15, 22, and 23.
> Black Pine 10 N 45 W 1.40 links
> Yellow Pine 20 S 40 E 180 links
> Surface rolling. Soil 3rd rate Timber Pine-brush".

The above quoted by FASSETT (1944) states that forty chains (804 meters) from the last corner, a quarter section post was set and witnessed with two trees of *Pinus banksiana* each six inches in diameter; the distance and direction from the corner are given, for example, "6° North 34° East 22 links" (link = 0.66 feet or 0.201 meters). At 80 chains, i.e. one mile, a section corner was set and witnessed by a ten inch *Pinus banksiana* and a twenty inch *Pinus resinosa*.

A township description from the same general area (T. 46 N. R 9 W.):

"This is a township of barrens that is almost worthless for agricultural purposes, or anything else, as there is but very little timber in it, and that is scrubby Black Pine; and there is hardly a drop of water in the township, in fact now, except small ponds in the South end of it, the Prairie that I have noted on the West side; can hardly be called a Prairie, as no very great time has elapsed since it was covered with small Pine which has been blown down, and burned up, remnants of which still lie on the ground." (from FASSETT 1944).

The description below of a line in Dodge County, Wisconsin gives not only the quarter section and section corner trees but also those falling along the survey line in an oak opening (NEUSCHWANDER 1957). Burr oak is *Quercus macrocarpa* and black oak is *Quercus velutina*.

North between sections 27 and 28 at
30.00 left prairie
40.00 quarter section corner
    bur oak 12″ N 64° W 1.31
    bur oak 14″ N 17° E 1.81
49.08 bur oak 27″
59.52 bur oak 24″
79.78 black oak 14″
80.00 corner to sections 21, 22, 27 and 28
    black oak 18″ N 77° E 70
    black oak 19″ S 65° E 58
    black oak 14″ S 32° W 85 marked
    black oak 12″ S 17° W 79 marked
    Land level first rate with bur and black oak timber. Undergrowth of hazel, grass and weeds.

The surveyor was required to note the vegetation found along the line between corners and in later surveys those trees which fell on the line. He often recorded changes in vegetation, tornado paths, lakes, streams, swamps and similar features (Fig. 2). Some surveyors were excellent observers and provided voluminous notes. Other reports are terse, while a few are completely false. BOURDO (1956) notes that the investigator should be aware that instructions varied among surveys as did the foibles of the individual surveyors. Although a surveyor may have shown a preference for a certain size class or species of tree as a witness tree, and may have estimated tree diameter and sometimes even estimated the distance between the tree and the corner, nonetheless he often had little choice but to use the nearest stem. Biases in the placement of trees with respect to a particular quadrant can be determined by statistical analysis.

BOURDO (1956) states, "Although true random is not an attribute of the General Land Office survey, the survey records do constitute a sample of the vegetation, recorded on the spot, according to a predetermined plan. Valid qualitative analysis of the vegetation the surveyors saw is possible when the bias apparent in the methods they used is small or has no effect on the inferences drawn, or if the bias itself can be used to advantage. If a worker fails to investigate the degree of bias beforehand, however, his conclusions are liable to contain error that could easily have been avoided."

7.3 Applications of Land Survey Information

Several workers made use of the early metes and bounds surveys common in the American colonies prior to 1785. Others have found valuable vegetation data in notes of explorations and geological surveys. However, the methodically gathered data from the rectangular township and section system has proven most useful.

Data from the surveyors records have been used for several purposes, initially to prepare vegetation maps. Maps of presurvey vegetation were drawn intuitively based on the tree species recorded at section corners or along lines of travel. In some cases these maps were compared with those of current vegetation to substantiate speculation on vegetational change. In other instances, they were compared with residual undisturbed vegetation to indicate growth over a period of 90 to 100 years. The data have also been used to prepare large scale soils maps based on correlation of vegetation types with soils.

Maps representing the original vegetation as it existed at the time of the land survey have been developed for large and small areas by many workers including DAVIS (1907), SEARS (1925), KENOYER (1929, 1933, 1939, 1942), LUTZ (1930), MARSCHNER (1930), SHANKS (1937, 1938), FASSETT (1944), ELLARSON (1949), POTZGER, POTZGER & MCCORMICK (1956), NEUSCHWANDER (1957), CURTIS (1959), LINDSEY, CRANKSHAW & QADIR (1965), MCANDREWS (1966) and finally JONES & PATTON (1966).

ROHR & POTZGER (1950), VEATCH (1953), and CRANKSHAW, QADIR & LINDSEY (1965) and others have examined the relationships between pre-settlement vegetation and soils and have developed soil maps. Others have used the survey information to assist in the interpretation of successional patterns and forest history. LUTZ (1930) made use of the DALE survey to compare the original (sensu 1800) forest with a modern stand in Pennsylvania, a procedure also used by STEARNS (1949) in northern Wisconsin, and by BURGESS (1964) in the forest-prairie border of North Dakota. THOMPSON (1940) employed survey records to locate prairie relicts in central Wisconsin. JONES & PATTON (1966) used the presence of trees or of grassland vegetation at section corners to demonstrate a pre-settlement black-belt in Alabama. BROWN (1950) suggested that, although the vegetation of central Wisconsin had not changed in composition since the 1850's, the density of the forest had increased. COTTAM (1949) made detailed and quantitative use of survey data by calculating distance between pairs of witness trees to determine density, dominance and frequency of tree species. His work resulted in the point quadrat method now used widely in ecological forest sampling (COTTAM & CURTIS 1949). LINDSEY and others have found survey records to be valuable in describing vegetation as an index of land value at the time of settlement. LINDSEY's work has been cited in legal disputes concerning the value of Indian claims (LINDSEY 1960). In Minnesota, SPURR (1954) utilized surveyors' records to interpret the relation of fire to the forests of Itasca Park. BOURDO (1956) compared stand composition as determined from land survey records with composition based on modern forest data for several large areas of hardwood forest in Michigan.

The reader will note that most work has been in the forested areas of the American middle west where the land survey usually preceded settlement. Little use has been made of the survey in the Great Plains where witness trees are few or non-existent or in the mountainous west where proportionately more area of undisturbed vegetation is available for study.

DIE BEDEUTUNG DER FORSTWIRTSCHAFTS-GESCHICHTE IN MITTELEUROPA UND FRANKREICH FÜR DIE SYNDYNAMISCHE VEGETATIONSKUNDE

F. Reinhold†

8 DIE BEDEUTUNG DER FORSTWIRTSCHAFTS-GESCHICHTE IN MITTELEUROPA UND FRANKREICH FÜR DIE SYNDYNAMISCHE VEGETATIONSKUNDE

Mit der Erstarkung eines biologisch ausgerichteten "Waldbaus" um die Jahrhundertwende gegenüber der Vorherrschaft der hauptsächlich, wenn auch nicht einseitig ökonomisch ausgerichteten "Forstbetriebseinrichtung" als Ordnungsplanung in Zeit und Raum, die Reinbestände (besonders der Nadelhölzer, aber auch von Buche und Eiche) im Hochwaldsystem zur Folge hatte, trat ein enger Kontakt zur damals blühenden Pflanzengeographie (DRUDE u.a.) ein, indem man die *"natürlichen" horizontalen Verbreitungsgebiete der einzelnen Nadelbaumarten* mittels forstgeschichtlicher Quellen zu umreissen versuchte (bes. DENGLER, VON WINDISCH-GRAETZ). Diese Forschungsrichtung wird bis heute fortgesetzt und auch detaillierter auf die standörtlichen Gegebenheiten bezogen (z.B. Neuentdeckung von autochthoner Kiefer im Ostmünsterland (HESMER 1958) und im Pfälzer Bergland).

Seit dem 16. Jahrhundert hatte sich ein grundsätzlicher Baumartenwechsel vollzogen, der aus folgenden Beispielen zu ersehen ist (in %, ohne Blössenangabe):

	Ki	Fi	Ta	Ei	Bu u.ä.	Bi u.ä.
1.) Land Sachsen						
1547/1591	7,5	21,2	26,1	9,3	25,8	10,1
1937	15,4	79,7	—	0,9	2,4	0,4
2.) Reichswald bei Nürnberg (Mitt. Staatsforstv. Bayern 37, 1968)						
1500	55	12	1	20	3	7
1959	82	8	1	3	1	4
3.) Nordrhein-Westfalen (nur 1883, 1937, HESMER 1958)						
1883	23	20		12	40	5
1937	20	39		16	23	2

(*Ki* = *Pinus sylvestris,* Fi = *Picea abies,* Ta = *Abies alba,* Ei = *Quercus petraea* et *robur,* Bu = *Fagus sylvatica,* Bi = *Betula pendula*)

Im ursprünglichen Laubwaldgebiet von Nordrhein-Westfalen ist ferner der Mittelwald von 17 % auf 2 %, der Niederwald von 40 % auf 23 % zurückgegangen, während z.B. in Frankreich Mittel- und Niederwald mit je rd. 30 % weiterhin vorherrschen (ein Ausfluss auch der abweichenden Besitzstruktur: 25 % Staatswald, je 37 % Kommunal- und Privatwald). Gewiss gab es auch umgekehrte, allerdings sehr seltene Entwicklungen (z.B. Sihlwald bei Zürich:

Mitte 16. Jahrhundert fast reiner Nadelmischwald, heute überwiegend Laubwald.)

Neben dieser *Erforschung* des Bestockungswandels auf Grund des forstlichen Unterlagenwandels trat die *der Mischungsverhältnisse* auf den einzelnen Forstorten, um Unterlagen für die Zusammensetzung der Baumschicht der natürlichen Waldgesellschaften zu erhalten, was nur mittels forstlicher Quellenforschung möglich ist. Hierfür liegen beispielsweise Karten für Kursachsen um 1550—1590 (Reinhold 1942), für Thüringen im 16. und 17. Jahrhundert (Jäger 1970) und den Schwäbisch-Fränkischen Wald ab 1650 (Jänichen 1956) vor. Diese Methode ermöglicht es, neue Kombinationen von Baumarten und Relikt-Vorkommen zu erkennen, z.B. Höhenkiefer-Tannen-(Fichten-)Wald ohne Buche im Vogtland, Ost-Thüringen und auf Buntsandstein im Südost-Schwarzwald, Relikt-Fichtenwälder in Nordost-Sachsen und im Elbsandstein-Gebirge.

Ein oft diskutiertes Problem, die Bedeutung des Buchenanteils in den Querceto-Carpineten, sei hier angeführt, da die starke Verbreitung von Hainbuche auf besseren Standorten ein Ergebnis der Niederwaldwirtschaft ist, die in den Laubwaldgebieten der niederen und mittleren Lagen in siedlungs- und gewerbenahen Bezirken schon ab ausgehendem Mittelalter die vorherrschende Betriebsart ist (in Frankreich mittelwaldartig mit Eiche überstellt), während in den Nadelwaldgebieten eine wilde "Blenderung" vorherrschte.

So muss es uns fraglich erscheinen, ob die seit dem 16. Jahrhundert stärker aussagefähigen forstlichen Akten uns wirklich den wahren "natürlichen Zustand" zu rekonstruieren gestatten. Denn diese Erhebungen, meistens durch "Amtsleute" z.T. mit Hilfe des Forstpersonals erstellt, waren ja gerade in den verschiedenen, soeben erst stabilisierten "Landesherrschaften" durch die mangelhafte Holzversorgung der "holzfressenden Gewerbe" (Bergbau, Salinen; in Frankreich auch Kriegs-, Marine- und Glasindustrie) veranlasst und wurden dann im 17. Jahrhundert zu Beginn der *autokratischen Staatsindustriewirtschaften* durch zahlreiche, nun durch Forstpersonal durchgeführte Beritt- oder Umrittberichte auf die einzelnen Forstbezirke detailliert und Grundlage der ersten, nicht nur der Brenn- und Bauholzversorgung der ländlichen und städtischen Bevölkerung dienenden Nachhaltsplanungsversuche [z.B. formuliert in der sächs. Generalbestallung von 1575 (Cod. August. 521): "Die Gehölze pfleglich und also angegriffen, dass Uns eine währende Nutzung, den Untertanen aber eine beharrliche Hülfe bleibe"]. Es ist dies die Zeit der mitteleuropäischen Forstordnungen der Flächenstaaten und der französischen Ordonnanzen (Colbert), die anstelle der (Grimmschen und Österreichischen) Weisthümer traten und im wesentlichen das "schlagweise" Hauen und den Schutz der fruchttragenden Bäume (also nicht Ulme, Ahorn) vorschrieben, Bodenstreunutzung verboten (z.B. Bayern, Salzburg) sowie Harz- und Pechnutzung (z.B. Bayern, Württemberg, Baden-Baden, Thüringer Wald) und Pottasche-Brennung für Glasfabrikation (z.B. Sachsen, Fichtelgebirge, Baden) einschränkten [Letzteres auf abgelegene Bezirke mit viel Fall- und Faulholz; während in Frankreich ganze "Glaserforsten gewidmet" ("affectées") wurden: insgesamt 197 Fabriken (PLAISSANCE 1961), im Norden, den Vogesen, in S. Frankreich und südlich der unteren Loire, die ausschliesslich auf Buchenschwachholzbasis mit kurzem Umtrieb arbeiteten]. Wie wenig aber die Forstordnungen effektiv waren, zeigte sich sehr bald, spätestens nach dem Dreissigjährigen Kriege, von dem ab die Walddevastierungen bis zum Ende des 18. Jahrhunderts in allen, den Industrien erreichbaren Waldungen grössten Umfang annehmen.

Zum Beispiel reichte das Einzugsgebiet der Saline Lüneburg bis nach Mecklenburg. Die Sole "wanderte" in Leitungen aus Holzröhren dem Brennholz nach, z.B. in Oberbayern von Reichenhall über Traunstein bis Rosenheim. In Frankreich gab es etwa 750 siderurgische Werke, besonders im Nordosten, Osten, Dauphiné, in den Ost-Pyrenäen, denen eigene Waldungen überlassen wurden; allein in Nivernais mit 138 Anlagen betrug der jährliche Bedarf 800 000 Stere (= Raummeter). In Deutschland wurden Bestände in der Oberpfalz zu Krüppelwald verwüstet. Entsprechende Beispiele lassen sich in grosser Zahl finden (z.B. HESMER 1958), wo selbst kleinste Landesherren dem Merkantilismus anhingen und so lokal die Waldungen verwüstet wurden.

Die kurze Übersicht soll zeigen, wie kritisch die forstlichen Quellen aus dem 16. bis zum 18. Jahrhundert hinsichtlich ihrer vegetationskundlichen Aussagekraft geprüft werden müssen. Aber gebietsweise ist schon für das 15. Jahrhundert Vorsicht am Platz, da zu dieser Zeit und vorher das natürliche Waldbild neben der Holzverkohlung durch *Waldweide* beeinträchtigt worden ist, z.T. auch durch mehr oder minder geordneten Waldfeldbau.

Das Recht der Waldweide, die nur bis zur Einführung des Kartoffelanbaus, der Kleesaat und der Stalleinstellung von grösserer Bedeutung war, dann aber gebietsweise von dem Recht auf Streunutzung abgelöst wurde, die eine verheerende Wirkung auf den Boden hatte, wie auch und schon vorher der Plaggenhieb in NW-Deutschland mit Verheidung als Folge, ist ein Ausfluss der Ordnung innerhalb der Organisation der Markgenossenschaft; in den landesherrlichen, auch gutsherrlichen Waldungen gewann die Mastnutzung, scheinbar merkwürdigerweise, erst einen grösseren Umfang, als nach dem 30-jährigen Krieg die Grundherren auf jeden zusätzlichen Geldbezug angewiesen waren und entgeltliche

Berechtigungen vergaben. Es ist also stets zu prüfen, *welche Besitzrechtsform* in einem bestimmten Waldbezirk vorlag. Es kann die frühere Sozialstruktur der berechtigten Bevölkerung auch heute noch, lokal je nach Stabilität der Standorte feststellbare Unterschiede an Bestand und Boden zur Folge gehabt haben; z.B. entwickelten sich rein landes- und gutsherrlich "reservierte" Waldungen (und deren Böden) im Hochwaldsystem ganz anders als benachbarte reine markgenossenschaftliche Waldungen, die im Lauf der Zeit zu Niederwäldern, besonders Stockausschlagwäldern absanken. Aber zwischen beiden Extremen gibt es alle Übergänge. Das trifft für Baumartenzusammensetzung, Holzqualität, Verjüngungsfähigkeit wie auch Bodenvegetation und Bodenzustand zu (Beispiele bei Hesmer 1958). Diese differenzierte Forschung scheint für künftige kombinierte vegetations-, standortskundliche und forstgeschichtliche Erkundungen von besonderer Bedeutung zu sein, was der Forderung K. Mantels im Programm (Punkt 6) der Sektion Forstgeschichte der IUFRO von 1963 entspricht: "Geschichte der Veränderungen des Waldes unter menschlichem Einfluss, . . . wie die Waldgebiete sich unter dem *Einfluss von Rechtsordnung, Agrarverfassung, wirtschaftlichen Ansprüchen und forstlicher Nutzungstechnik* entwickelt haben".

Die wirtschaftsbedingten Ursachen der Änderungen der Waldvegetation sind mit Hesmer (1958) in folgendem Katalog zusammenzufassen: "Brennen, Holzschlag, Streunutzung und Plaggenhieb, Vieheintrieb und landwirtschaftliche Zwischennutzungen", zu denen noch die Zeidelweide (klassisch im Nürnberger Reichswald; in Ostpreussen sogar bis in unser Jahrhundert), die Kriegseinflüsse (z.B. Pfalz, Ostpreussen; totale Waldverwüstung in England in der Zeit der Kämpfe Karls I. mit dem Parlament und zu Cromwells Zeit; 1. Weltkrieg in Frankreich) sowie die Jagd hinzukommen (in Deutschland waldkonservierend: Spessart als Musterbeispiel; in Frankreich dagegen öfters waldfeindlich: z.B. Fontainebleau, Rambouillet mit den "tirés" = Niederwald, auf 1—1,50 m "abgeschoren", um darüberschiessen zu können, mit vielen Anstellschneisen).

Die Anzahl und Ergiebigkeit der forstgeschichtlichen Quellen sind in den einzelnen Jahrhunderten sehr unterschiedlich. Die mittelalterlichen "Weisthümer" sind wenig ergiebig und enthalten nur wenig Baumartenangaben.

Aufschlussreich sind die Inventuren des 16. u. 17. Jahrhunderts mit zahlreichen Aktenunterlagen über erste Beschreibungen der einzelnen Forstbezirke. Kritik ist am Platz: z.B. Bedeutung des Wortes "Tanne" je nach Gegend für alle drei, in Deutschland einheimischen, forstlich wesentlichen Nadelbaum-Arten. Dazu kommen Grenzbeschreibungen mit Markbäumen; hierbei ist allerdings fraglich, inwieweit diese für die Holzarten-Zusammensetzung repräsentativ sind. Das gilt auch für die ersten Landeskarten mit Holzarten-Signaturen. Wesentlich ist die Zunahme von Einzelbeschreibungen (z.B. Beritt-Berichte) und von Angaben in Feststellungen von Forstberechtigungen, Akten über Besitzaufteilungen und Holzverkaufslisten (besonders für Holzhöfe an Floss-Strassen).

Ab Mitte des 18. Jahrhundert liegen Waldbeschreibungen mit Altersklassenaufteilung, Massenschätzungen, Bodenbeurteilung, Bodenvegetation vor, die die Grundlagen für die ersten Forsttaxationen zu Ende des Jahrhunderts und Beginn des 19. Jahrhunderts werden. Um diese Zeit (etwa ab 1750) erfolgen systematische Kulturnachweisungen über Nadelholzeinbringung, die vereinzelt bis 1368 zurückgeht (erste Kiefernsaat in Nürnberg; später in Frankfurt, Pfalz, Baden, Hessen, Brandenburg, Mecklenburg, Holland u.a.; "neu erfunden" in Thüringen 1564) und Nachweisungen über Samenverkäufe (besonders wichtig für Herkünfte der Lärche, aber auch bestimmter Fichten- und Kiefernprovenienzen).

Ab beginnendem 19. Jahrhundert liegen periodische Forsteinrichtungswerke vor mit Beurteilung der einzelnen Holzarten, Bodenbeschreibung, Exposition und Inklination, dazu mit entsprechenden Holzarten-Altersklassen nach neuem Forstvermessungswerk und jährlichen Hiebs- und Kulturnachweisungen. (Bis heute gleiches System mit laufender Verbesserung, besonders über Ertragsleistung und Vorrat, aber auch Standorts- und Bodenvegetationsbeschreibung.)

Bei jeder vegetationskundlichen, insbesondere syndynamischen Beurteilung von Wäldern sollte die Bestockung des Vorbestandes und die Entstehung (künstlich oder natürlich) des jetzigen Bestandes, ferner auch die Meliorierung (Düngung, Entwässerung, Bodenbearbeitung u.a.) bekannt sein (Beispiel: Olberg 1943).

Zum Abschluß seien einige spezielle, forstgeschichtlich erfassbare Tatbestände genannt:

1.) Ertragsleistung über das ganze Bestandesleben hin: Der "*ertragsgeschichtliche Zuwachs*", als Mass-Stab für die *Trockensubstanzproduktion* bestimmter Standorte oder bestimmter Wald-, bzw. Forstvegetationseinheiten geeignet, ist nach Wirtschaftsfiguren getrennt (Abteilung, Unterabteilung, Jagen) erfassbar, besonders wenn Vorrat und Nutzung nach Baumholz (Baden, Schweiz) erhoben wurden. Langfristige Probeflächen der Versuchsanstalten erlauben Ableitungen der Ertragsdynamik bestimmter Standorte und Bestandestypen (Vegetationseinheiten). (Beispiel: Drescher 1965.)

2.) "*Fluktuationen*" *in der Bestandesmischung* innerhalb einer gleichen Behandlungsweise können abgeleitet werden: z.B. beim langfristigen Femelschlag im Fichten-Tannen-Buchenwald am Westabfall des Schwarzwalds Wechsel zwischen Grundbeständen von Buche, bzw. Tanne (HOCKENJOS 1956; im Hochschwarzwald nur modifizierte Verschiebungen, DRESCHER 1965).

3.) *Naturwaldreste* können nach deren Bestandesaufnahmen zu verschiedenen Zeiten analysiert werden: z.B. Eichen-Buchen-Naturwaldsfluktuation, bedingt durch verschiedene Langlebigkeit der Baumarten auf der Baar bei Donaueschingen (REINHOLD 1949).

4.) Mittels *Bohrspahnanalysen* kann die *Entwicklungsdynamik* bestimmter naturnaher Einheiten analysiert werden: z.B. Höhenkiefer-Tannen-Fichten-Wald des SO-Schwarzwalds auf Buntsandstein im langfristigen Femelschlag (KWASNITSCHKA 1955).

5.) Das Problem des *Fichtenvorstosses* in der "Waldbauzeit" (nach dem 15. Jahrhundert, nach HORNSTEIN 1958) wird untersucht: z.B. Hornisgrindegebiet im N-Schwarzwald (HAUSBURG 1967).

8.1 SUMMARY

The Importance of Forestry History in Central Europe and in France for the Syndynamical Vegetation Science

Since the 16th century, the tree species composition in forests of Central and West Europe is fundamentally influenced by economical activities in connexion with changes in agricultural methods and with the development of industry and forestry. Simultaneously, numbers of written and printed documents with more and more detailed records on forests were increasing; they afford studies on the course of changes in tree species composition and on its economical reasons. Historical work in forestry on the basis of these records is essential for the recognition of the natural characters of tree species composition on different sites and for the investigation of natural relic stands. But it is necessary to evaluate these records in a critical way.

Many examples of early industrial influences (e.g. mining activities, glass factories, salt-works factories) are mentioned. In agriculture, introduction of cultivation of potatoes, *Trifolium pratense* and other crops diminished or stopped the necessity of grazing and browsing of domestic animals in the forests. Agricultural use of forest litter became highly important (in many cases only temporarily).

Some facts detectable by means of studies in forestry history are (1) productivity in the whole developmental course of a forest stand; (2) fluctuations in the percentages of the tree species during long periods; (3) changes in structure and composition in natural forest reservations; (4) syndynamics of certain forest associations, additionally by means of stem bore analysis; (5) problems of increasing percentages of *Picea abies* since the 15th century.

9 CYCLIC SUCCESSIONS AND ECOSYSTEM APPROACHES IN VEGETATION DYNAMICS

R. KNAPP

Contents

9 CYCLIC SUCCESSIONS AND ECOSYSTEM APPROACHES IN VEGETATION DYNAMICS

9.1 Definition and Examples of Cyclic Succession

Seres regarded by many authors as most typical and "classic" should develop via a more or less great number of stages to a selfperpetual terminal equilibrium (climax community, permanent community = Dauergesellschaft). Only natural events of catastrophic dimensions (e.g. hurricanes, earth quakes, landslides etc.) or actions of man (e.g. felling of trees, artificial burning, ploughing etc.) should be reasons of destructions of these terminal communities. Such successional seres are also called "*linear*" or "*directional*".

"*Cyclic*" successions are distinguished from these seres by certain stages (often terminal stages) which are destroyed with certainty by any immanent properties of their plant composition, mainly by attributes of their dominant species. After destruction of these stages, the succession begins again with an initial stage or a certain intermediate stage.

Cyclic successions are highly important in certain forests. The work of Watt (1925, 1947) on beech forests (*Fagus sylvatica*) initiated an approach to these phenomena. In the gaps originating from death of old trees, light-demanding species can invade or spread temporarily (gap-phase). With regeneration of beech, nearly no herbs can live in the dense shade under the close-growing saplings. An herb layer with shade-tolerant species like *Oxalis acetosella* can develop when the canopy opens out somewhat. 10—20 years later best conditions develop for the majority of forest herbs and forest grasses, until the development of new gaps by death of dominant trees. Then, the cycle begins again with the gap-phase.

All these different phases and stages of a cyclic vegetational change are summarized also under the name "*regeneration complex*". Forests, bogs, and other plant communities with such cyclical regeneration are mostly more or less mosaics of different phases or stages.

Regeneration cycles and regeneration complexes are often highly obvious in certain tropical forest communities.

The composition of tropical and subtropical rain-forests by patches of different species combinations can be observed in many regions. This structure results from regenerational successions ("microseres" or "serules", Daubenmire 1968) in the gaps origi-

Fig. 1. Stages of a cyclic succession in montane beech forests (Fagion sylvaticae) of the conservation area on the Schafstein, Rhoen, Germany (832 m above sea-level, basalt). Anthropogenic influences have ceased in these forests since long periods. In the foreground: stage with high percentages of grasses and herbs spreading since the death of the large beech tree, whose decaying stump and branches are visible (gap phase). Behind it: mature old beech forest with herb layer consisting mainly of species moıe shade tolerant. The photograph is taken in late autumn; the branches of the beech trees have lost already most of their leaves. Or.

nating from death of old and tall emergent trees (AUBRÉVILLE 1938, RICHARDS 1952, JONES 1955). Critical studies and considerations revealed that this patchiness of rain-forests is also occurring on sites with primarily uniform conditions of soil, climate and relief (KERSHAW 1964, AUSTIN & GREIG-SMITH 1968, WILLIAMS et al. 1969b, WEBB et al. 1972, KNAPP 1973). Differences in microclimate, in properties of the upper substratum horizons and in other environmental attributes can be induced by influences of the different plant species composition of the patches. The areas of the patches are mainly dependent on the gap sizes, resulting from the extent of the trees whose death produces the gaps. In rain-forests in Queensland, the area of patches is 100—150 m² (WEBB et al. 1972). But in other regions, e.g. in certain African tropical rain-forests, these areas are much larger.

CAJANDER (1909, 1926) described clearly stand developments referring also to cyclic changes. He emphasized that his forest types (Waldtypen) include all regeneration stages and phases. Thus, the forest types in the definition of CAJANDER are attributable in some way to regeneration complexes.

On xerothermic sites of central and south-eastern Europe, cyclic vegetational changes were observed in connexion with polycormous spreading of certain perennial herbs and woody species (within the vegetational complexes and mosaics of Festuco-Brometea, Quercetalia pubescenti-petraeae and related plant communities, JAKUCS 1972). Polycorms are aggregations of sprouts connexed by (mainly subterranean) stolons belonging to the same species and (at least originally) to the same individual; examples of species forming such polycorms are *Prunus spinosa*, *Polygonatum odoratum*, *Brachypodium pinnatum* etc. In the centres of the polycormous aggregations, other species can get dominant by different reasons (e.g. alterations of microclimatic and edaphic conditions by influences of the polycorm-forming species). Eventually, the newly dominant species can again form a polycorm centrifugally spreading. In certain cases in connexion with progressive successions (starting from xerophilous grass and herb vegetation), a dense woodland patch can develop in the centre. In the course of its spreading, the dense woody canopy can be opening. These woodland openings have sometimes such an extreme microclimate, that highly drought resistant grass and herb vegetation can exist only; thus, the progressive succession just mentioned can start eventually again at these places.

Examples of cyclic successions were supposed often within a development of vegetation mosaics of certain peat bogs. Plant communities connexed with vigorously growing peat (e.g. communities with dominance of *Sphagnum rubellum*, *S. magellanicum*, *S. fuscum*) are replaced by vegetation units with dominant dwarf shrubs (e.g. *Vaccinium uliginosum*) in parallelity with decreasing availability of water (new reports OVERBECK 1950, ELLENBERG 1963, CASPARIE 1969). In consequence of retarded accumulation of peat in such dwarf shrub communities, neighboring areas get a relatively higher surface. Thus, by reason of improved water conditions, again the communities with the *Sphagnum* species mentioned can replace the dwarf shrub vegetation. But this type of cycles seems to be not so generally occurring and important as it was often assumed until recent times.

The existence of this type of cyclic successions in peat bogs got really doubtful in many areas. In certain cases, the constance of mosaics of Caricetum limosae and Erico-Sphagnion growing side by side during very long periods could be demonstrated (JAHNS 1969, CASPARIE 1969).

9.2 Relations between Cyclic Changes, Linear Successions, and Fluctuations

However, it remains disputable in many cases whether a successional sere is cyclic or linear. Certainly, many conifer forest associations and various shrub communities of winter rain areas (macchias, fynbos etc.) are so inflammable that they will be nearly inevitably destroyed by natural or man-made burns. But theoretically a further successional development or a selfperpetuating status can be assumed. Examples of such syndynamic processes and of more or less selfperpetuating stands can be found sometimes also in some places in consequence of many special investigations in this vegetation; but they are so rare that emphasizers of cyclic successions regard them often as exceptions.

In other cases, it can be problematic whether the forces destructive for a certain stage can be regarded as catastrophic or as inevitable by biological attributes of plant species. An example is the destruction of particular grassland stages by rodents attracted by the palatability of the species composing the plant community. It could be claimed that the rodent invasion is a catastrophic event for the vegetation. But it could be also emphasized the attraction of the rodents by attributes the plant species.

In prevalence of theoretical considerations in connexion with climax theories or related ideas, a tendency exists to neglect cyclic characters of syndynamical processes. All successions are preferently subordinated to linear seres culminating into terminal climax or similar vegetation in such theories.

On the other hand, emphasis on the actual status of vegetation demonstrates more and more the importance of cyclic processes. In recent times, it is increasingly realized for instance that action of indigenous animals, insects, and other small species as well as big mammals and spontaneous fires are important natural factors in ecosystems. These factors are most important causes releasing cyclic successions. The frequent neglection of these factors until few years ago in vegetation science or in case of fire influences their attachment to human (anthropogenic) factors were a primary reason in underestimation of the importance of cyclical nature in successional processes in earlier periods and often still in present times.

It can be also disputable whether cyclic changes should be subordinated under the conception of succession. This depends highly on the quantity of change involved. On the one end of this quantity gradient, the species composition within a cycle is not changing at all and even quantitative differences can be only detected by most sophisticated methods. This applies to nearly ideal climax-like communities. In such cases syndynamical processes are in no way obvious.

On the other gradient end in certain tropical rain forests for instance, a great number of stages has to be passed within a complete cycle on greater gaps originating from death of highly shading emergent trees. Centuries can be necessary in such forests until the terminal stage is attained again, primarily by reason of the long life span of the trees composing the late intermediate stages. Most authors agree to denominate such a long lasting sequence of stages a succession.

Cyclic changes were also suggested to be divided in *inter-community cycles* and *cycles within a community* (CHURCHILL & HANSON 1958, KERSHAW 1964, MORAVEC 1969). A problem of this division is the definition of the differences necessary to separate a certain community from other vegetation units. In case a cycle includes stages or phases differentiated qualitatively by species composition (e.g. by groups of differential species) it seems well referable to phenomena mostly regarded as successions.

This is well in conformity with definitions of BARKMAN (1958) and BRAUN-BLANQUET (1964). They define all syndynamical changes with qualitative species differentiation (i.e. stages or phases distinguished by differential species) as successions, whereas they regard syndynamical processes with only quantitative differences in the representation of species (differences in coverage, in frequency, in constance, in abundance etc.) as fluctuations. RABOTNOV adhers to the same definition on the basis of the ample Russian experience on such changes (p. 21). A considerable part of the cyclic changes is to be classified under fluctuations in consequence of this reasonable definition.

A division of periodical changes caused by environmental factors or by internal structure (CHURCHILL & HANSON 1958) seems to be highly diffuse in many cases. The interrelations and interferences between environmental and biotic factors are so complicated within a biocoenosis with inclusion of all the influences of animals and microbial populations that this theoretically seemingly clear division becomes often arbitrary in nature.

9.3 Considerations on Ecosystems Approaches in Vegetation Dynamics

Cyclic successions are appropriate to introduce an ecosystems approach into studies on vegetation dynamics. Studies on problems connexed with ecosystems (without using this term) started already in the beginning of this century and even earlier, but were concentrated often on lakes and other aquatic environments (limnology, THIENEMANN 1925, RUTTNER 1940). Scopes to other biotopes were mostly of a general or more theoretical nature (e.g. MOROZOV 1912, 1928). In recent years, ecosystem research was extended in a planned and detailed way to a great variety of habitats (e.g. DUVIGNEAUD 1961, 1967, WATT 1966, ODUM 1969, VAN DYNE 1969, 1972, HEAL 1970, PATIL, PIELOU & WATERS 1971, PATTEN 1971, 1972, ELLENBERG 1971, 1973, moreover also SUKACHEV & DYLIS 1964 using the term biogeocoenology).

The projects are emphasizing until now mainly the interactions within a single ecosystem. Such studies on a single ecosystem are meaningful to minor vegetational changes, e.g. cyclic or fluctuational changes not connexed with fundamental qualitative changes of species composition. Namely, the causal explanation of such changes seems to be highly promoted by ecosystem views, since they are dependent mostly on attributes of organisms or on fluctuations of weather conditions.

Successions connexed with fundamental quantitative changes of species composition, with disappearances or new occurrence of species in the course of the seres, are sequences of a more or less great number of ecosystems. Therefore, the understanding of such successions can be promoted by results of studies on the mutual interrelations of syndynamically connexed ecosystems.

The interactions between the organisms in a stable ecosystem afford permanent life possibilities for individuals of species which are components of the ecosystem concerned. However, the growth possibilities of an individual are mostly not maximal in natural ecosystems, but limited by various effects of associated organisms (competition etc.). These limitations are the price for the maintenance of an environment permanently appropriate for the species concerned. But important exceptions are existing, namely in certain terminal stages. E.g. trees in the highest layers of forests of terminal or sub-terminal stages can get gradually immense and maximal dimensions, e.g. *Sequoia sempervirens* and *Sequoiadendron giganteum* in North America, *Eucalyptus regnans* in Victoria, Australia, and certain specimens of *Abies alba* in Europe.

Accumulations of litter, toxic substances and other products, which could change fundamentally the equilibrium between the species, are decomposed by members of stable ecosystems. Moreover, the expansion of species, which would be a reason of extermination of associated species, is limited by organismic or physicochemical influences in stable ecosystems. All these attributes are characteristic for ecosystems, which are syndynamically terminal stages of successional seres viz. steady-state communities (Dauergesellschaften) and climax vegetation.

In the ecosystems of initial and intermediate stages, a high percentage of organismical interactions are also mutually supplementary. But in a more or less large percentage, organisms in these ecosystems change fundamentally in the course of their development the life conditions of certain associated species. This can occur by accumulation of products inhibitory and toxic to other organisms or by alteration of the physical environmental situation (e.g. by shading or excessive consumption of water). It can also occur positively (in the sense of ecosystem productivity): e.g. by enrichment of the substratum with plant nutrients (nitrogen in connexion with bacterial symbiosis etc.), certain species can grow better, and new species (often more pretentious and more productive) can invade the changed site, resulting ultimately in a new ecosystem.

Ideas on interactions in ecosystems can be attained by comparative studies and observations. But evidences for the actions of certain physico-chemical or organismical influences in ecosystems can be gained most conclusively by experimental work. By controlled alterations of certain physico-chemical factors or by exclusion of certain organisms and species, the ways of the actions can get apparent within the members of ecosystems and within the successional relations of different ecosystems. Such complex experimental work is done for instance in some projects of experimental plant sociology (e.g. KNAPP 1954, 1967, WOIKE 1958, BECHER 1963, BERAN 1964, PFLÜGER 1966, GOLF 1973). The effects of certain plant nutrients within syndynamical ecosystem changes become obvious in fertilizing experiments in pastures, meadows and related vegetation (e.g. KIRSTE & WALTER 1955, STÄHLIN 1959, KLAPP 1962, VAN BURG & ARNOLD 1966, GEERING, FREI & LANINI 1966, SPEIDEL 1966, KÜNZLI 1967, TYLER 1967, GOODMAN & PERKINS 1968, BODROGKÖZI & HARMATI 1969, KNAUER 1969, THURSTON 1969) and in forests (e.g. ZÖTTL 1964, SCHLÜTER 1966, BAULE & FRICKE 1967, LANZ 1969, HAUSSER 1971) under the condition that whole species combinations are surveyed and considered in these studies.

9.4 Computer Models and Numerical Methods in Ecosystem Studies Relevant for Vegetation Dynamics

New numerical methods and sophisticated computer programs (computer simulation) developing in connexion with ecosystem research and other system orientated projects could be used for predictions on the course of successions and of the species composition and interaction in future stages. Since the stages of successional seres are composed of various species, an intricate network of competitional and other mutual influences has to be considered in the computer programs, besides additional effects and changes. Therefore, these programs and models will be very complicated. As an example of an result surpassing facts already well proved by other methods, a computerized evidence of fluctuational changes in percentages of tree species (basal areas per plot) may be mentioned, consequential only to competitional relations, differences in longevity, in growth curvatures and reactions to main site factors of the various tree species (independent of weather fluctuations and periods of abundance of certain animals) (Botkin et al. 1972a, b, also Leak 1970, Bledsoe & van Dyne 1971).

In cases of fundamental changes in environmental factors during successions, models of regressions with dynamic scalars (e.g. Austin 1972, Fitzpatrick and Nix 1969) are applicable. These methods are preferently feasible in vegetation with plant communities dominated by one species. Examples of successions including such plant communities and simultaneously with fundamental environment changes are certain hydroseres (e.g. from aquatic plant communities to reed and sedge swamps and ultimately to woody swamp vegetation).

Computer models treating vegetational changes by heavier or lower grazing pressure of animals are developed in connexion with studies on grassland ecosystems (Goodall 1967, 1969, Bledsoe & Jameson 1969, van Dyne 1969b, 1972, Jones 1970, Milner 1972). The complexities of interactions in these vegetational changes and successions are often extremely high, since plant communities rich in species can be involved, and a great number of competitive and other mutual influences have to be considered consequently. Therefore, considerable difficulties are often existing in attempts of computer modelling (e.g. critical discussion by Milner 1972).

The numerical and computer-modelling methods elaborated for the dynamics of animal populations (e.g. Schwerdtfeger 1968, recent bibliographies: Schultz 1971, Conway & Murdie 1972) are important for the analysis of fluctuational changes in vegetation. These methods can be applied on population dynamics of dominant species essential for fluctuations of certain plant communities. Moreover, dynamical changes in animal populations are often reasons of vegetational fluctuations.

D. CYTOGENETICAL, COMPETITIONAL, ALLELOPATHIC AND SIMILAR CAUSES OF VEGETATION DYNAMICS

10 GENETICAL AND CYTOLOGICAL CONDITIONS FOR SYNDYNAMICAL VEGETATION CHANGES

R. KNAPP

Contents

10 GENETICAL AND CYTOLOGICAL CONDITIONS FOR SYNDYNAMICAL VEGETATION CHANGES

10.1 Genotype and Phenotype in Relation to Vegetation Change

A fundamental reason of vegetation changes is the differentiation of plant taxa in their physiological behavior concerning biotic and abiotic environmental factors. These differences are primarily manifestations of the genotype, particularly of certain genes or special combinations of genes. The specific genetical properties induce a particular physiological constitution of a plant taxon. The genetical processes (gene action, mutations etc.) in properties important for vegetation dynamics are apparently not fundamentally different from those in other attributes. This is evident in the many cases of well analyzed properties equally important for economic cultivation of crops and for survival or competitiveness in syndynamical processes. Such properties are studied in details by applied geneticists and by plant breeders in the purpose of development of crop varieties better adapted to modern cultivation methods and to higher productivity. Drought resistance, survival in extreme temperature, ability to withstand to parasites or virus infections, productivity of seeds may be mentioned as examples.

Additionally to the genotypic constitution, the phenotype is of immense importance in vegetation dynamics. By differences of phenotypic modifications induced by influences of associated plants and directly or indirectly by abiotic environmental conditions, plants with the same genotype can be highly different.

Additionally, lasting phenotypical influences (Dauermodifikationen) may be of considerable influence. In experiments with controlled conditions (e.g. in phytotrons), obviously properties of parent plants acquired phenotypically under certain environmental conditions can be transferred to the next generations in a lower degree. The effect is decreasing from generation to generation in the way typical for permanent modifications (Dauermodifikationen) (Knapp 1956a, 1962b, Stearns 1960).

These results can contribute to the explanation of some observations on deviating properties of progenies of more or less homozygote crop varieties cultivated in climatically different areas (Boguslawski 1961, Quinby, Reitz & Laude 1962, Weinberger 1965). Also these deviations cannot always be totally explained by genotypic differentiations.

10.2 Genotypic Pattern of Plant Populations in Relation to Successions

Successional vegetation changes often are mainly the reason of shifting dominance of species. Species represented only in small percentages in preceding stages can become dominant in later parts of successional seres or vice versa. Dominants of pioneer stages originate also occasionally from single plant individuals in terminal communities. E.g., immense numbers of plants of *Betula pendula* in pioneer tree stages in Central Europe can develop from seeds shed from few birches or even a single birch tree persisting in climax-like forests with dominant oaks (*Quercus robur*, *petraea*) and beech (*Fagus sylvatica*).

A parallel behavior can occur in the genotypic pattern of species populations within stands of plant communities. A well documented case is the behavior of *Agrostis tenuis* in pastures and on heavy metal soils in Great Britain (Bradshaw 1971, McNeilly 1968, McNeilly & Antonovics 1968). The populations of *Agrostis tenuis* on heavy metal sites are mainly or nearly exclusively composed by a genotype highly resistant to adverse influences of Zn and other heavy metal ions. This genotype is differentiated also by earlier flowering (McNeilly & Bradshaw 1968) and an higher self-fertility rate (Antonovics 1968). Other genotypes of *Agrostis tenuis* are dominant in pastures on rich productive soils. They suppress the heavy metal resistant genotype on good pasture sites, being more competitive by higher vigor of growth. But the heavy metal genotype is not totally exterminated on good pasture sites in consequence of this competition effect; considerable numbers of its seeds can germinate and a small percentage of its plants can even develop flowers and fruits. In case a new heavy metal site is created in areas of good pasture sites by mining deposits, seeds of these few plants can be the source of a rapid establishment of a new heavy metal *Agrostis* stage.

Thus, species populations in stands of plant communities can be a mixture of various genotypes, differentiated only weakly in morphological properties, but well adapted to the special conditions of particular successional stages and site conditions. A genotype occurring sparsely can get dominant in the course of successional changes or on newly created pioneer stage sites.

10.3 The Importance of the Genetical Penetrance for Vegetational Changes

The genetical term "penetrance" describes the quantity of

phenotypic realization of an attribute potentially induced by a gene. The penetrance of genes for attributes important for syndynamical behavior can be highly dependent on environmental conditions (e.g. Bateman 1959, Stern 1959). Since the environmental conditions become very different during the course of vegetational changes and in various successional stages, great morphological and physiological differences in plants of the same species may be induced often by alterations of the penetrance of particular genes. Plants highly different for this reason are uniform genotypically. Thus, differences in the penetrance can be an important reason for the adaptability and for the variability of genotypically uniform plant populations in successional seres.

10.4 Cytological and Other Features of Therophytes in Connexion with Syndynamical Behavior

The percentage of species well adapted to initial successional stages is higher in therophytes than in most of the other life form classes. Therefore, many cytological and reproductive attributes of therophytes probably are connexed with syndynamic behavior and promote adaptation to the special situations in initial stages.

Tendencies can be stated in therophytes reducing possibilities of hybridization by differentiation of the chromosome apparatus. Different chromosome numbers, dysploidy, structural cytological changes, pre- and postzygotic barriers are effective (Ehrendorfer 1970). The genetical flexibility and variousness is limited by this way; the heterozygosity of therophytes is often reduced in consequence. Cytotaxonomic work on *Crepis* (Babcock 1947), *Clarkia* (Lewis & Raven 1958 et al.), *Viola* (Schmidt 1964), *Helianthus* (Heiser 1965), *Spergularia* (Ratter 1969), *Vicia* (Hanelt & Mettin 1970) and *Dipsacaceae* (Ehrendorfer 1965) are examples of studies highly informative on therophytes.

Many therophytes produce enormous numbers of seeds, have most effective dispersal mechanisms, and germinate rapidly in conditions favorable for growth of the species concerned. Versability and survival is promoted often by the ability of extreme phenotypic modification and of the production of germinable seeds also in adverse conditions (Knapp 1957a, c). All these properties can be deciding for prevalence of therophytes in initial successional stages.

10.5 Cytological Specialities of Perennial Herbs and Grasses of Early Successional Stages

Perennial herbs and grasses (mainly hemicryptophytes and

geophytes) are not uniform in their syndynamical behavior. Many species of these groups are constituents of terminal stages (e.g. species of herb layers in many forests). Other species, mainly certain hemicryptophytes, are most prominent in early intermediate stages or even in initial stages. Recently EHRENDORFER (1970) has compiled cytogenetic patterns of perennials which in parts refer particularly to early settlers and colonizing groups:

(1) Polyploids: Divergent from diploids, more or less isolated by postzygotic barriers, mostly of hybrid derivation and often connexed by further hybridization. Examples in *Agropyron*, *Anthoxanthum*, *Ranunculus* subg. *Batrachium*, *Fragaria*, *Lotus corniculatus* agg., *Epilobium angustifolium*, *Mentha*, *Galium mollugo* agg., *Campanula* sect. *Heterophyllae*, *Leucanthemum* etc.

(2) Agmatoploids: Holo- or polycentric chromosomes, consequently strong tendency for fragmentation and for numerical change, often coupled with polyploidy. Examples in *Juncaceae* (e.g. *Luzula campestris* agg.) and *Cyperaceae* (e.g. *Eleocharis*, *Carex*).

(3) Agamic groups: Anortho- and aneuploid hybrid derivatives more or less apomictic, new apomicts as result of occasional hybridization. (Diploids of these groups are mostly sexual and more or less isolated.) Examples in *Poa*, *Bothriochloininae*, *Potentilla* "*verna*" agg., *Alchemilla*, *Hieracium*, *Taraxacum*, *Antennaria*, etc.

Taxa highly competitive and important in early stages of anthropogenic successions are apparently often of rather recent hybrid origin (partially amphiploids). In some cases, this origin is well studied, e.g. in *Tragopogon* in Northwestern North America (OWNBEY 1950, STEBBINS 1971), in a taxon intermediate between *Viola alba* and *V. odorata* in California (BAKER 1972), and in weedy *Sorghum* forms (QUIMBY et al. 1958, BAKER 1972).

10.6 Cytological and Reproductive Specialities of Temperate Phanerophytes in Connexion with Differences in Syndynamical Behavior

The temperate phanerophytes (trees and shrubs) can be divided in connexion with syndynamic properties in two groups: (1) The majority of these tree species occur in later or terminal stages of successions and in climax vegetation. A part of the species (forming a special subgroup) can establish already in early successional stages (e.g. some *Pinus* species). (2) The species of the second group are prevalently confined to early and intermediate successional stages. With exception of extreme sites and of timberline vegetation, they are suppressed or eliminated mostly in terminal stages and in climax vegetation. These two groups are differentiated rather drastically in their cytological and reproductive specialities.

In the first group (e.g. *Quercus*, *Ulmus*, *Platanus*, *Tilia*, *Acer*, *Fraxinus*, *Picea*, *Abies*, *Pinus*) cytological differentiation in chromosome numbers and chromosome structure is low. Thus, crossing barriers are not frequent, and hybridization can occur easily. Wind pollination, wide dispersal of seeds and fruits, monoecious and dioecious flowers in many groups promote gene flow, recombination, heterozygosism and genetical polymorphism. Different taxa are the result of geographic isolation in high percentage and in some genera even in most cases. These taxa can mostly hybridize freely. Clear differences between species are breaking down in consequence often as soon as geographical or ecological isolation is ending.

On the other hand in the second group with species prevailing in early and intermediate stages, polyploid patterns of young origin (e.g. *Betula*, *Salix*, *Rhododendron*) or agamic complexes (e.g. *Rubus*, *Pomoideae*) or heterogamy (permanent hybridity based on anorthoploidy, e.g. in the *Rosa canina*-group) are prevalent (EHRENDORFER 1970).

10.7 Cytological Basis of Special Attributes of Tropical Successions

In tropical successions woody phanerophytes (shrubs, certain trees) are highly important mostly already in early stages of seres. The special syndynamical behavior is not only induced directly by tropical climate. Many tropical phanerophytes are apparently different from taxonomic groups of high latitude trees and shrubs (e.g. groups of *Pinaceae*, *Salicaceae*, *Fagaceae*, *Betulaceae*, *Aceraceae* etc.) in speciation, reproduction in the widest sense, and cytology. In parallelity with the richness in species, the range of chromosome numbers is greater in tropical taxonomical groups in average due partially to dysploidy and polyploidy (mostly ancient polyploidy).

Particularly in the humid tropical areas, hermaphrodite flower types are dominant, fruit or seed dispersal and ranges of pollination are prevalently limited, and breeding interchange more or less restricted to relatively few and scattered individuals. In consequence, gene flow and recombination rates are relatively low. But possibilities for genetical drift are rather important. Local populations tend to be relatively uniform and narrowly specialized. In consequence of this specialization, hybrids are mostly suppressed by competition. (Examples for special data and further considerations:

BODARD 1962, ROBYNS 1963, CUATRECASAS 1964, FEDOROW 1966, FARRON 1968, EHRENDORFER 1968, 1970, GHOSH 1968, ASHTON 1969, WHITEMORE 1969, VINK 1970).

Comparisons of African, European and North American areas imply that diversification rates are not alway greater in the tropics than in high latitude temperate areas (KNAPP 1973). The great number of species in certain tropical areas is apparently mainly a consequence of rather undisturbed evolutionary processes since ample geological periods and an expression of coexistence of many ancient and recently generated taxa for reasons mentioned above.

Fig. 1. A mosaic of various stages of regenerative successional seres in a tropical lowland rain-forest area near Kumba, Cameroon. Foreground: A shrub stage with banana plants as remnants of former cultivation (successions from shrub stages of Rauvolfio-Cnestidetalia to intermediate forest stages of Musangetalia smithii to sub-terminal forest stages of Lophiretalia alatae. Details on species composition of the vegetation units: KNAPP 1966b, 1973). Or.

The many species living in certain areas are specialized only partially to particular conditions of soil and macroclimate. The majority of the taxa often is adapted to certain syndynamical and stand structural conditions. Specialization on certain strata or layers and to various kinds of epiphytism may be mentioned as examples. For this reason the numbers of stages differentiated by characteristic species is much higher in most tropical humid areas than in other parts of the world. (Figures 1, 2.)

Fig. 2. Tree fern brake in the mountains of Eastern Puerto Rico, a stage highly differentiated physiognomically within the multiform successional seres of the Caribbean tropical montane rain-forest complex. Or.

10.8 Polyploidy and Vegetation Dynamics

Polyploidy was sometimes emphasized to be advantageous in certain successional stages. In various Centraleuropean areas, the percentage of diploids is high in initial stages. The percentage of polyploids is increasing in the following stages. But in terminal stages again, the percentage of diploids is raised.

This phenomenon is mainly connexed with life form composition and proportion of certain plant families in the different stages (KNAPP 1953b). Therophytes are generally diploid in a rather high percentage, whereas polyploids are often prevailing among hemicryptophytes. Phanerophytes are intermediate compared with the life form classes just mentioned. All these comparisons apply to species belonging to a certain particular family. Thus, the relation of diploids versus polyploids is already explained partially, since therophytes abound in initial stages, hemicryptophytes in intermediate stages, and phanerophytes in terminal stages.

Additionally, the percentage of polyploids is highly different in different families. It is high in grasses (*Gramineae*, *Poaceae*) for instance. Since grasses are most important in many intermediate stages, this special attribute of the *Gramineae* is a further reason of the high percentage of polyploids in the intermediate parts of successional seres.

Syndynamically advantageous species with multiplied genome numbers are obviously mainly allopolyploids (= heteropolyploids).

This is shown by experimental and cytotaxonomical work. Examples for allopolyploid species much more competitive and propagative in initial and intermediate successional stages compared with the diploid taxa concerned are *Poa annua* (2n = 28, derived from *Poa supina* Schrader, 2n = 14, and *Poa infirma* H.B.K., 2n = 14, Tutin 1954), *Arenaria serpyllifolia* (2n = 40, derived from *A. leptoclados* (Rchb.) Guss., 2n = 20, and *A. marschlinsii* Koch, 2n = 20, v. Woess 1941), and *Anthoxanthum odoratum* (2n = 20, derived from *A. alpinum*, 2n = 10, and *A. ovatum*, 2n = 10, Jones 1964).

11 MUTUAL INFLUENCES BETWEEN PLANTS, ALLELOPATHY, COMPETITION AND VEGETATION CHANGES

R. KNAPP

Contents

11 MUTUAL INFLUENCES BETWEEN PLANTS, ALLELOPATHY, COMPETITION AND VEGETATION CHANGES

11.1 Nature of Mutual Influences between Plants

The vegetation changes are not only caused directly by different reactions of the plant species on environmental factors. In case a plant species of a community is promoted by an environmental change, the possibilities of its better development are influenced by the associated plants. The expansion of an individual of a species favorized by a changed environmental factor is different

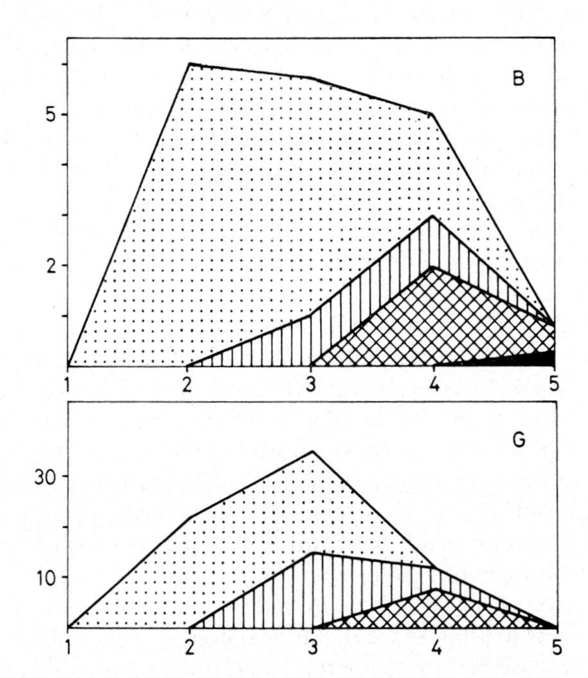

Figs. 1 and 2. Numbers of seedlings (Nr.) of different ages per 10 m² on areas (in the Taunus, Germany) clear-cut before periods of different length (abscissa: 1 = immediately after felling of the trees; 2 = after 6 months; 3 = after 18 months; 4 = after 44 monthes; 5 = after ten years). Ages of the seedlings: stippled = less than 1 year; vertically hatched = 1—2 years; cross hatched = 2—4 years; black = more than 4 years. Above: *Betula pendula* seedlings. Below: *Sarothamnus* (= *Cytisus*) *scoparius* seedlings. Young seedlings occur only in the first years after clear-cutting. Most of the young plants of *Betula* and *Sarothamnus* die within a few years, mainly by reason of competitional and partially also by allelopathic influences of associated plants. Or.

in plant communities as compared with isolated experimental cultivation. In extreme cases by influences of associated plants, the advantage of a changed environment can be totally obscured. Thus, in vegetation the environmental factors act mostly not directly on the individuals of a plant taxon; their effects are more or less modified by mutual influences between the plants.

11.2 Competition and Repression

In former times and also until today in many institutions, mutual influences are mostly studied under the concept of "competition" (Konkurrenz, Wettbewerb). This term implies that necessarily the advantage of a plant must be the consequence of a repression of other species or other individuals. In many cases, such competition models are really applicable. But it should not be neglected that in considerable numbers of cases also mutual promoting influences occur between species and individuals of plants. Additionally, mutual inhibiting influences can be observed in vegetation not referable to the concept of competition.

Competition and its implications have been treated in comprehensive considerations (e.g. Clements & Hanson 1929, Caputa 1948, Ellenberg 1953, 1954, Knapp 1953a, 1953c, 1955a, 1960, 1961a, 1964a, 1967a, d, 1971d, Harper 1961, Walter 1962, Bornkamm 1963, Donald 1963, Gigon 1971). One of the main objectives of these considerations is the difference between the reactions of a plant taxon on various conditions of soil and climate growing in pure stands or in monocultures versus living in plant communities of several species or in mixed cultures ("physiological behavior" versus "ecological behavior", Ellenberg 1953).

Mutual influences on which the term competition can be applied often occur most pronouncedly and most exclusively in intermediate successional stages.

The repression of other species and the consequential expansion of plants of the next stages can be induced in very various types. These can be classified in connexion with successional or fluctuational changes partially in the following way (G. & R. Knapp 1955):

1. *Overgrowth* (Überwachsung): Species attaining a greater height grow over the assimilation organs (leaves) of other plants; these are weakened consequentially by decreased possibility of photosynthesis. This effect can be attained by high herbs and grasses or more pronouncedly by shrubs and trees.

2. *Undergrowth* (Unterwachsung): In the course of the process of undergrowth, shade resistant species invade or occupy the bare

space near the soil surface in stands of relatively tall plants. The low species can develop by germination or can invade by means of runners (Ausläufer) from neighboring stands. The formerly dominant tall species can disappear by influence of root competition or by impossibility of development of seedlings in the shade of the dense stands of low plants. The process of undergrowth occurs rather often in successions on old fields and on road sides. For instance, high stands of communities with *Artemisia vulgaris* or *Melilotus albus* can be replaced by low grassland of *Poa pratensis* and *Festuca rubra* in the course of successional vegetation change.

3. *Interpenetration* (Durchwachsung): This type of expansion occurs in cases of presence of species with long runners or with ability of development of root sprouts. The runners or root sprouts penetrate into stands of other species (which can be very dense partially). During the first phases of interpenetration, only few stems of the expanding species appear sporadically in the invaded stands. Later on, these stems become more and more numerous and are expanding simultaneously repressing the plants of the former stage. Examples of species expanding under favorable conditions by this way are *Convolvulus arvensis*, *Petasites hybridus*, *Phragmites communis*, *Equisetum* div. spec., *Prunus spinosa*, *Populus tremula*, *Populus tremuloides* etc.

4. *Marginal repression* (seitliche Verdrängung): Expansion by marginal repression can be observed in species growing in ample bunches, cushions, mats, or other dense aggregations (e.g. *Galium odoratum*, *Convallaria majalis*, *Majanthemum* div. spec., *Mercurialis perennis*, *Dryas octopetala*). Marginal repression is effective by expansion of these aggregations and consequential starving and disappearance of the stems of other species with diminished competition power growing formerly on the invaded places.

5. *Outlasting* (Überdauerung): One of the most effectful ways of ultimate or later dominance of a species is a lifespan longer than those of the associated plants. Successional stages often are mainly the consequence of such differences in lifespans of species which actually germinated together in the period of the start of successional seres. Species of greater longevity outlast annual, biennial, and later also shortliving perennial plants. Another consequence of a kind of outlasting is the dominance of a species in early successional stages whose seeds remained in living state in the soil during long periods under the vegetational covers of later stages (also chapter 4.6). These seeds can germinate as soon as environmental conditions get favorable, for instance after destruction of dense vegetation by hurricanes, clearcuts, fires, ploughing etc.

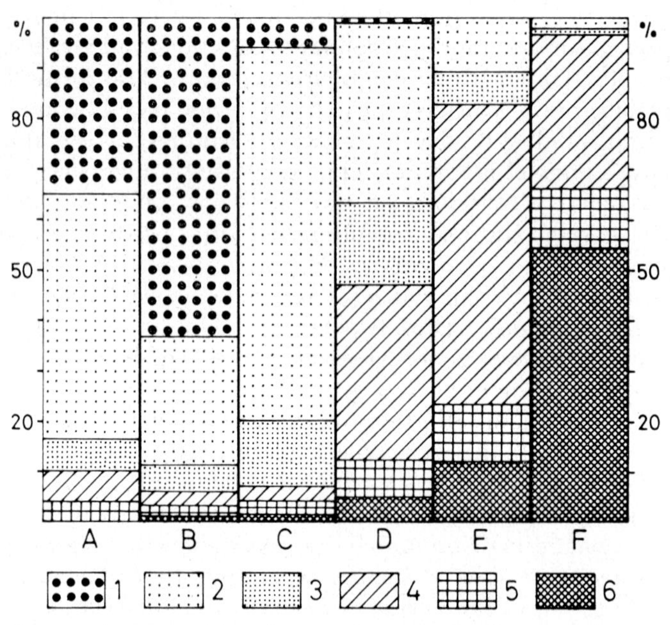

Fig. 3. Tree species composition of stages of a successional sere starting on cleared forest areas (moderately rich upland sites with "Braunerde" = brown forest soils) in the environments of Wetzlar, Germany (averages of 10 studied plots for each stage). A = two years after start of the sere; B = after 4 years; C = after 8 years; D = after 46—54 years; F = after 180—223 years (A = percentage of the numbers of individuals; B—F = percentages of surfaces covered by the particular species.) 1 = *Salix caprea*; 2 = *Betula pendula*; 3 = *Populus tremula*; 4 = *Quercus petraea* et *Qu. robur*; 5 = *Carpinus betulus*; 6 = *Fagus sylvatica* (sum of the data for these species in a particular stage = 100). The dominance is shifting from *Salix caprea* (in B) to *Betula pendula* (in C) to *Quercus* (in E) and finally to *Fagus sylvatica* (in F). But already after 4 years, all the tree species are present. The changes in dominance are caused by different growth rates and shade tolerance of the various species within this sere. Syntaxonomically, the succession developes (from B to F) from Sambuco-Salicion capreae to Carpinion and finally to communities of intermediate character between Fagion and Carpinion. Or.

11.3 Promoting Mutual Influences between Plants

Promoting mutual influences between plants often increase in importance and in percentage in the later and terminal stages of successions. The rich stratification and the abundance of species in these stages partially are explainable by such promoting influences.

Herb layers and strata of mosses in forests, shrub communities and in tall grassland mostly are dependent on influences of the plants with greater height. This is most obvious in plants avoiding sites with high light intensity and consequently dependent locally on the shade casted by tree layers and other high strata. But several other promoting effects can be effective. On the other hand, species of herb layers can have favorable influences on the forest ecosystems with inclusion of the trees.

The percentage of plant species and of plant individuals promoted by other plants is extremely high in many forests of the humid tropics. The promotion is so intensive that even direct or indirect dependence on other plants in nature is apparent in many species. The immense numbers of epiphytes and lianas, but also many low story trees and undershrubs belong to these manifold group of dependent plants. New experiments established the strict dependence of such species from the shade and other microclimatic conditions created by the high trees of the upper layers (KNAPP 1967d).

Examples of action of substances liberated from living plants promoting components of associated vegetation are effects in gibberellins (KNAPP 1961c, 1962a, CRUDEN 1969) and in some amino-acids (LINSKENS & KNAPP 1955).

11.4 Allelopathy and Related Influences

Most mutual influences and competition effects can be referred to modifications of various environmental factors of fundamental importance on plant growth, e.g. nutrient contents of the soil, light conditions, availability of water etc. Additionally, mutual influences originate from chemical substances exuded or leached from living parts of plants (e.g. roots, leaves etc.). The term allelopathy was introduced for such influences by MOLISCH (1937). These phenomena were studied in the last decades in an extensive and critical way (e.g. BODE 1940, BONNER & GALSTON 1944, BONNER 1950, KNAPP 1953a, 1954, 1960, 1967a, d, MULLER 1953, 1965, 1969, GRÜMMER 1955, BÖRNER 1956, 1960, RADEMACHER 1957, EVENARI 1961, RICE 1967, GRODZINSKIJ 1967, WHITTAKER 1970, references and discussion of earlier precursory work: KNAPP 1954, 1967a).

Other mutual chemical influences between plants can be exerted by substances liberated from dead parts, mostly decomposing leaf litter or roots (e.g. KNAPP 1953a, 1954, 1958, 1961b, 1966, 1967d, WINTER & SCHÖNBECK 1953, LOSSAINT 1959, LEMÉE & BICHAUT 1971). These influences are not to be included in the term allelopathy sensu stricto. But in practice, the separation of influences by dead plant material and by living parts of plants is sometimes difficult.

Mutual chemical influences between higher plants can act directly or via microorganisms of the soil. For instance, substances from higher plant species may primarily inhibit nitrogen fixing microorganisms (KNAPP & LIETH 1952, PARKS & RICE 1969, BLUM & RICE 1969, BORMANN 1969). Secondarily, the growth of associated higher plants can be retarded by consequential nitrogen deficiency. Thus, the disappearance of nitrophilous species from abandoned cultivation areas is accelerated apparently by substances from plant species of the early stages of old-field successional seres which inhibit the development of nitrogen-fixing bacteria and *Cyanophyceae* (RICE 1965, RICE & PARENTI 1967, also BORMANN 1969).

The effect of substances exuded from parts of evergreen sclerophyllous shrubs living in winterrain regions is an example of apparently direct allelopathic effects between higher plants. These substances inhibit the therophytes represented in large numbers of individuals and species in preceding stages (e.g. on recently burned areas). Thus, nearly no therophytes grow often in dense communities of these sclerophyllous shrubs, and the differentiation of species composition is highly accentuated in successional series (in the Mediterranean area: DELEUIL 1950, 1951; in California: NAVEH 1961, MULLER 1965, MULLER & DEL MORAL 1966, HANAWALT 1969, MCPHERSON & MULLER 1969).

The quantity of chemical mutual influences is dependent on other environmental factors. Colloidical properties of the soil (quantity of interior adsorptive surface etc.) and simultaneous presence of other substances are highly modifying allelopathic effects for instance (KNAPP & FURTHMANN 1954, KNAPP 1960). Therefore, experimental results under certain laboratory or field conditions can be generalized only very critically. But causal explanations of dynamical processes in successions are remarkably promoted by combined laboratory and field work on allelopathy (e.g. KNAPP 1963).

Allelopathic and related chemical effects act often synergistic with other factors. This is very obvious in modification of the equilibrium between grassland and forest vegetation (KNAPP 1965a, 1969a, 1973). For instance, effects of phenolic compounds from litter of the grass *Festuca rubra* and oxygen deficiencies in the upper soil horizons inhibit simultaneously and synergistically germination and growth of woody plants in regeneration stages of deciduous forest in the West German hill country (KNAPP n.p.).

Allelopathy and influences of chemical substances from dead

parts of plants are effective apparently in nearly all stages of successions and fluctuations. But obviously, allelopathic effects are most important in intermediate successional stages with dominant grasses or perennial herbs. Many dominants of such stages were found inhibiting associated plants by allelopathic influences in experimental work (e.g. *Bromus inermis*, *Agropyron repens*, *Aristida*, *Melilotus*, *Artemisia*, *Helianthus*).

Species of these genera can be important in the course of successions on old fields in eastern and central United States for instance. The development of certain stages of these successions is influenced by substances from *Helianthus annuus*. These substances inhibit species of the preceding stages, but also seedlings of the next generation of *Helianthus annuus* (autotoxy). *Aristida oligantha* dominant in the next stage is resistant against the inhibiting substances from *Helianthus*. Thus, the dominance of *Aristida oligantha* in the next stage is promoted indirectly by creation of its superiority over the competitors weakened by inhibiting substances from *Helianthus* (PARENTI & RICE 1969, WILSON 1969).

11.5 Modifications of Mutual Influences between Plants and of Syndynamical Equilibria by Indigenous Animals

Many earlier assumptions and considerations on successional possibilities neglect the influences of indigenous animals on vegetation. The indigenous animals are members of the biocenosis or of the ecosystems and have to be included consequently in the natural factors determinative on syndynamical processes.

Fig. 4. Grassland vegetation (with *Pennisetum mezianum*, *P. massaicum*, *Themeda triandra* etc., Penniseto-Themedetalia) in central Kenya, promoted by influences of indigenous wild ungulates; visible on the figure: zebra (*Equus burchellii* resp. *boehmii*) and kongoni (*Alcelaphus buselaphus cokii*). In the background (dark tints), Acacia woodland (dominant *Acacia xanthophloea*, Carisso-Acacietalia) on sites with more water available. (More explanations KNAPP 1965d, 1973). Or.

Local enrichment of nitrogen in soil (e.g. in the environment of rest and nest places) and creation of spots of bare, highly porous soil (excavations by burrowing animals) can induce vegetational changes and successions with stages similar in species composition with communities anthropogenic in connexion with agriculture and pasture management (e.g. KNAPP 1959b, 1964b, 1971c, DINESMANN 1968, LAWS 1970, LAWTON 1971). Mainly in climatical and edaphical areas marginal for tree growth, the syndynamical competition equilibrium of forests versus grassland or savannah can be shifted in direction to the open formations with dominant gramineous species by influence of herbivorous indigenous animals (e.g. TALBOT 1965, KNAPP 1967b, 1969a, 1971a, c, 1972, DE VOS & JONES 1968). But also the development of species composition of tree canopies can often be decided by animal influences.

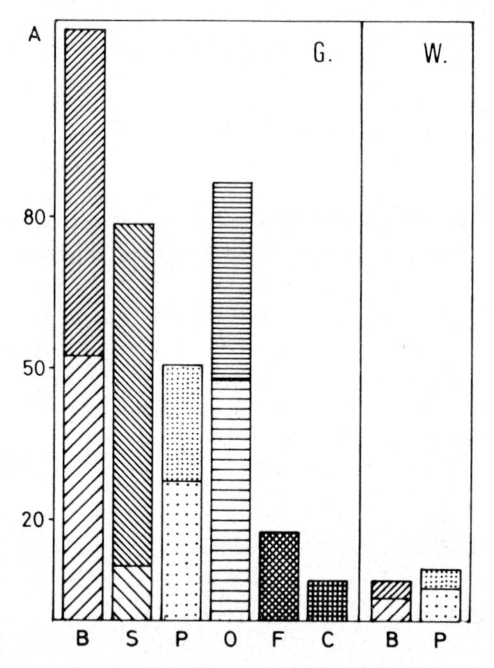

Fig. 5. Average numbers of young plants of tree species on plots of 100 m² with originally uniform site conditions. B = *Betula pendula*; S = *Sorbus aucuparia*; P = *Populus tremula*; O = *Quercus petraea*; F = *Fagus sylvatica*; C = *Carpinus betulus*. Columns left of the vertical line (G) = protected against game browsing; right of the vertical line (W) = exposed to influences of game (mainly *Carpreolus capreolus* and *Cervus elaphus*). Narrowly hatched or stippled (upper parts of the columns) = plants 3—4 years old; signatures more distant (lower parts of the columns) = plants 5—10 years old. Areas of studies in Hessen, Germany. Development of high layers with many individuals of several woody species in the protected plots; few scattered woody plants (of *Betula pendula* and *Populus tremula*) only in the plots exposed to game browsing. Or.

The stability of many plant communities is endangered by accumulation of layers of more or less decomposed litter and raw humus under interrelated influences of certain climates and plant species. These accumulations of dead organic material can be avoided by consumption of plant substances by herbivorous animals. For this reason, such plant associations can be stable permanent communities (Dauergesellschaften) only under the influence of indigenous herbivorous animals (e.g. WATT 1960, 1971, BARADZIEJ 1969, BORISOVA & POPOVA 1971, WELLS 1971, REMMERT 1972).

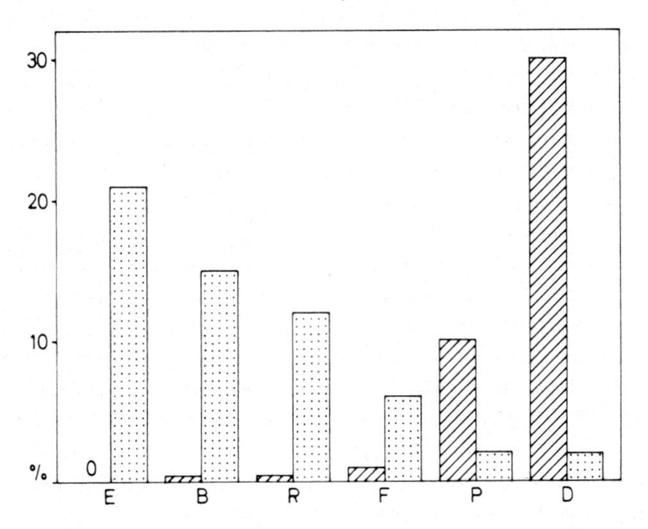

Fig. 6. Surface parts (% = coverage) covered by 6 plant species in plots (average of 10 plots with 25 m² size, Hessen) of originally uniform sites conditions: (a) exposed to moderate game browsing, mainly by *Capreolus capreolus* and *Cervus elaphus* (hatched columns); (b) protected against such animal influences since 10 years (stippled columns). E = *Epilobium angustifolium*; B = *Betula pendula*; R = *Rubus idaeus*; F = *Rubus fruticosus* agg.; P = *Poa chaixii*; D = *Deschampsia flexuosa*. The first 4 species (E, B, R, F) are mainly repressed directly by game browsing of their sprouts. The two grasses (P, D) are promoted by reason of the reduced competition of the woody species (B, R, F etc.) and of *Epilobium* in consequence of the game influences. Or.

11.6 Stability of Species Composition as a Result of Mutual Influences

The species composition of plant communities is mainly congruent to certain combinations of climatic and edaphic environmental factors. Therefore in most cases, the species combination is changing in parallelity with alterations of the climatic and edaphic conditions. But several observations state cases of lasting survivals of plant species and communities, formerly growing in an adequate optimal environment, after radical changes of the site conditions, e.g. by drainage or by eutrophication (e.g. KNAPP 1967a, SUMMER-

FIELD 1972). Experimental studies proved, that this survival is promoted by the presence of a dense vegetation cover and by isolation of the stand from possibilities of vegetative spreading (namely by stolons) of species in adjacent vegetation better adapted to the new site properties (KNAPP 1955b, 1967a). Under these conditions, a species combination of small sedge communities (Caricion canescenti-fuscae), normally growing on wet acid soils, with *Agrostis canina*, *Carex echinata*, *C. panicea*, *C. fusca*, *Eriophorum angustifolium* etc., persisted more than two years after transfer to drained, well aerated, and approximately neutral pH properties of the substratum. In the same way, a species combination of heath vegetation, normally growing on highly acid and poor sites, including even *Vaccinium vitis idaea*, survived in a relatively vital status after transplantation in loamy soil rich in plant nutrients and with pH 6.0.

This stability of vegetation is promoted by mutual influences between plants. The competitive power of the established plant community can prevent at a certain degree the invasion and settlement of species of other associations, even when they are adapted better to the new site conditions after environmental changes. This effect is much more obvious in cases of invasions by seedlings than by stolons. This is partially a result of allelopathic influences. Since allelopathic effects are often most pronounced in germination and in seedling stages, the different behavior just mentioned is well understandable.

E. CLASSIFICATION OF SUCCESSIONS AND OF THEIR TERMINAL STAGES

12 TYPES OF SUCCESSION

P. Dansereau

Contents

12 TYPES OF SUCCESSION

12.1 Introduction

The phenomenon of succession cannot readily be separated from the whole of vegetation dynamics, which is the object of this section of the handbook. The complexity of vegetation change has been revealed to us in ever-greater detail since the turn of the century when COWLES (1899) and later CLEMENTS (1936) formally defined the phenomenon of community replacement and suggested that the combined forces of the environment were conducive to a sort of convergence which they called the climax.

Detailed critical research oriented towards a verification of this hypothesis was not forthcoming for some time. Dogmatic incorporation of the climax theory in many textbooks served the purpose, for awhile, of deadening the issue on the one hand, and of lending excessive vulnerability to a workable theory that had not been sufficiently challenged. There followed a period when, in the eyes of some ecologists, succession could be witnessed everywhere and, in the eyes of others, virtually nowhere (EGLER 1951, 1967). It had long been observed by such management-minded men as foresters (SPURR 1952). Studies on rightsofway, "abandoned" land, exploited forest revealed how often an expected stage could be "skipped" or cornered, and how surprisingly stable many pioneer or consolidation communities could be (KENFIELD 1966).

It does not seem to be the purpose assigned to me in the present context to discuss the climax hypothesis. I shall therefore attempt to separate the issues involved in succession from their possible culmination in a state of equilibrium. I am nevertheless somehow bound to state a personal position on this still surprisingly controversial topic by referring to my former paper (DANSEREAU 1956b). In the intervening fifteen years I have had occasion to revise my judgment of many ecological phenomena which I had witnessed and then interpreted within a hopefully coherent framework (see 1957), and I trust this has been apparent in my publications of recent years and that it will be evident in the present text. However, I am bound to say that I honestly see little reason to doubt that the allogenic and autogenic forces of the ecosystem *tend towards change* on the one hand and *are subject to a great variety of early or late inhibitions*, on the other hand.

This means that the double task of the investigating ecologist still is:

1) to decompose the forces that promote succession (or change), and to identify their strength and orientation;

2) to locate the counter-forces of inhibition, arrest, and deviation that re-orient change, set it back, or induce intermediate and often durable equilibria.

2.2 Gradients: Smooth and Bumpy

The first line of approach is geared to the continuity-discontinuity phenomenon. Change in vegetation, whatever its tempo, is continuous, but certain turnovers are more likely self-perpetuating than others because of built-in forces or of repeatedly active outside pressures.

It is inevitably a subjective urge and possibly a cultural hang-up that gives greater emphasis to continuity (e.g., the Wisconsin School) or discontinuity (the S.I.G.M.A. School). The "mind and the eye" (Arber 1954) are congenitally geared to each other very differently in the body and spirit of different investigators and this inner matrix is in turn variously influenced, not to say impregnated, by the schools. The requirements of synecological study do not preclude the assumption of this subjective approach, even in highly sophisticated research (Williams 1967a, b). They do, however, imply a constant recognition of the duality of alternating continuity and discontinuity.

The ordination of plant-communities along a gradient is a *constat* of the first order. Many shorelines with a regularly pulsating flood-level have belt after belt of vegetation dominated by different species (typical example in Dansereau 1956a, Fig. 16). Whereas we know that all zonation is *not* indicative of succession, we can also point to many instances where it is, and where, precisely, it is the displacement of the underlying gradient itself that induces it. Thus, the gradual progress of sedimentation on a shoreline displaces the quantities and periods of water-availability in such a way as to make the ecological conditions of the position originally occupied by belt A so very similar to those of belt B that the typical (and eventually the dominant) species of A yield to the invasion of B-adapted species.

Smooth gradients lend themselves to an investment by a gamut of species (and communities), but this may occur in four different ways.

1) Where each of the segments of the gradient is large enough, this gamut fully develops into well-marked belts.

2) When the critical bands ABCDEFG either contract or expand, the gradient can be reduced to ACDFG, where the B and E positions are too narrow to allow minimal Lebensraum (they have been cornered).

3) Where the values undergo a reversal, the spatial arrangement will favour an ordination such as ABCBABCDEBEF.

4) Where there occurs an intrusion of a quite different influence, it will cause a spatial sequence such as AB′CDE′FG.

These four kinds of gradient are, therefore: 1) smooth, open, continuous; 2) contracted, somewhat discontinuous; 3) sequentially reversed; 4) bumpy and heterogeneous.

12.3 Sere and Ecosystem

A sere is a group of plant-communities related in a predictable linear order that replace one another in time on a given site. Thus, on a dry sandy plain in the Montreal Lowlands (see DANSEREAU 1956a, Fig. 8), the following succession of communities is observed: Oenotheretum, Danthonietum, Festucetum, Solidaginetum, Crataegetum, Pinetum strobi. (Fuller description of these plant-communities, their structure and dynamics, will be found in DANSEREAU 1959.) In the course of this replacement, organic content and moisture-retaining capacity of the substratum have increased; so has the differentiation of soil horizons; the vegetation has become more stratified and this has buffered the macroclimate considerably. On the other hand (see also DANSEREAU 1956a, Fig. 8), on a silting floodplain in the St. Lawrence River Valley, from the permanently submerged to the regularly flooded land the following are apparent: *Nupharetum*, *Scirpetum*, *Calamagrostetum*, *Spiraeetum*, *Alnetum*, *Acereto-Ulmetum*. In the course of this movement the water-content of the substratum decreases; exposure to air increases; likewise, vegetation mass stratifies and increases.

Virtually all landscapes, all over the world, can be mapped as to their driest and wettest parts, leaving the well-drained (mesic) areas in-between. Commonly, therefore, one reads of a hydrosere and of a xerosere, with a mesosere in-between (where water availability *in the soil* is not critical).

However, this recognition of three main segments in the landscape is not enough since it points only to one factor. Important as water availability may be, it frequently happens that other forces in the landscape are more critical because of the stress induced by

TABLE I. Huguet del VILLAR's (1929) ecological, non-geographical classification of the physiological regimes that characterize ecosystems

Substratum	Habitat	Relative harmony	Nature of control	Quality of control	Regime	Typical ecosystems
	Totally or partly aquatic	Harmony of factors		Constant	1. *Limnophytia*	Lakes, ponds, streams
				Sub-constant	2. *Helophytia*	Marshes, temporary ponds
			Chemical	Alcalinity	3. *Halohydrophytia*	Sea, salt lakes
				Acidity	4. *Oxyhydrophytia*	Acid lakes
	Hydrophytia	Dominant discrepancy of one factor	Thermic	Excess	5. *Hydrothermophytia*	Warm springs
				Deficiency	6. *Cryophytia*	Arctic seas, ice, snow
geophysical			Biotic	Mephitic accumulations	* (*Hydrosaprophytia*)	
	emerged	Harmony of factors	*Mesophytia*	Constant	7. *Hygrophytia*	Tropical rainforest
Ecophytia	*Pezophytia*			Sub-constant	8. *Subhygrophytia*	Subtropical and temperate rainforest
				Discontinuous	9. *Tropophytia*	Monsoon and deciduous forest
		Dominant discrepancy of one factor	Water scarce: *Xerophytia*	Moderately	10. *Mesoxerophytia*	Mediterranean forest
				Extremely	11. *Hyperxerophytia*	Desert

TABLE I (continued)

geophysical:		emerged *Pezophytia*		Temperature extreme	Excess	12. *Subxerophytia*	Savana
					Deficiency	13. *Psychrophytia*	Tundra
				Reaction diverging from neutral	Alcalinity	14. *Halophytia*	Seashore
					Acidity	15. *Oxyphytia*	Bogs and needle-leaf forest
Ecophytia			Dominant discrepancy of one factor	*Edaphophytia* Physical condition unfavorable substratum excessively:	Loose	16. *Psammophytia*	Dunes
					Dry	17. *Chersophytia*	Shallow gravels
					Petrophytia	18. *Chasmophytia*	Crevices
					Compact	19. *Lithophytia*	Rocks
				Perturbing biotic factor	Putrescible accumulations	** (*Pezosaprophytia*)	
				Biogenophytia	General transformation of the environment	20. *Biogenophytia* (s.str.)	Bird cliffs
						21. *Paranthropophytia*	Buildings yards, railways
organic:	*Saprophytia* (dead)	Aquatic Emerged	Harmony of factors	Texture of substratum	Constant	22. * *Hydrosaprophytia*	Logs under water
						23. ** *Pezosaprophytia*	Rotting logs
	Biophytia (living)	Supporting Harboring	Harmony of factors	Texture of substratum	Constant	24. *Ectobiophytia*	Bark of trees, sheaths of bromeliads
						25. *Endobiophytia*	Intestines of animals, living wood

their scarcity or excessive abundance, or immediate impact. Thus the prevalence of salt induces a *halosere*, of acidity, an *oxysere*, of hard rock, a *lithosere*, of moving sand, a *psammosere*, etc.

If we were to propose a classification of "types of succession," should not these determining factors serve well? Are they not the basic cause of any observed discontinuities between ecosystems?

A classification of *types of succession* could therefore be delineated on *types of ecosystems*. In several publications from 1952 onwards (see especially 1956a, 1957, 1959, 1966b, 1970), I have suggested that HUGUET DEL VILLAR's (1929) scheme was very appropriate for this purpose. In a very slightly modified way I have used it as a frame for regional studies in the St. Lawrence Valley (1959), in Puerto Rico (1966a), in the Azores (1970).

DEL VILLAR's classification (Table I) is based upon *physiological regime* and therefore upon the *equipment that all forms of plant life which are present in a given ecosystem must have to meet a definable kind of stress*: hardness or acidity of substratum, flooding, extreme permeability, etc. Admittedly the 25 resulting categories may be contested, and I shall not discuss their inclusiveness or exclusiveness at this time, but I do believe this kind of scheme to be highly useful, both as a means of description and as a functional framework.

Thus, successions will be: (1) limnophytic, (2) helophytic, (3) halohydrophytic, etc. (see Table I, sixth column). In any given landscape, the mosaic of plant-communities will be such that ecosystems will border upon each other. Thus, a dune (psammophytia), a bog (oxyphytia), and a marsh (helophytia). As long as vegetational change progresses within each ecosystem (dune, bog, marsh), the physiological regime is invariable: the various plant-communities (Caricetum, Chamaedaphnetum, Ledetum, Piceetum ericaceum in the bog, for instance) essentially effect a similar cycling of similar resources, similarly accessible and transformable. The boundary between the dune and the bog may be permanent or shifting, but it rather *breaks the continuity of succession*. The physiology of dune plants is so utterly different from that of bog plants that virtually no *binding* species will mark the transition. This *hiatus* is definitely lesser in the transition from bog to marsh where several species are likely to carry over, although in a reduced role. However, the relay in quality, quantity, and availability of resources involves a major qualitative change and a new ecological background has developed and a potential *new succession* has begun.

12.4 Trends and Inhibitions of Succession

The causes of succession are numerous and very unevenly influential under different ecosystematic regimes. It is easier to detect the effects of a progressive succession and to list them, as follows.

In the soil:

1) moderation of drainage (deficient or excessive to regular);
2) addition of organic material;
3) redistribution of minerals;
4) improvement of structure (loosening or cohesion);
5) tapping or liberation of buried or bound elements.

On the microclimate:

6) attenuation of temperature extremes;
7) reduction of temperature and humidity fluctuations;
8) decrease of radiation and evaporation, at least near soil surface;
9) increase of shade;
10) buffering of wind.

In the vegetation:

11) increase of coverage and of biomass;
12) changes in light exposure with stratification;
13) greater utilization of airmass and soil layers;
14) changes in composition and diversity of exploiting species;
15) structural complexification;
16) niche multiplication;
17) fuller utilization.

In a field study of the strategies involved in succession, it will appear that critical factors undergo a regular shift that taxes the physiology of the participating plants, first in one way and then in another. Thus, such plants as *Oenothera biennis* and *Danthonia spicata* are excellently adapted to great temperature and humidity fluctuations and to low water and organic content of the soil, but poorly equipped to compete with *Festuca rubra* or *Poa pratensis* once the above-mentioned conditions have "improved", and not at all able to withstand either the crowding or the shade created by invading *Solidago*.

What is it, then, that favours the progress of succession? What stops it? What sets it in reverse?

These questions should be posed by looking at the cycling of resources in the ecosystem and by testing the efficiency of cycling. Figure 1 shows the now classic triangle (or pyramid) which has served to illustrate the relative position of the agents and the resources in the ecosystem.

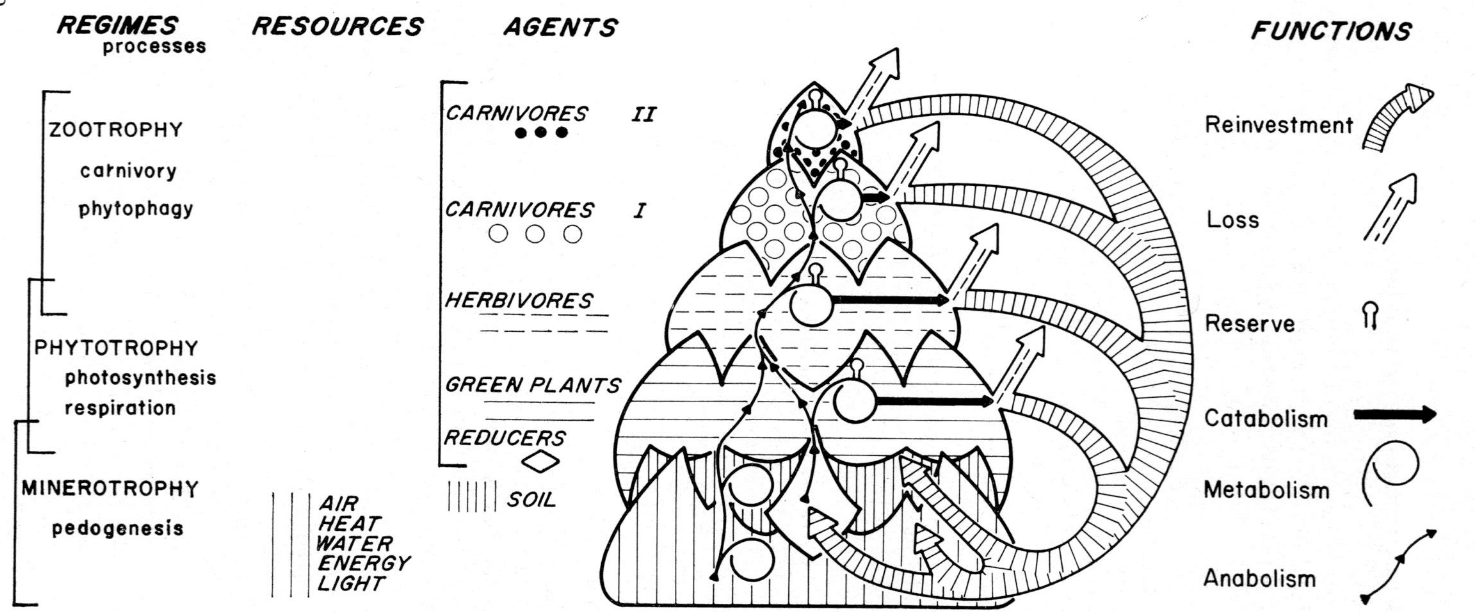

FIGURE 1. *Structure of the ecosystem.*

It is possible to considerably improve this diagram (published in DANSEREAU 1969) as I have done in another context. It will serve the present purpose well enough, however, since our concern is with primary productivity.

Three principal regimes, involving three basically different metabolisms, are involved in the cycling of resources:

minerotrophy takes place through several pedogenetic processes that transform parent-rock under the impact of climate and vegetation and make available certain qualities and quantities of resources (light, energy, heat, gases, liquids, solids in the air and soil);

phytotrophy involves the metabolism of plants which operate a transformation of *mineral products* and transform them into vegetable tissue;

zootrophy results from the further elaboration of *mineral and vegetable products* by animals.

Each turnover, at each level, makes for *storage*, *reduction*, *reinvestment*, and, of course, *loss*. It is largely the balance between loss and storage that controls succession. Considerable loss of water (through erosion or excessive evaporation) at the minerotrophic level, of plant-exploiters (through a sudden, widespread disease) at the phytotrophic level, of animal consumers (through emigration) at the zootrophic level may very well set back the successional trend and thereby cause a retrogression.

On the other hand, an accumulation of a large capital of resources, creating a surplus, invites invasion. The latter may be temporary or cyclic, such as the passage of migrating birds in the tundra or in the tropical savana (see MOREL & BOURLIÈRE 1962), or it may be permanent, such as the colonizing of an "old field" by birches or pines.

What is it, then, that *stops* succession? One force is the capacity of many species, especially dominants of a community, to maintain their spatial position after their habitat and even the climate itself have ceased to favour full vitality ("law of persistence", DANSEREAU 1966b). Another is the impact of allelopathy (MOLISCH 1937, EVENARI 1961, MULLER 1966, KNAPP 1954, 1967), the chemical power of some plants to exert an antibiotic effect. These are autogenic forces, inherent to the flora.

Among the inhibitions caused by a permanent or recurring feature of the site are numerous edaphic and physiographic conditions affecting pedogenic or microclimatic processes, such as frost-pockets, abruptness and hardness of substratum, in fact any of the 25 ecological situations enumerated in Table I.

12.5 Autogeny, Allogeny, Biogeny

These considerations lead naturally to TANSLEY's (1935) recognition of three basic types of succession: *autogenic*, *allogenic*, and *biogenic*.

Autogenic succession is the replacement of one plant-community by another which is primarily due to the better adaptation of the invaders to ecological conditions that result from the residents' accumulation of resources and "improvement" of the site. The residents, therefore, cannot fully utilize conditions that are essentially "of their own making" or at least cannot compete with better-equipped invaders, which could not have withstood the adversities that prevailed at an earlier time. It thus appears that it is the prevailing resident *agents*, through their metabolic power, that have induced a significant change.

Allogenic succession, on the contrary, implies only a minimum of resource accumulation by the resident community of plants and animals, whereas materials *from other ecosystems* more or less drastically *alter the resource-basis* and thereby allow new agents to prevail, inasmuch as the residents are poorly adapted to the new conditions.

Biogenic succession supposes the rather catastrophic, or at least sudden, interference of a living agent which is capable either of altering some significant resource (light, heat, and evaporation in the grassland, on the tracks of buffalo herds) or of destroying or reducing the mass of a major agent (chestnut blight in Eastern North American forests).

Thus, the three "types of succession" differ fundamentally in their ecosystematic strategy:

autogenic succession results from the accumulation of *excessive resources* that permit the access of new primary-producing agents;

allogenic succession results from the neutralization (sedimentation) or the ablation (erosion) of existing resources and the liberation (erosion) or substitution (sedimentation) of new resources which lend themselves to a new occupancy;

biogenic succession is the result of the excessive utilization by a primary consumer that changes the composition and/or structure of the community and thereby the existing balance.

Succession, therefore, as an ecological phenomenon, involves a shift in the interlocking mineral, vegetable, and animal cycles. It obeys the uneven pressures of more or less rapid turnovers, more or less abundant storages, and is at the mercy of both resources and agents that originate in other ecosystems.

There are many types of succession, and they imply different kinds and degrees of primary productivity. They may be classified as:

a) showing greater or lesser continuity where they are geared to coordinated environmental gradients;

b) being characteristic of certain identifiable ecosystematic controls (or physiological regimes);

c) showing discernible trends towards progression, retrogression, stagnation, inhibition;

d) as being essentially inner-controlled (autogenic) or outer-controlled by incoming resources (allogenic) or by primary consumers (biogenic).

13 CLIMAX CONCEPTS AND RECOGNITION

R. H. Whittaker

Contents

13 CLIMAX CONCEPTS AND RECOGNITION

13.1 Introduction

Ecologists seek to capture with concepts the meaning of vegetation change. The idea that this change could be described and understood was a major development in the history of ecology, that may be traced back to KERNER (1863), HULT (1885, 1887), WARMING (1891, 1896), and others [see preceding articles and reviews of LÜDI (1921, 1930), FURRER (1922), CAIN (1939), WHITTAKER (1953), and ALEKSANDROVA (1964)]. In counterbalance to the idea of vegetation change it was natural to recognize the condition of relative vegetational stability that came to be known as "climax". Early sources of the concept of climax include HULT (1885, 1887), WARMING'S (1896) *Schlussverein*, the chief associations of MOSS (1907, 1910), the stable formations of CRAMPTON (1911, 1912), and the American work of COWLES (1899, 1901, 1910, 1911), and CLEMENTS (1904, 1905, 1916).

The work of COWLES and others emphasized the manner in which successions in a given area converge toward the same, or similar, climax communities. The end-point of this convergence was seen as the major community of a geographic area. One may on this basis state an early conception of the climax as: (1) a stable, self-maintaining, mature plant community, (2) toward which plant successions of an area converge, and which terminates these successions, and (3) which consequently characterizes a geographic area as its principal, prevailing (undisturbed) plant community and expresses that area's climate. One must note, however, that the three components of this conception — stability, convergence, and regional prevalence — bear no *necessary* relation to one another. From different interpretations of the manner in which they relate to one another result the varied approaches to the climax discussed below.

13.2 Monoclimax Theory

One of the most widely influential treatments of climax theory was stated in a monograph and a later article by CLEMENTS (1916, 1928, 1936). CLEMENTS accepted a direct and necessary relation among the three components of the climax conception. It was assumed that all successions of an area converged toward a single stable, mature plant community or climax. Characteristics of this climax were determined solely by climate; given sufficient time, successional processes and modification of environment by the community would overcome effects of difference in topographic position and parent material. In principle, at least, the climax was the same for all habitats in a given climatic area. There was in consequence a direct, one-to-one correspondence of climax communities with climatic regions. For its assumption of a single climax for a geographic area, the interpretation has been termed the "monoclimax" theory. The climax communities may be recognized on two levels: formations or biomes defined by community structure or physiognomy and characteristic of larger geographic areas, and within these "associations" defined by dominance of plant species or genera and characteristic of smaller areas. The deciduous forests of the eastern United States represent a climax formation or biome; the three or more subdivisions of these forests that characterize regions within the eastern forests are "associations" (CLEMENTS 1916, WEAVER & CLEMENTS 1929, BRAUN 1950). [These American "associations" are collective dominance-types (WHITTAKER 1962), and they are very much broader units than the associations of European phytosociologists.]

CLEMENTS' interpretation was strongly influenced by analogy of the community with the individual organism, though this analogy is not a necessary part of the monoclimax theory. CLEMENTS (1916) considered that the unit of vegetation, the climax formation, is a complex organism that arises, grows, matures, and dies. The formation is able to reproduce itself, repeating the stages of its development in a life history comparable to that of an individual plant. The climax formation is the adult organism; as a reproductive process succession can no more fail to end in the adult form than it can in the case of the individual plant. Retrogression, as a development of the community backwards, is just as impossible for a succession as for a plant's growth (CLEMENTS 1928: 125—6, 147, cf. COOPER 1926, GLEASON 1927, WHITTAKER 1953). The organismic analogy was accepted in more qualified form by TANSLEY (1920, 1935) and other authors but has been widely criticized by others (see WHITTAKER 1957, 1962: 52) and finally left aside from the concerns of the science.

In actuality the vegetation of a given area is complex and includes a number of types of stable communities. Stable communities other than the true climax must be accommodated in a monoclimax interpretation; they are recognized and interpreted as proclimaxes. A number of types of proclimaxes are distinguished (CLEMENTS 1936, WEAVER & CLEMENTS 1938). Serclimaxes are successional communities that are arrested, held more or less permanently in an early developmental stage by special conditions of their environments. Preclimaxes may occur in the less favorable (e.g. drier) habitats of an area, and postclimaxes in the more favorable (e.g. more mesophytic) habitats of an area. Preclimaxes and postclimaxes may be relicts of past climatic change and vegetation migration, and may occur as climaxes of other areas with less or more favorable climates. Subclimaxes are late successional communities, the maturation of which to the climatic climax is long delayed. Disclimaxes are communities that are held in stable conditions different from the climatic climax by effects of animals or disturbance by man.

CLEMENTS' monoclimax formulation has been widely criticized for its assumption that all successions in an area must converge to the same climax (GAMS 1918, DU RIETZ 1921, DOMIN 1923, BOURNE 1934, CAIN 1947, CROCKER & WOOD 1947, BEADLE 1948, BEADLE & COSTIN 1952, RICHARDS 1952, WHITTAKER 1951, 1953, 1956, 1957, SJÖRS 1955). The proclimaxes are the means of fitting to an over-simplified ideal, the single climax community, an actual complexity of stable communities in an area that violates the ideal (WHITTAKER 1953, 1957). Other authors have offered monoclimax interpretations that may be less vulnerable to criticism than Clements:

1) WALTER (1943, 1954, 1962) has sought to resolve the difficulty of relating successional convergence and climax regions by considering the climax independently of succession. Climaxes are conceived as communities of geographic areas with uniform climatic conditions free from disturbance by man. Effects of local peculiarities of habitat must be set aside, only in habitats with "typical" climatic relations can the climax appear. Hence climaxes are to be recognized in large areas of relatively level terrain with mature soils; the steep topography of mountains must be left out of consideration. The climax concept is then identical with zonal vegetation of Russian soil science; there is direct correspondence between climatic region, zonal soil, and climax community. Climax areas will include islands of extrazonal and azonal vegetation of special habitats (corresponding to some of the proclimaxes of CLEMENTS) (cf. GAMS 1936).

2) DANSEREAU (1954, 1956, 1957) considers that succession involves releases of communities from local controls (by topography, parent material, biotic interactions, etc.) that determine the characteristics of successional communities; release from these

controls eventually permits the community to develop to true climax status. Local controls and available flora may, however, inhibit succession and keep communities more or less permanently short of climax status. Certain areas consequently lack true climaxes and are occupied by successional communities of varying degrees of instability or relative stability. Only certain landscapes permit the development of a true climatic climax, to be recognized as a widespread community on upland sites. On very flat or very steep topography the vegetation is under lasting edaphic or topographic control, and one can only surmise what the climax of the uplands would be.

3) Among American ecologists BRAUN (1950, 1956) and OOSTING (1948, 1956) have stated monoclimax conceptions derived from CLEMENTS in some respects, but freed from some of CLEMENTS' assumptions. Both authors recognize the existence of a climatic climax community, but also accept the occurrence of other stable communities in an area (topographic climaxes, edaphic climaxes, subclimaxes). In theory these other communities should develop to the monoclimax given sufficient time; in practice they are stable communities or climaxes of habitats other than those which support the climatic climax. BRAUN's (1956) later article considers climaxes as regional, rather than simply climatic, because history, including erosion cycles and past climates, may along with climate have much to do with composition and extent of present climaxes. The concept of regional climax does not imply uniformity of vegetation in an area. In old areas of mature topography, a mosaic or pattern of intergrading climax communities is seen; in young areas of immature topography succession is active, and most of the vegetation is not climax. In this later conception BRAUN (1956) has departed so far from CLEMENTS' monoclimax that her view is convergent with polyclimax and climax pattern interpretations.

4) In European phytosociology the interpretation of BRAUN-BLANQUET (1928, 1932, 1951, 1964) may be regarded as monoclimax in its recognition of a single climatically determined climax or terminal community (*Schlussgesellschaft*) toward which all successions ultimately converge. BRAUN-BLANQUET recognizes also other communities that are stabilized in their local habitats (*Dauergesellschaften*). The grouping of successional communities related to the same climax forms the *Klimax-Komplex*, which corresponds to a geographic territory, the *Klimax-Gebiet*. Though the concepts resemble those of CLEMENTS, the practice in the school of BRAUN-BLANQUET has been different. In the earlier work of the school of CLEMENTS successional interpretation was primary, and classification of vegetation was based upon it. In the school of

BRAUN-BLANQUET, in contrast, vegetation of an area is likely to be classified first, and successional relations among the resulting community-types then interpreted. Developments towards polyclimax or climax pattern concepts in the school of BRAUN-BLANQUET are discussed below.

13.3 Polyclimax Theory

WALTER, DANSEREAU, BRAUN, and others have qualified the monoclimax conception; while retaining a monoclimax perspective they have in various ways departed from the assumption of convergence to a single climax determined by climate alone. A direction of departure from the monoclimax conception that is common to many authors may be formulated:

1) Although convergence of successions is a significant phenomenon, this convergence is partial, incomplete.

2) From incomplete convergence of successions there result several different stable or climax communities in different habitats in a given area.

3) Among these one community may be most widespread and most directly expressive of climate. This is the climatic climax, but there is no need to assume that other stable communities in the area will ever develop into this climatic climax.

4) Stability, and regional prevalence or expression of climate, are thus to be considered separately. Stability defines climax communities, but among the several stable communities of an area the climatic climax community is the one which is also regionally prevalent.

5) Since existence of a number of more or less equally stable communities in an area is granted, the interpretation may be termed the "polyclimax" theory.

Although many ecologists have stated polyclimax viewpoints, TANSLEY is often given principal credit. TANSLEY (1920) regarded a formation as a grouping of successional communities around a mature community for their habitat, a concept related to CLEMENTS' climax formation and BRAUN-BLANQUET's climax-complex. Mature or climax communities can be determined by factors other than climate, however (TANSLEY 1935, 1939, 1941). Successions do not lead toward a single climax but toward a mosaic of climax communities determined by a corresponding mosaic of habitats (TANSLEY 1939: 216). In addition to subclimaxes arrested at stages before the climatic climax, TANSLEY (1935, 1939) recognizes plagioclimaxes resulting from deflection of successions and stabilization of successional communities by disturbance.

In the school of BRAUN-BLANQUET, TÜXEN (1933, 1935) has advanced the concept of paraclimax for a widespread stable community the characteristics of which are determined by soil factors and not climate alone. Oak-birch forests of nutrient-poor sandy soils of low relief in northwestern Germany are regarded as an edaphically determined paraclimax in a climate capable of supporting other forest types (TÜXEN 1933, ADRIANI 1937, VLIEGER 1937, ELLENBERG 1963: 245). TÜXEN & DIEMONT (1937, DIEMONT 1938, RICHARD 1961) further suggest recognition of climax-groups (of more than one climax community on different parent materials in an area) and climax-swarms (of more than one climax community determined by topographic positions). A paraclimax might be termed an edaphic subclimax by a follower of CLEMENTS' terminology; but the climax-group and climax-swarm represent polyclimax interpretations. ELLENBERG (1953, 1959, 1963) has rejected the convergence toward a single climax, for alpine meadows of different habitats as stated by BRAUN-BLANQUET & JENNY (1926) and for tropical rainforests on different parent materials. ELLENBERG (1963) and KLAUSING (1956) have related patterns of forest community-types to environmental gradients, and KRAUSE (1952) has developed an interpretation of climax vegetation through community mosaics and community complexes and considered the manner in which the complex of one geographic area gives way or changes into that of another. SCHMITHÜSEN (1950) rejects the monoclimax assumption of identity of vegetational and climatic areas and convergence toward a theoretical climatic climax or *Schlussgesellschaft*. Topographic and parent material differences remain significant despite succession; various factors are effective in determining climax characteristics in different parts of a landscape's pattern of habitats. To the spatial pattern of habitats, then, must correspond a pattern of climax communities.

13.4 Climax Pattern Hypothesis

The landscape interpretation of SCHMITHÜSEN (1950), TÜXEN (1956) and KNAPP (1949, 1971a) and the regional climax of BRAUN (1956) are very close to a third viewpoint, formulated by WHITTAKER (1953) as the climax pattern hypothesis.

Two significant differences in perspective underlie the emphasis of the concept of "pattern" over that of "mosaic" as the latter appears in the polyclimax viewpoint. It is considered, first, that plant communities predominantly intergrade with one another continuously (though some discontinuities must result from topographic and soil contrasts and disturbance). From American work in gradient analysis it has appeared that plant communities are parts of community continua along environmental gradients (WHITTAKER 1951, 1956, 1960, 1962, 1967, CURTIS & McINTOSH 1951, BROWN & CURTIS 1952, CURTIS 1955, 1959). Second, it is considered that the species in a landscape respond "individualistically", each in its own way, to environmental factors including interactions with other species (RAMENSKY 1925, GLEASON 1926, 1939, LENOBLE 1926). Species are consequently combined in many different ways into the communities, and shared in many different ways by the communities, of a landscape; the species do not simply fall into distinct groups corresponding to particular communities. The vegetation of a landscape thus consists not so much of a mosaic of distinct pieces, as of a complex and subtle pattern of communities interrelated by continuous intergradation and by sharing of species (WHITTAKER, 1954b, 1956, 1962, cf. MEUSEL 1940, SCHMID 1942, 1950, 1955).

The climax pattern hypothesis may be formulated thus (WHITTAKER 1952, SHIMWELL 1971):

1) The environments of a landscape form a complex pattern of environmental gradients (or of "complex-gradients" as directions of variation in environment that involve many individual factor-gradients, WHITTAKER 1967).

2) At each point in the landscape pattern communities develop toward a climax. Because of the intimate functional relation of community and environment in the ecosystem, characteristics of the successional communities and the climax are determined by the actual environmental factors at that point, not by abstract, regional climate. The climax is to be interpreted as a steady-state community adapted to the characteristics of its own, particular environment or habitat. Differences in environment — different topographic position or marked difference in parent material, etc. — normally imply difference in composition of climax communities.

3) Differences in environment along a continuous environmental gradient usually imply a continuum of intergrading climax communities adapted to different positions along that gradient. To the complex pattern of environmental gradients of the landscape will correspond a pattern of climax communities (where these are not replaced by successional communities). Plant species have "individualistically" scattered population centers in the climax pattern of communities, which may be interpreted also as a complex population continuum. The community-types (associations, etc.) into which the population pattern is divided are largely arbitrary (though essential for some research purposes).

4) Among the climax community-types, one is usually most widespread in the landscape. This regional type may be termed the prevailing climax community of the landscape.

13.5 Climax Recognition

The difficulties of defining and recognizing climaxes that have plagued ecologists are partly solved if the climax is identified with a self-maintaining, steady-state condition in relation to a particular habitat. There are then two separate questions to be asked of a community: (a) Is it climax — is it stabilized in relation to its environment? (b) If so, does it also represent the most widespread stable community-type of the landscape, the regional (or climatic or prevailing) climax? The various criteria that have been offered for climax recognition (Cooper 1913, 1923, Weaver & Clements 1938: 479—80, Braun 1950: 13, Whittaker 1953, Shimwell 1971) may be considered in relation to these questions.

1. Observation of successions and terminal communities. In many cases there is no substitute for careful study of successional sequences, interpreting these with all available evidence on replacement of plant populations by other populations. The more disturbed the vegetation of a landscape, the more essential is careful research on succession to determine what communities are stable and what is their relation to the climax pattern. Techniques of successional study have been discussed by other authors, above. For our present purposes the problem is recognition of which communities are terminal and stable, and the following points contribute to this recognition.

2. Population structure of the community. In climax communities the populations should be in steady-state; loss from the populations by death should be equalled by gain to the populations by germination and seedling growth. In some cases an all-age population structure can be recognized in the dominant species; in a forest one may look for seedling and sapling numbers sufficient to replace the canopy trees as these die. Such correspondence of the young-tree undergrowth, from which the canopy is recruited, with the canopy itself is termed *accordance* by Braun (1950) and is reflected in characteristic J-curves of age or size distribution (Whittaker 1956). Forests which, in contrast, have young-tree undergrowths of species that differ from the canopy species, but are capable of reaching canopy size in that environment, are in most cases successional. Application of accordance as a criterion is limited, however, by the fact that in many communities the replacement of the

canopy or dominant population is not continuous. Forests may reproduce only at irregular intervals, or only in gaps formed by destruction of the canopy (JONES 1945, KNAPP 1963); and in other climax communities replacement of the dominants is infrequent, or involves cyclic processes (WATT 1947). There are cases also in which pioneer and successional communities are dominated by species from the climax community; such is the case in the condensed or telescoped successions of the desert (SHREVE 1942, 1951, MULLER 1940), the arctic (MULLER 1952, cf. RAUP 1941, 1951), and aquatic communities (WAUTIER 1951, BLUM 1956).

3. Maxima of successional trends. The climax represents a steady state of energy flow and materials cycling, as well as of population replacement. A number of trends or progressive developments toward the climax characterize succession (WHITTAKER 1953, MARGALEF 1963, ODUM 1962, 1969). Through the course of succession community productivity, biomass, stature, stratal complexity, and species diversity tend to increase. The ratio of total community respiration to gross primary productivity increases toward 1.0 in the climax, when it is 1.0 the biomass accumulation ratio (of biomass to net annual productivity) reaches a stable and maximum value (WHITTAKER & WOODWELL 1969). Plants that are taller, longer-lived, and more tolerant of growth in the shade of an existing community tend to replace lower, shorter-lived, less tolerant species. There may consequently be reduction in the rate of population change and increase in relative stability of the community up to the self-maintaining climax. The pattern of population distribution within the community in some cases becomes less irregular, and the community more homogeneous (PICHI-SERMOLLI 1948a, 1948b, GREIG-SMITH 1952, PIELOU 1966). In comparing a number of communities representing a successional sequence, the community which is maximal in some of these trends may be thought the climax, or closest to the climax. The trends may, however, reverse themselves in succession. Productivity and species diversity may decrease from late successional into climax communities. Replacement of a successional community by a climax community of lower biomass and plant stature is uncommon, but occurs in some circumstances (WHITTAKER 1953: 57, DRURY 1956, JENNY et al. 1969).

4. Consistency by habitats. A particular kind of habitat (e.g. southwest-facing open slopes on granite at a given elevation in a mountain range) should support a particular kind of climax community adapted to the environment of that habitat-type. If the vegetation is mature and undisturbed, its climax condition should be expressed in consistency by habitats — occurrence of

similar climax communities in similar environments. If the vegetation has been in various ways disturbed, different successional communities will appear in similar habitats. If the vegetation is a mixture of climax and disturbed communities, the climax communities may be recognized by their consistency (together with their population structure and maxima of successional trends) in a given kind of habitat, contrasted with the more diverse successional communities also occurring in habitats of this kind.

5. Successional convergence. The preceding point reflects the fact that in a given kind of habitat different successional communities, resulting from different kinds and timings of disturbance, converge toward similar climax communities. Convergence is also, with somewhat more qualification, a basis of recognizing regional climax communities — the second question to be asked. Convergence of successional communities in the different kinds of habitats of a landscape is only partial. The communities of significantly different habitats (north-facing vs. south-facing slopes, granite vs. limestone or serpentine parent material) cannot be expected to converge to climax communities of the same composition. Convergence in species composition of climax communities is to be sought within classes of habitats, not for the landscape as a whole. For the landscape as a whole, however, successions in many kinds of habitats (not necessarily all) may lead to climax communities that are convergent in relation to a more broadly defined community-type (a formation, or a collective dominance-type represented by somewhat different species combinations in different habitats).

The following points refer primarily to recognition of regional climaxes (question b).

6. Maximum mesophytism, or occupation of intermediate habitats. Maximum mesophytism has often been stated as a characteristic of the climax (COOPER 1913, WEAVER & CLEMENTS 1938, OOSTING 1948). Related to successional convergence is the fact that many successions lead from communities that are xeric (e.g. rock faces) or hydric (e.g. ponds or marshes) to communities that are mesophytic for the area (e.g. the different climax forests that may develop on an open slope and on a valley floor). [There are, however, cases (LÜDI 1923, BOURNE 1934, DAHL 1957, WHITTAKER 1953, JENNY et al. 1969) in which leaching of nutrients causes replacement of a late-successional community by another more xerophytic in character.] Among the different climax communities in a landscape it may be difficult to say which one is most mesophytic. It may also be doubtful that the most mesophytic community (in the sense of adaptation to most favorable, moist

but not wet environments) is the regional climax; this community may be a "postclimax" in the sense of CLEMENTS. It may be better to specify that the regional or climatic climax occurs in intermediate habitats for the area (cf. KUJALA 1945) — neither valley bottoms nor southwest-facing open slopes nor unusual parent materials, but on "normal" parent materials in topographically intermediate situations (e.g. open east- and west-facing slopes).

7. Occurrence on the uplands. Closely related is the view that the climatic climax is to be sought on the uplands (NICHOLS 1923, DANSEREAU 1954). Such interpretation excludes valley vegetation, whether or not this is regarded as most mesophytic for the area, from designation as climatic climax. It may also imply that a climatic climax is not present in areas of extensive lowlands or flat terrain. Distinctive climaxes occur in areas of low relief, such as the oak-birch paraclimax in Germany (TÜXEN 1933), the soil- and fire-affected savannas in forest climates in South America (BEARD 1953), the extensive swamp forests of the Gulf Coast and Mississippi Valley (WEAVER & CLEMENTS 1938), the different vegetation patterns of outwash plains or bajadas from those of mountain slopes or uplands in desert areas (WHITTAKER & NIERING 1965). It may be somewhat arbitrary to say that the vegetation of uplands is a more climatic climax than that of level lowlands. For study of the vegetational expression of climate, however, it may be convenient to use as a basis of comparison the vegetation of uplands (and, especially, intermediate topographic positions in these).

8. Regional prevalence. As an alternative to definition of a climatic climax by occurrence in intermediate topographic positions on the uplands, the regional climax may be defined as the climax occupying (at least potentially) the largest share of the habitats in an area. The regional climax may then best be termed prevailing climax, rather than climatic climax. The edaphically affected climax communities of lowlands mentioned in the preceding paragraph are prevailing climaxes for their lowland areas; they need not be termed "climatic climax". Extensive areas of contrasting parent materials support different climax patterns (WHITTAKER 1954a, 1960, WHITTAKER & NIERING 1968); in each of these patterns the climax community-type that is most widespread in the landscape represents its prevailing climax type. Essential similarity over a wide area (BRAUN 1950) is a characteristic not of the climax condition as such, but of the regional climax community.

9. Soil maturity. Development of the soil accompanies development of the community in succession; of the communities in a successional sequence the climax should possess the most mature

soil. Climatic climaxes should in general correspond to mature, zonal soils. The definition of soil "maturity" is, however, subject to almost as many difficulties as the definition of the climax community. Soil characteristics may contribute to recognition of climax communities, however, when the two aspects of soil maturity (steady-state condition for a given habitat, vs. possession of zonal profile characteristics) are clearly distinguished. These two aspects apply as evidence to recognition of the climax condition, and of the regional climax community, respectively.

10. Climatic correspondence. Physiognomy of vegetation shows convergent response to similar climates, in different parts of a continent and on different continents. In a disturbed landscape in a climate corresponding to those supporting broadleaf deciduous forest in other areas, it is reasonable to consider that deciduous forests are most likely to be climax among the various communities observed. The physiognomic convergence will not, however, indicate which of the deciduous forests in the area are part of the climax pattern and which are successional. The physiognomic convergence also has its limitations. Convergence of physiognomy of vegetation in similar climates on different continents is imperfect (BEADLE 1951); Australian woodlands may occur in climates that would support grasslands on other continents. The soils effects referred to in paragraphs 7 and 8 in some cases imply that physiognomically different climax communities and climax patterns occur on different soils in the same climate. Differences in occurrence of fire can produce alternative, physiognomically different, stable communities — such as deciduous forest, versus prairie subject to frequent fire, in parts of the eastern United States. Some climax communities, once destroyed, are replaced by other stable communities, and the effect may be apparently irreversible (GAUSSEN 1951).

13.6 Conclusions

These conclusions are suggested on climax concepts, climax recognition, and the value of climax interpretations.

The diversity of climax concepts — monoclimax, polyclimax, and climax pattern, and variations within these — may seem at first confusing. The climax is an interpretive concept that cannot be directly proved from nature. The concept has been created by ecologists on the basis of observed differences in relative stability and instability of communities, as a needed means of abstraction, inference, and generalization about communities. The forms of climax interpretation are to a degree free or underdetermined, in the sense that the same vegetation may be plausibly interpreted (using many of the same terms) by monoclimax, polyclimax, or pattern concepts.

It is sometimes suggested in consequence that the differences among these concepts are largely semantic (CAIN 1939, 1947, EGLER 1951). It need not be thought that semantic differences are unimportant. The three viewpoints differ significantly as perspectives on the interpretation of vegetation and in the research approaches these perspectives suggest. The monoclimax theory emphasizes the unity of vegetation in terms of successional convergence and climatic expression; the polyclimax theory grants the diversity of successions and climax types related to different habitats of a landscape; and the climax pattern emphasizes the coherence, as a landscape pattern of interrelated communities, of this diversity of climax types. It may not be surprising if a biogeographer concerned with broad generalizations is drawn to a monoclimax viewpoint, a phytosociologist concerned with detailed classification and mapping of local vegetation to a polyclimax viewpoint, and an ecologist concerned with gradient analysis and interpretation of vegetation patterns to a climax pattern viewpoint.

The three viewpoints have, however, features in common, and there may be basis of preference among them. All three viewpoints must allow for (1) diversity of successions in an area and partial convergence among these successions, (2) occurrence of a number of relatively stable terminations of successions, or climaxes, in an area, and (3) the need for recognizing from among these a major community, a regional climax. Point 2 can be allowed for by means of the proclimaxes of the monoclimax theory, the topographic or edaphic climaxes of polyclimax theory, or the intergrading communities that form the pattern in climax pattern interpretation. Diversity of types of stable vegetation in relation to the different habitats of a landscape is essentially universal, however. The polyclimax and climax pattern viewpoints have advantage in rendering this climax diversity explicit, while also retaining concepts by which the needs of point 3 may be met.

Between the polyclimax and climax pattern viewpoints a question of theory is involved: Is vegetation "really" a mosaic of distinct communities, or a pattern of intergrading communities? Vegetation patterns variously combine continuity and discontinuity, but the results of gradient analysis support continuity as a widespread condition, probably the general condition in landscapes not strongly disturbed. Individuality of species distributions and complex and multidirectional sharing of species by communities are also well established. There is thus basis for preferring the climax pattern hypothesis for realism in accepting complex and relatively continuous population patterns as part of the character of vegetation. The preference of the author naturally lies with this hypothesis, on the grounds that: (1) It is more realistic for, or appropriate to, the population structure of vegetation, (2) it provides for the needs met by the other theories, for the pattern can very well be classified into climax types, and a prevailing (or a climatic) climax type can be recognized among these, and (3) it provides better conceptual basis for certain types of research than the other climax theories. It is appropriate to gradient analysis and makes possible the quantitative comparisons of climax patterns between different parent materials, elevation belts, and climates discussed by WHITTAKER (1956, 1960).

There can be no single full solution to the problems of climax recognition. The difficulties are, however, much reduced when the complexities of climax vegetation are made explicit. It is then possible to distinguish clearly the two questions — of stability or steady-state condition, versus regional prevalence or climatic expression. In dealing with communities in a particular kind of habitat, observation of population structure and disturbance evidences, combined with successional trends and consistency of climax composition will generally, even without historic records of disturbance, permit recognition of the climax for a habitat-type if that climax type is still present. For the recognition of regional climaxes there are alternative approaches through occurrence on the uplands and in topographically intermediate conditions, or prevalence as a type in a majority of topographic positions. Either of these may be combined with inferences from soil maturity and climatic correspondence. Choice of the regional climax type will in some cases be unmistakable, in others somewhat arbitrary; but choices that are effective bases of vegetation description and relation to climate can be made. The concepts of climatic climax and prevailing climax will in many cases coincide. They will not necessarily coincide, however; for example the prevailing climax of an extensive area of serpentine soils may differ from the prevailing (and also climatic) climax of an adjacent area of more typical soils.

There are, finally, those who consider that the climax concept has outlived its usefulness (EGLER 1947, MATHON 1949, 1952); this author does not. The climax concept has these continued values:

1) It expresses a real and important (though relative) difference in stability between successional and self-maintaining communities.

2) It offers by this means a basis for reducing research variables. Effects of climate, topography, parent material, etc. on community composition and function may be compared on the basis of climaxes, without complication of the comparison by different successional stages.

3) It provides a standard of comparison for study of the characteristics, function, and practical management of successional communities.

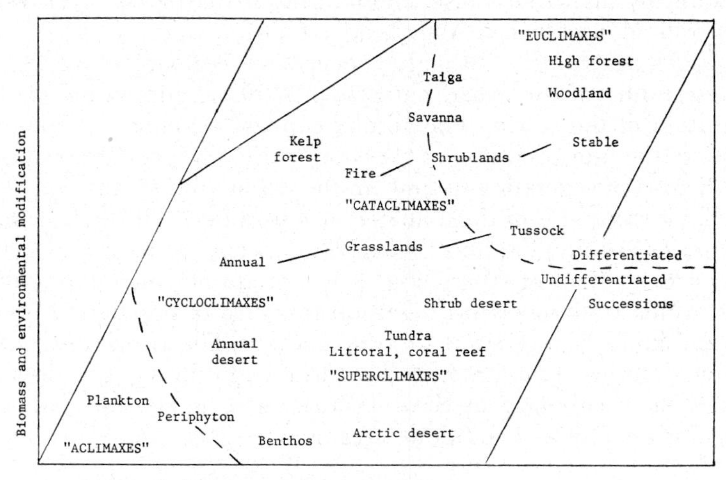

Fig. 1. A scheme relating characteristics of climaxes and successions to one another and other community characteristics. Effectiveness of the distinction of climax from succession, and utility of the climax concept, increase obliquely upward to the right.

Climax types:

"Aclimaxes" — generation times of dominants are short relative to environmental change, community fluctuation is incessant; climax and succession are not distinguishable.

"Cycloclimaxes" — generations are timed to annual environmental fluctuations; "climax" can be characterized by principal annual dominants, but succession is undifferentiated.

"Cataclimaxes" — generations correspond to irregular intervals between destructions (by fire, blow-down); the climax is characterized by the mature dominants, succession is weakly or well differentiated.

"Superclimaxes" (MULLER 1940) — generations are long relative to environmental fluctuation with replacement in dominant populations more or less continuous, but biomass and environmental modification are small; the climax is characterized by stable populations, but succession is undifferentiated.

"Euclimaxes" — generations are long relative to environmental fluctuation with replacement in dominant populations more or less continuous, biomass and environmental modification are large; the climax is characterized by relatively stable populations, and succession is clearly differentiated (classic successions of well-defined stages with different dominants in sequence).

4) It provides means of ordering and understanding in relation to one another, the diverse stable and unstable communities of an area.

5) It thus makes possible comprehension of the vegetation of a landscape through a climax pattern (or mosaic) and the successional communities related to parts of this.

6) It provides a standard of comparison by which may be measured the increasingly widespread effects of disturbance and pollution by man that cause communities to retrogress (DYKSTERHUIS 1949, WOODWELL & WHITTAKER 1968).

7) It contributes through concepts of regional climaxes to understanding of the broad patterns of climatic adaptation of the vegetation of the world. The climax concept remains an essential means to the interpretation of the complexity of vegetation in space and the flux of vegetation in time, in the world around us.

Some criticisms of the climax concept result from the fact that successions are simpler, and the climax concept less useful, in some environments than in others. Fig. 1. is a rough scheme relating different kinds of climaxes (as based on manners of population function and kinds of successions) to one another. The various prefixed "-climax" terms are not intended as formal additions to the already formidable vocabulary of these (WHITTAKER 1953), but only as labels for the climax conditions suggested.

14 DIFFERENCES IN DURATION OF SUCCESSIONAL SERES

J. MAJOR

14 DIFFERENCES IN DURATION OF SUCCESSIONAL SERES

Judging by species composition we can quote some times of plant succession to a more or less equilibrium state. IVES (1941) believed the fire scar in Colorado which showed a plant succession from *Salix* to *Populus tremuloides* to *Picea engelmanii—Abies lasiocarpa* would vanish in some 300 years. VIERECK's (1970) *Picea glauca* forest along the Chena River of interior Alaska was 220 years old, and presumably the *Picea mariana* climax muskeg which developed from the more productive white spruce forest was only a few hundreds of years older. The succession in the subalpine belt of the Swiss National park described by BRAUN-BLANQUET, PALLMANN & BACH (1954: 43, 153—164) and BRAUN-BLANQUET (1964: 669—673) probably developed a climax Rhododendreto-Vaccinietum cembretosum in somewhat less than 2000 years. COOPER's classical studies (1923, 1931, 1939) of plant succession at Glacier Bay, Alaska have been followed up by LAWRENCE (1958), DECKER (1966) and UGOLINI (1966).

The qualitative descriptions of vegetation indicate the near-climax spruce-hemlock forest is attained in something over 200 years. At Mt. Robson in the subalpine belt of the Canadian Rockies (1680 m elevation) TISDALE, FOSBERG & POULTON (1966) found little change in the forest growing on a 160 year old recessional moraine during the 49 years since COOPER (1916) had first studied the area, but they noted the still older, undisturbed forest was different from the 160 year-old one with pioneer plants absent and replaced by more typically forest plants. A rudimentary podsol was replaced by a well-developed one. An increase in cover by the growth and crown closure of the dominant *Picea engelmannii* would probably bring about this change in a matter of a few hundred years, so only a few centuries are evidently required for development of a mature forest over a well-developed podsol in this region. HEUSSER (1956: 276—7) describes both the mature forest and the one on the terminal moraine which was 170 years old at the time of his visit. The mature forest is evidently climax and, although its minimal age is four centuries, the time required for its development may be less. The maritime, coastal Alaskan climate results in more rapid plant succession than the continental, drier Rocky Mountain climate.

A still slower rate of succession to climax occurs on the north side of the Alaska Range at the upper limit of the forest-tundra. VIERECK (1966) believes that on outwash of various ages from the Muldrow Glacier climax *Betula glandulosa—Eriophorum vaginatum* tundra would form after 5000 years or so (see below).

A summary of vegetation development before retreating Swiss glaciers is given by LÜDI (1958). Before the low altitude terminus (1250 m) at the lower edge of the spruce belt at the Upper Grindelwald glacier *Alnus incana* and *Picea abies* had formed forests in only a little more than 100 years. Before the more continental Aletsch glacier at 1870 m elevation an open *Larix decidua-Betula pendula* woods with *Pinus cembra* reproduction (LÜDI 1945) had developed in less than 100 years. Maturity of the pine will establish the climax vegetation here.

In PLOCHMANN's (1956) study of spruce-fire (*Picea glauca—Abies balsamea*) forests of northwestern Alberta, Canada which have followed forest fires, succession is related to replacement of successive generations of different species of trees. Only a few life spans of these short-lived trees are therefore involved, and climax vegetation must be achieved in a few hundred years on most sites—although sites which differ ecologically also differ in species and rates of succession. However, PLOCHMANN recognizes that this mature forest breaks up, so the descending curve is followed. Also, fire is on at least as short a cycle as 150 years, so this length of time usually limits the appearance of the spruce-fir.

Successions in regions of more temperate climates are exemplified by OOSTING's (1956: 250—1) illustration and discussion of the succession from abandoned agricultural fields in the Piedmont of North Carolina through loblolly pine (*Pinus taeda*) to an oak-hickory (*Quercus-Carya*) climax forest in "200 years or more". In central New England CLINE & SPURR (1942: 40) thought that 400 years was a minimum time for development to climax vegetation. Farther north in southern Ontario MARTIN (1959: 215) believed 800 years would be necessary. In both cases *Tsuga canadensis* would be a very prominent member of climax vegetation in this region of mixed conifers and hardwoods.

There are few documented studies of long-time grassland succession, but SHANTZ (1917) believed that bared areas in eastern Colorado would return to the regional shortgrass vegetation of *Bouteloua gracilis* and *Buchloë dactyloides* in 20—50 years.

The recent study by WILLIAMS *et al.* (1969) of secondary succession initiated by bulldozer clearing within a state "Beauty Spot" in a moist subtropical forest in Australia is interesting from several stand-points: 1) By applying various analyses based on the computer's ability to handle large masses of instructions and data no rates of vegetation change were produced. In fact, the authors plotted their observations, which were made at intervals of 24 to 529 days, as equally spaced. 2) Instead of rates of change, highly interesting qualitative distinctions were made; a) Temporal variability separated after passage of only 1/5 of the period of 7 years from spatial, b) Spatial variability was marked and would condition the further course of succession. 3) It is unfortunate that the real paucity of quantitative studies of plant succession was amplified by attention only to work published in English. 4) The study illustrates plant ecologists' tendency to study non-quantitative characteristis of vegetation. 5) Is diversity a useful parameter with which to describe plant succession? Could simple observational knowledge of species' ecologies substitute for diversity measures?

TAGAWA (1964: 195) does use an index of diversity to illustrate primary succession from fresh lava to a climax forest of *Machilus thunbergii* in southwestern Japan off the island of Kyushu in 700 years. Rate of succession had a maximum in mid-course (*op. cit.* p. 224,). TAGAWA further found that secondary succession was 5—7x as fast as primary succession (*op. cit.* p. 209).

15 SPECIAL RUSSIAN METHODS IN CLASSIFICATION OF SUCCESSIONS

V. D. Aleksandrova

15 SPECIAL RUSSIAN METHODS IN CLASSIFICATION OF SUCCESSIONS

The dynamic approach to the study of vegetation is traditional in Russian Geobotany. This idea was already widely represented as early as in the end of the last and in the beginning of this century in works of DOKUCHAYEV (1885, 1892, 1899) and a number of eminent Russian botanists (KRASNOV 1888, 1894, PACHOSKY 1891, 1896, 1908, 1910, TANFILYEV 1894, TALIEV 1897, 1901, 1904, GORDYAGIN 1901, MOROZOV 1912, POPOV 1914, VYSOTZKY 1915, et al.). Very important for classification of successions were at that time works of VYSOTZKY who formulated several concepts associated with the effect of anthropogenic factors on vegetation and proposed a number of terms most of which are still used by Soviet geobotanists. Thus, he introduced the concepts of *digression* (degradation) of vegetation (including the *pasqual digression*, as result of grazing, the *phenisectial digression* as result of mowing, etc.) and *demutation* (re-establishment) of vegetation after cessation of the action of the destructive factor; the concept of *catastrophic digression* (corresponding to "katastrophale Sukzession" of GAMS 1918 and "catastrophic change" of TANSLEY 1935) including the *excisional digression* (caused by felling), *exarational digression* (caused by ploughing) etc.

The theoretical basis of classification of successions, adopted most widely in the U.S.S.R., was developed by SUKACHEV in his works published in the period from 1915 to 1964 (SUKACHEV 1915, 1928, 1938, 1942, 1954, 1964 et al.); its elaboration was also promoted by other Soviet geobotanists. The vegetation changes distinguished are primarily *secular changes* (RAMENSKY 1938, LAVRENKO 1940; otherwise: paleogenic, SUKACHEV 1928; general, YAROSHENKO 1961; phylocoenogenetic, SHENNIKOV 1964, etc.) and *short-term changes* (LAVRENKO 1940, otherwise: rapid — RAMENSKY 1938; particular — YAROSHENKO 1961, etc.). Secular successions are the changes of the plant cover over vast areas involving the changes of plant communities types as a result of the *phylocoenogenesis* process (SUKACHEV 1942) or the evolution of plant communities (YAROSHENKO 1961). BYKOV (1957, 1966) distinguishes three processes taking place at the time of *Phylocoenogenesis*: *speciogenesis*, constituting the evolutionary transformation of species; *ecogenesis*, constituting the change of phytocoenotic role of species; and *transgenesis*, associated with incorporation of new species into coenosis and with consequential exclusion of some species.

Short-term successions (Ramensky 1938, Lavrenko 1940, neocoenogenesis, Sochava 1944; local or particular changes—Yaroshenko 1961) are the changes of concrete phytocoenoses at the same place. According to Sukachev, they involve four processes: The first process is the *syngenesis*, i.e. "process of colonisation of the area by plants, struggle between them for space and for the means of existence, their accustoming to one another and formation of their interrelations" (Sukachev 1950: 460). The second of these processes is *endoecogenesis* or "the change of vegetation in consequence of change of environment by plants themselves by means of their vital activity" (ibid.: 460). The third process is *hologenesis* or "the change of the plant cover in consequence of change of a more wide-scope unity, a part of which is the given biogeocoenosis, say, of all the geographical environment or its parts such as athmosphere, lithosphere etc." (ibid.: 394). The fourth is the process of *heitogenesis* or "the process of succession of vegetation caused by the action of some factors external to the given phyto- or biogeocoenosis as a whole due to development of other natural phenomena, either more nearby or more remote, such as the expansion of some animals into the given phytocoenoses, its destruction by an avalanche, by the action of man" (ibid.: 305). In the same way *syngenetic*, *endoecogenetic*, *hologenetic* and *heitogenetic successions* are distinguished depending on the prevalence of one or the other of the four processes mentioned above. In the earlier works of Sukachev the latter two types of successions were regarded as one type designated as *exogenetic*.

Other characteristics are also considered in the classification of successions. Thus, Ramensky distinguished successions associated with disturbance of the plant cover (occasional or systematic) and successions not involving any disturbances of the existing equilibrium of the plant cover (series of balance-changed coenoses), while among the "exodynamic" successions two types are distinguished, viz. natural and cultural caused by the systematic activity of man (Ramensky 1938: 326—327). Lavrenko (1940) considered it expedient to distinguish among *short-term exodynamic successions* such categories as *pyrogenic*, *climatogenic*, *edaphogenic*, *zoogenic* and *anthropogenic* (the latter being defined as only those that are caused by the direct action of man's implements on the vegetation, otherwise designated as *laborogenic* successions sensu Aleksandrova 1964: 308). Shennikov (1964) supplements the classification of succession by the category of *geomorphogenic* successions belonging according his terminology to exogenetic successions. Yaroshenko (1961) subdivides *particular changes* into *natural* and *anthropogenic* ones distinguishing among the former ones *successive changes* (*endodynamic* and *hologenetic*) and *abrupt changes* (*climatogenic*, *edaphogenic* and *biogenic*); anthropogenic changes are also subdivided by this author into successive and abrupt changes.

The division of successions into *progressive* and *regressive* is also practiced in the Soviet geobotany. According BYKOV progressive successions are characterized by the relatively more complicated structure of coenoses, by the maximal utilization of space, by the most efficient utilization of the resources afforded by environment, by the increase of the coenoses productivity, by the presence of young endemic forms, by the recent speciation involving the adaptation to the phyto-environment, by the prevalence of the struggle for existence with the phyto-environment, by the mesophytization of phytocoenoses, by a conspicuous transformation of the coeno-environment. On the contrary, regressive successions are characterized by the relative simplification of the structure of coenoses, by the incomplete utilization of the environmental resources, by the decrease of coenoses productivity, by the presence of relic endemic forms and the speciation with adaptation to the coeno-environment, by the prevalence of struggle for existence with the coeno-environment, by either xerophytization or hydrophytization of coenoses and by slight transformation of the coeno-environment (BYKOV 1957: 273). Besides the characteristic features of progressive successions formulated by BYKOV, ALEKSANDROVA also takes into consideration the increase of the amount of matter and energy, involved by the phytocoenosis into its biological turnover, the decrease of the average entropy per individual, the increase of the average negentropy per individual, the higher degree of organization of phytocoenosis, the increase of the amount of information it contains (ALEKSANDROVA 1961, 1964: 311).

The views of the Soviet authors on the methods of classification of successions are rather similar to TANSLEY's (1920, 1929, 1935 et al.) that had developed independently. Thus, *syngenetic succession* sensu SUKACHEV can be compared to TANSLEY's *autogenic succession* (TANSLEY 1929: 680); like the Soviet authors, TANSLEY asserts that succession is not always progressive, as CLEMENTS insisted. The classification of successions developed by the school of CLEMENTS (1916, 1928, et al.) based on the characteristics of the site at which the succession begins, was but rarely and only partially used by Soviet geobotanists. Likewise CLEMENTS' concept of *climax* in its classical form was never too popular among Soviet geobotanists. The concept of "*aboriginal*" or "*completely formed*" association developed by SUKACHEV (1928), the same as "*stabilized coenosis*" of RAMENSKY (1938: 283—284) etc. are similar to the concept of climax sensu TANSLEY (1920, 1935) and WHITTAKER (1953: 54—57), but not in the sense of CLEMENTS. However the concept of climax-formation sensu CLEMENTS has its analogue in the Soviet Geobotany (the zonal vegetation of placors = sites of upland positions, VYSOTZKY 1909, 1927, 1930, ALEKHIN 1936, LAVRENKO 1947, 1957, et al.), excepting the requirement of successional connections with non-placoric formations, as it was already pointed out in the literature (WALTER 1942: 16, 1954: 146, WHITTAKER 1953: 42, LAVRENKO 1957: 1933, ALEKSANDROVA 1964: 317).

Classification of dynamic categories in connection with the large- and medium-scale geobotanical mapping is being developed at present by SOCHAVA and his collaborators (SOCHAVA 1962, 1963, 1968, GRIBOVA & SAMARINA 1963, GRIBOVA & KARPENKO 1964, KARPENKO 1965, ISACHENKO 1964, 1965, 1967, et al.) and also by some Estonian geobotanists (EILART & MASING 1961, MARVET 1964, 1967, 1968, MASING 1968, et al.). SOCHAVA operates with the concept of *aboriginal vegetation* (closely similar to climatic climax) and of *quasi-aboriginal* vegetation (closely similar to Dauergesellschaften, BRAUN-BLANQUET 1964, et al.); the concepts of *actual* (*autochtonic* and *anthropogenic*) and *potential* vegetation (SOCHAVA 1962: 15) are closely similar to TÜXEN's concepts (TÜXEN 1956, 1957 et al.). Long-term and short-term derivative communities are distinguished (after SUKACHEV); the term "*serial*" is used by SOCHAVA only for the designation of spontaneous communities; the anthropogenic successions are designated as "*series of transformations*" (SOCHAVA 1962: 17, 1963: 8 et al.).

A new approach to the more precise definition of the concepts associated with the classification of successions was applied by KARAMYSHEVA & RACHKOVSKAYA (1962), ISACHENKO (1964, 1965), GURICHEVA, KARAMYSHEVA & RACHKOVSKAYA (1967). This approach pertains to the establishment of successional relations between the combinations of phytocoenoses (the units closely similar to "Fliesen" sensu SCHMITHÜSEN by principles of their discrimination).

Very important for the classification of successions are the experimental studies on dynamic of plant communities carried out by Soviet investigators such as experimental studies in meadows (RABOTNOV 1958, 1960, 1967 et al.), and experiments in forest phytocoenoses (KARPOV 1961, 1969) and others.

Recently a direction in the Soviet Geobotany initiated using mathematical methods for studying dynamic features in the plant cover (LOPATIN 1956, 1967, RABOTNOV 1965, VASILEVICH 1968, BOCH 1968, et al.), which affords a possibility of the development of classification of successions on a new methodical basis.

16 SOME PRINCIPLES OF CLASSIFICATION AND OF TERMINOLOGY IN SUCCESSIONS

R. KNAPP

Contents

16 SOME PRINCIPLES OF CLASSIFICATION AND OF TERMINOLOGY IN SUCCESSIONS

16.1 Stages and Seres

The vegetation changes in the course of successions can be subdivided in stages. A "*stage*" (or "*stadium*") is distinct from the preceding or following successional steps in any way by attributes of species composition. This differentiation may be only a slight diversity in dominance or coverage (Bedeckungsanteile) of certain species, but can also be a great qualitative difference by abundant occurrence of species totally absent in the preceding or following stage.

Fig. 1. Various stages of a successional sere resulting in regeneration of tropical lowland rain-forests in Selangor, Malayan Peninsula. In the foreground, stage with dominant grasses; behind it, shrub stage. In the background: three secondary forest stages. Or.

A "*sere*" (or "*series*") includes all stages of a succession from earliest initial settlement or colonization of plants on bare areas until attaining the equilibrium of syndynamically terminal vegetation (terminal stages, climax vegetation with all its implications). A sere can include many stages belonging to several plant communities defined by differential and characteristic species. On the other hand in certain cases, seres can be observed only with stages differentiated weakly in their species composition. Such short and simple seres occur often in extreme habitats, e.g. in desertic, arctic or high alpine areas.

16.2 **Phases**

The terminus technicus "*phase*" refers to a syndynamic status within a particular plant community defined by characteristic and differential species, mainly to an association in the sense of the Zürich-Montpellier school (Braun-Blanquet 1928, 1964).

Fig. 2. Optimal phase of the *Ipomoea pes-caprae-Canavalia*-association (Hydrophylaci-Ipomoeetalia) on coral sands at the beach of southeastern Ceylon (Knapp 1965d). Or.

The stage within a sere best representing the association concerned (most rich in characteristic species or most approaching the total characteristic species composition) is the "*optimal phase*". It is preceded by the "*initial phase*" (or in case of occurrence in plurality by two or more initial phases). In the initial phase the characteristic properties of the association are already prevailing; but plants of the associations preceding in the successional development and absent in the optimal phase are occuring still in more or less great numbers.

The "*terminal phase*" (or terminal phases) supersedes the optimal phase in the course of succession. Also in the terminal phase, the characteristic attributes of the association continue to be dominant; but it is differentiated by pioneer plants of the following stages belonging syntaxonomically already to another association. Characteristic species of the association are mostly decreasing in number or in constance (respectively in presence, Stetigkeit) in the terminal phase.

Early Stages of Successional Seres

Fig. 3. Initial stage on sandy deposits of the Yamuna river in the plains of Hindustan, India. Successfull immigration and ecesis of *Tamarix* (*T. dioica*, *T. indica*). Or.

Fig. 4. Early successional stages on gravelly river deposits near Portage, Southern Alaska. Shrub stage with dominant willows (*Salix sitchensis*, *S. alaxensis*, *S. longistylis*) (Salicion sitchensis) and the subsequent cottonwood stage with dominant *Populus balsamifera* = *tacamahaca* and its *ssp. trichocarpa* (Populetalia trichocarpae). On the slopes in the background, conifer forests (dark tints) with dominant *Picea sitchensis* (Piceion sitchensis) and *Tsuga mertensiana* (at higher elevations, Vaccinio-Tsugion, Abieto-Tsugetalia mertensianae). Or.

Examples of early stages of successional seres were studied already by the first American and British authors specialized in syndynamics partially in several details (e.g. COWLES 1899, 1901, CLEMENTS 1905, 1916, WEAVER 1917, TANSLEY 1920). The terminology concerned is still useful. Certain special properties of plants able to settle bare areas are discussed already in another paragraph (pp. 105 110).

The *settlement* of a plant species on a bare area (*invasion*, *colonization*, *immigration*) has to be preceded by successful *migration* of a *propagule* to the place concerned. The propagules can be seeds (*disseminules*) or spores, but also mobile vegetative parts able to regenerate to an adult plant. Examples are parts of lichen thalli (isidium etc.) and certain small bulbs of higher plants (e.g. in *Allium* div. spec., *Ficaria verna* = *Ranunculus ficaria*, *Dentaria bulbifera* = *Cardamine b.*). Such propagules are transported by various means. In many plant species one of these means of transport is prevailing (anemochory, hydrochory, zoochory etc.).

Bare areas are often colonized by marginal invasion of pioneer plants from immediately neighboring areas, particularly by means of long stolons. An example of an invader species effective by this way is *Carex arenaria* on European sand dunes.

Immigration is finally successful when individuals of the immigrated species develop seeds, spores and other propagules. Such successful immigrations are denominated by the terms "*establishment*" or "*ecesis*".

The plants cover often only a small part of the soil surface in the initial stages. Competition and other mutual influences are consequently not essential in such cases.

During the next stages, seedlings develop often in the immediate neighborhood of the parent plants. Certain plant species of the initial stages are also able to spread extensively by stolons (development of polycorms). Thus, the vegetation replacing the open initial stages often is characterized by highly uneven distribution of individuals or stems of species (underdispersion). The area is a micromosaic of facies (in the sense of local dominance of a particular species, microfacies, KNAPP 1959a, 1960, 1967a).

The development of these *aggregations* of plants of one species originating from a single immigrated individual is named "*multiplication*". During the process of multiplication, a dense vegetation cover is often developing. Thus, mutual influences including competition and allelopathy become more and more important.

Fig. 5. Intermediate stage of an old-field succession in the "Gladenbacher Bergland" (Hessen, Germany). Aggregation of *Chrysanthemum leucanthemum* (in the central part of the figure, recognizable by the white flowers) surrounded by aggregations of the grass *Poa trivialis*. The aggregations (= micro facies) were originating from multiplication of single plants of the initial stage. Or.

16.4 Classification of Seres

Various views exist for the classification of seres. Most known is the classification on the basis of environmental conditions. Already COOPER (1913) and CLEMENTS (1916, 1928) delimited "*hydrarch successions*" or "*hydroseres*" beginning in water (in saline areas

Fig. 6. Subsequential to a woodland fire, a subsere (= secondary sere) is starting. Area in the Darling Range, West Australia, with *Eucalyptus* woodlands (e.g. *Eu. marginata*, *Eu. calophylla*) and sclerophyllous scrubs with *Proteaceae* (*Banksia div. spec.*, *Hakea div. spec.*, *Stirlingia latifolia*), *Myrtaceae* etc. Or.

"*haloseres*") and "*xerarch successions*" or "*xeroseres*" beginning on dry substrates, on bare rock ("*lithoseres*") or wind-blown sand ("*psammoseres*"). This type of classification can be more subdivided and related to life forms of plants following the ideas of Del Villars (1929, pp. 128 129).

Former influences of vegetation on the substrate colonized by the initial stages or its absence are essential for other classifications. Thus, a "*primary sere*" or "*prisere*" starts with the colonization of material not occupied by vegetation at any time before (e.g. on glacial morains after retreat of glaciers or on volcanic lava). The term is not correlated with degrees of human influences. Moravec (1969) distinguishes for this reason additionally naturally and anthropogenically influenced primary successions.

"*Secondary seres*" start on substrates already formerly covered and modified by vegetation after its destruction by any catastrophe, e.g. by fire, hurricanes or anthropogenic deforestation. The name "*subsere*" is also applied to such seres (Weaver & Clements 1938). The term "secondary sere" is used by some authors exclusively for successions released by man-made vegetation destructions.

The both groups are named by Swiss authors "*complete sere*" (Vollserie, = primary sere) and "*partial sere*" (Teilserie, = secondary sere) (Furrer 1922, Braun-Blanquet 1928, 1964, Lüdi 1930). These names exclude advantageously confusions with primary and secondary vegetation (in the sense of natural and anthropogenic vegetation).

The seres can also be classified by vegetational properties of their terminal stages (climax or permanent communities = Dauergesellschaften). This is the principle of the following classification by Braun-Blanquet (1964):

A. Terminal stages, with one layer (Schicht)
- I. Plankton seres. Competition for space only.
- II. Fungal, algal and bacterial seres. Competition mainly for space, light and nutrients
- III. Lichen seres. Competition mainly for space and light.
- IV. Moss seres. Competition dto.

B. Terminal stages with two or more layers. Competition for space, light, and nutrients
- V. Therophyte seres (e.g. on special sites in semideserts)
- VI. Grassland seres (e.g. in natural grass steppe, in natural prairie)
- VII. Eu-chamaephyte seres.
- VIII. Shrub seres
- IX. Forest seres

This classification is well in conformity with the length and differentiation of seres. In the first groups (I—V), the seres include only few stages of rather similar structure. The seres of the last groups (VI—IX) are beginning mostly with simply structured stages dominated by therophytes or cryptogams. They proceed to stages of higher organization and structure. The differentiation and number of stages is increasing usually from group VI to group IX.

Other classifications emphasize typical intermediate stages or initial stages. For instance, JENNY-LIPS (1930) distinguishes among other successions an Oxyrietum digynae-sere on siliceous screes (with initial stage belonging to the Oxyrietum digynae), a Stipetum calamagrostidis-sere on montane calcareous screes, a Thlaspietum rotundifolii- sere on high-alpine dry calcareous screes.

16.5 Terminology of Stages

The dominant species can be used for the denomination of a stage (e.g. *Sida cordifolia*-stage, *Baccharis salicifolia*-stage). In cases of dominance of a single species in different stages, another species should be added in the name to the dominant one differentiating the stages concerned (e.g. *Sarothamnus scoparius-Agrostis tenuis*-stage, *Sarothamnus scoparius-Betula pendula*-stage).

When a stage is identical with a plant community defined by any syntaxonomical classification system, this vegetation unit can be used for its name (e.g. Crucianelletum-stage, Scirpo-Phragmitetum-stage). In these cases the addition "-stage" sometimes is ommitted in the names (e.g. in certain successional schemes in BRAUN-BLANQUET 1928, 1964, HORVAT 1962).

16.6 Dynamic-constructive Evaluation and Classification of Species

Plant species are differentiated in the quantity of their syndynamical influence. Many species can attain not much influence on the progress of successions even under conditions most favorable for their growth. Reasons for these restrictions can be: (1) occurrence in relatively small numbers per area unit or in relatively low coverage (Bedeckungsanteile) also under best conditions (e.g. many species of *Orchidaceae*, *Gentiana* div. spec) and (2) growth forms or other properties which exclude important influences on associated plants or on soil and microclimate.

Certain classifications consider these properties in connexion with aspects whether a species is most influential for the initiation of a new stage of succession, for deterioration or conservation of the present stage. All these attributes important on the course of vegetation changes are summarized under the term "constructiveness" (Bauwert).

Species can be deciding for the development of the subsequent successional stages. Such species are called "*pioneers*" or "*constructive species*" (aufbauende Arten).

The reasons for the importance of pioneers can be diversified.

One group can alter the substrate conditions in a way that the soil becomes convenient for the next stage. E.g. mobile dunes are stabilized by the pioneer *Ammophila arenaria* and can be invaded consequently by vegetation not or only slightly resistant against covering by blown sand. The soil surface can be raised under stands of the pioneer species *Phragmites communis* in connexion with intensive peat accumulation in a way that plant communities less tolerant against flooding can develop. Account of its high pioneer importance, *Phragmites communis* is studied in details in its syndynamical behavior (e.g. BITTMANN 1953, HÜRLIMANN 1951, OREKHOVSKY 1969).

Another group of pioneer species influences the microclimate near the soil surface in a way that species of the next stages can grow up which are rather sensitive to irradiation influence, extreme

Fig. 7. Area in the southern Yukon Territory, Canada, near Whitehorse. The primary forests with dominant conifers (*Picea glauca*, Abieto-Piceetalia albae, on drier sites *Pinus contorta ssp. latifolia*) are destroyed by lumbering or by fires, except some remnants (visible in dark tints). Consequently, deciduous trees of intermediate stages dominate mostly in the area (mainly *Populus tremuloides*, on the figure highly contrasting to the conifers by the bright yellowish autumn colour of the leaves).Or.

temperatures and periodical desiccation. E.g. on old fields or on abandoned pastures in West German hill country on acid soils, *Sarothamnus* (= *Cytisus*) *scoparius*, *Betula pendula* and *Populus tremula* can be pioneers for beech forests (*Fagus sylvatica*). The beech seedlings are nearly unable to grow up in open areas; they are promoted by the microclimate in the slight shade of the pioneer shrubs and pioneer trees with less extreme temperatures and smaller water saturation deficits of the air.

Other species are deciding for the replacement of a stage. They destroy gradually the environment conditions indispensable for the stage concerned in the course of their invasion and spreading. They are called *destructive species* (abbauende Arten). The destruction can be attained also by different means: e.g. by shading of light demanding groups of plant species, by accumulation of litter with properties unbearable for certain species. Plants destructive for the former stage can be often simultaneously pioneer species with highly positive edificatory value for the following stage.

A third group of species is important for the consolidation and conservation of a stage (*consolidating* and *conserving species*, festigende und erhaltende Arten). The last group contains *indifferent* (*neutral*) *species* without special importance for successions.

The groups of species different in constructiveness were emphasized already by PAVILLARD (1919, 1928, cit. in BRAUN-BLANQUET) and later applied by BRAUN-BLANQUET (1928, 1964, B. and JENNY 1926) and other authors.

The constructive abilities of certain species is studied most detailed in coastal sand dunes of Europe and North Africa (e.g. KUHNHOLTZ-LORDAT 1923, BUROLLET 1927, CHRISTIANSEN 1927, VAN DIEREN 1934, BRAUN-BLANQUET & DE LEEUW 1936, PAUL 1944, 1953, WESTHOFF 1947, STEUBING 1949, SALISBURY 1952, LAMBINON 1956, GIMINGHAM, WILLIS et al. 1959, RANWELL 1960, FUKAREK 1961, ADRIANI & VAN DER MAAREL 1968, LUX 1969, GEHU 1969, TÜXEN 1970, GORDIYENKO 1970, RAZUMOVA 1970, KNAPP 1973). In connexion with projects on land reclamation, constructiveness is also considered extensively in early successional stages of coastal salt marshes (e.g. CHAPMAN 1960, WOHLENBERG 1969a, 1969b).

Also the highly detailed studies on successions in river valleys contain informative data on constructiveness of several species (e.g. in Central and Western Europe: KLIKA 1936, VOLK 1939, SAUBERER 1942, TCHOU 1948, 1949, SCHRETZENMAYER 1950, KÁRPÁTI, PÉCSI & VARGA 1952, KNAPP 1962c, 1967c, 1970a, SEIBERT 1962, WENDELBERGER-ZELINKA 1952, ELLENBERG 1963, HELLER 1969, MOOR 1969). The successions in fresh water are highly various in connexion with constructive abilities of the plants and the properties of the water (examples of recent work in Central Europa: HEJNÝ 1960, MÜLLER & GÖRS 1960, KNAPP & STOFFERS 1962, ELLENBERG 1963, NEUHÄUSL 1965, SEGAL 1965, KNAPP 1967c, HILD 1968, BORHIDI 1970, FIALA & KVĚT 1971).

17 VEGETATIONSENTWICKLUNGSTYPEN

E. Aichinger

17 VEGETATIONSENTWICKLUNGSTYPEN

Diesem System ist der Vegetationsentwicklungstyp als Einheit zugrunde gelegt. *Zu demselben Vegetationsentwicklungstyp fasse ich alle diejenigen physiognomisch einheitlichen Pflanzenbestände zusammen, welche sowohl in ihrem floristischen und soziologischen Merkmalen als auch in ihrem durch die Standortsverhältnisse bedingten Haushalt übereinstimmen und demselben Stadium einer Entwicklungsreihe angehören.*

Damit erfasse ich die Vegetationsentwicklungstypen in folgender Weise:

I. *Physiognomisch-floristisch.* Alle Vegetationseinheiten mit dem gleichen Erscheinungsbild werden zur selben Obergruppe gestellt. Z.B. erfolgt Zusammenfassung aller natürlichen Fichtenwälder zur Obergruppe "PICEETUM".

II. *Ökologisch-floristisch.* Die Vegetationseinheiten der Obergruppen werden nach ihren Umweltbedingungen zu ökologischen Gruppen vereinigt. Z.B. erfolgt Trennung der physiognomisch-floristisch erfassten Obergruppe "PICEETUM" in folgende Gruppen:

1. Gruppe der mehr oder weniger bodentrockenen, bodenbasischen Fichtenwälder: PICEETUM basiferens.
2. Gruppe der mehr oder weniger bodentrockenen, bodensauren Fichtenwälder: PICEETUM acidiferens. Diese können weiter in solche, deren Böden schon ursprünglich sauer waren (PICEETUM silicicolum acidiferens) und solche, deren Böden erst nachträglich oberflächlich versauerten (PICEETUM calcicolum acidiferens) untergliedert werden.
3. Gruppe der Auenwald-Fichtenwälder: PICEETUM inundatum.
4. Gruppe der bodenfrischen Unterhang-Fichtenwälder: PICEETUM superirrigatum.
5. Gruppe der Bruchwaldboden-Fichtenwälder: PICEETUM paludosum.
6. Gruppe der Hochmoor-Fichtenwälder: PICEETUM turfosum.

III. *Syngenetisch-floristisch.* Die Vegetationseinheiten innerhalb der einzelnen Gruppen werden als Glieder einer Vegetationsentwicklungsreihe betrachtet. Einige Beispiele mögen unsere Bezeichnungen in diesem Zusammenhang verdeutlichen.

↗ = zeigt die Entwicklung hinauf zu anspruchsvolleren Gesellschaften an (progressive Vegetationsentwicklung),
↘ = zeigt die Entwicklung herunter zu anspruchsloseren Gesellschaften an (regressive Vegetationsentwicklung).

— "etosum" = bezeichnet die Zugehörigkeit zu einem syngenetisch besonderen Untertyp, z.B.: Ericetum carneae ↗ PINETUM sylvestris ericetosum carneae ↗ Piceetum (also ein *Pinus sylvestris*-Wald, welcher der *Erica carnea*-Heide nahe steht) oder Ericetum carneae ↗ PINETUM sylvestris piceetosum ↗ Piceetum (also ein *Pinus sylvestris*-Wald, der ebenfalls in der *Erica carnea*-Heide aufgekommen ist, aber dem *Picea*-Wald schon nahe steht).

— "osum" = bezeichnet eine besondere fazielle Ausbildung (osus = reich an).

Bodentrockener, heidelbeerreicher *Picea abies*-Wald, welcher ein Verwüstungsstadium des Rotbuchen-Tannen-Mischwaldes auf ursprünglich basischem Boden ist und sich bei pfleglicher Wirtschaft wieder zum Rotbuchen-Tannen-Mischwald entwickelt = Abieto-Fagetum ↗ PICEETUM calcicolum acidiferens myrtillosum ↘ Abieto-Fagetum.

Pinus sylvestris-Wald, welcher in der *Erica carnea*-Heide aufgekommen ist und sich weiter zum *Picea*-Wald entwickelt = Ericetum carneae ↗ PINETUM sylvestris ericetosum carneae ↗ Piceetum.

Pinus sylvestris-Wald, welcher in der *Calluna vulgaris*-Heide eines Hochmoores aufgekommen ist und sich weiter zum *Picea*-Wald entwickelt = Callunetum vulgaris ↗ PINETUM sylvestris turfosum ↗ Piceetum.

Sekundärer *Vaccinium myrtillus*-reicher *Pinus sylvestris*-Wald, welcher ein Verwüstungsstadium des bodensauren Eichenwaldes ist und sich zum *Picea*-Wald entwickelt = Quercetum roboris acidiferens ↘ PINETUM sylvestris myrtillosum ↗ Piceetum.

Alnus glutinosa-Bruchwald, welcher im *Salix cinerea*-Buschwald hochgekommen ist und sich weiter zum Fichtenwald entwickelt = Salicetum cinereae ↗ ALNETUM glutinosae paludosum ↗ Piceetum.

Caricetum firmae, welches im *Dryas octopetala*-Bestand aufgekommen ist und sich zum Elynetum myosuroidis entwickelt = Dryadetum octopetalae ↗ CARICETUM firmae ↗ Elynetum myosuroidis.

Brachypodietum ramosi, welches nach Vernichtung des *Quercus coccifera*-Buschwaldes aufgekommen ist und sich wieder zum *Quercus coccifera*-Buschwald aufwärts entwickelt = Quercetum cocciferae ↘ BRACHYPODIETUM ramosi ↗ Quercetum cocciferae.

Sekundäre *Erica carnea*-Heide, welche nach Abhieb des streugenutzten *Erica carnea*-reichen *Fagus sylvatica*-Waldes aufgekommen ist und vom *Pinus sylvestris*-Wald abgebaut wird = Fagetum sylvaticae ericosum carneae ↘ ERICETUM carneae ↗ Pinetum sylvestris.

Die ökologischen Bezeichnungen basiferens und acidiferens werden der Einfachheit halber dort weggelassen, wo der erfahrene Pflanzensoziologe aus der Bezeichnung der Entwicklung oder der Dominanz ohnehin sieht, ob die Vegetation im basischen oder sauren Boden wurzelt.

Ökologische und syngenetische Differentialarten ermöglichen die Trennung der ökologischen Gruppe und der syngenetischen Vegetationsentwicklungstypen. (Hierzu und über weitere Einzelheiten der mitgeteilten Methoden: Aichinger 1951, 1954, 1957, 1967.)

Es ist mir klar, daß der primären Vegetationsentwicklung (primäre Dynamik) nur unter bestimmten Umweltbedingungen entscheidende Bedeutung zukommt, z.B. in der Verlandung der Seen, Bewaldung der Alluvialgebiete der Auenwälder und Schuttkegel, in Bergsturz- oder Flugsandgebieten; ebenso klar ist es auch, daß die vom Menschen ausgelöste Vegetationsentwicklung nach Kahlschlag, Niederwaldbetrieb, Streunutzung, Weidenutzung, Düngung, Mahd, Entwässerung, Bewässerung usw. eine viel grössere Rolle spielt (sekundäre Dynamik).

Dies gilt für die Alpentäler ebenso wie für die Mittelgebirge, Flachländer und nordischen Länder; denn selbst, wenn kein neuer Boden mehr besiedelt wird, keine Seen verlanden und die Flüsse keine Alluvionen aufschütten, verändern alle menschlichen Eingriffe die Umwelt und lösen damit eine neue Vegetationsentwicklung aus.

SUMMARY

Vegetation Development Types

A "vegetation development type" comprehends all stands similar in physiognomy, conform in floristic and sociological properties and in ecologically essential site conditions, and belonging to the same stage of a successional sere. Properties of the vegetation development types are expressed by their nomenclature. The first part of the names refer to the physiognomy and to the floristic composition (e.g. "PICEETUM"). The site connexions are dominated by adjectives (e.g. "PICEETUM basiferens", "PICEETUM turfosum"). The position in successional seres is characterized by mentioning the preceding and following stages; ascending or descending arrows indicate whether the connexions of the three stages refer to progressive or regressive successions (examples in the German text). Most successional changes in the Central European vegetation are of anthropogenic origin (in connexion with forestry, agriculture etc., secondary syndynamics). Primary vegetational successions occur only in special environmental conditions (e.g. in certain alluvial areas, successions in connexion with increasing natural organic deposits in lakes, on mountain screes, on sand dunes etc.).

F. PRODUCTIVITY AND CHEMICAL CHANGES IN SUCCESSIONAL STAGES

18 PRIMARY PRODUCTIVITY OF SUCCESSIONAL STAGES

H. LIETH

Contents

18 PRIMARY PRODUCTIVITY OF SUCCESSIONAL STAGES

18.1 Introduction

The primary productivity of successional stages is of interest for various reasons. The climax communities in the different biomes of the world operate with different levels of biomass and humus per unit area. The contribution of the successional stages before the climax equilibrium is reached must be of primary importance for the early establishment of the climax community.

The question of whether successional stages are more productive than the climax community [or close to climax (tenure)] is still unanswered for most successional series of the world. This also includes the comparison of different management forms of farms and forests with the potential primary productivity of the climax community of that same location.

The primary productivity is usually a good indicator for the functional capacity of any ecosystem. Its characteristics in natural communities, however, have not been emphasized very much until recent years. Only since the International Biological Program was organized around 1962 did productivity become a focal point for ecologists around the world, and this naturally includes the productive capacity of the successional stages.

This paper can touch only a selected number of viewpoints and also only a limited number of papers on the subject. The literature listing is, however, greatly expanded over the number of papers cited in the text. This will give interested readers the opportunity to consult further journals and books. The field is just now rapidly expanding, and contacting the listed authors for papers published after this summary is bound to be more informative than using this paper itself. A selection of criteria of interest in ecologically oriented successional studies is compiled in Table I and II.

Much of the earlier literature on productivity is compiled in LIETH (1962), RODIN & BAZILEVICH (1966), WESTLAKE (1963) and REICHLE (1970). These include aspects of productivity in successional stages.

18.2 The Concepts of Succession and Productivity

The traditional viewpoint in successional studies is to observe and explain the sequence of species that change completely from bare ground to climax vegetation. This is called primary succession

TABLE I. Tabular model of ecological succession: trends to be expected in the development of forest ecosystems (selected from ODUM 1966).

Ecosystem attributes	Developmental stages	Mature stages
Community energetics		
Gross production/community respiration (P/R ratio)	Greater or less than 1	Approaches 1
Gross production/standing crop biomass (P/B ratio)	High	Low
Biomass supported/unit energy flow (B/E ratio)	Low	High
Net community production (yield)	High	Low
Food chains	Linear, predominantly grazing	Weblike, predominantly detritus
Community structure		
Total organic matter	Small	Large
Inorganic nutrients	Extrabiotic	Intrabiotic
Species diversity—variety component	Low	High
Other attributes		
Nutrient exchange rate, between organisms and environment	Rapid	Slow
Role of detritus in nutrient regeneration	Unimportant	Important
Internal symbiosis	Undeveloped	Developed
Nutrient conservation	Poor	Good
Stability (resistance to external perturbations)	Poor	Good
Entropy	High	Low

if the development occurs for the first time on a given newly generated surface (outcrops, bare rock, new islands, fresh lava flow, melting glaciers). If the climax vegetation is accidentally removed (fire, lumbering, tornado) and the remainder regenerates, we speak of secondary succession. Productivity considerations were mostly all but forgotten in the classical studies. Productivity figures were only of interest when crops with or without fertilizers were compared with close-to-climax vegetation in similar habitats. In order to generate models one can think of different management types like fields, pastures, or certain tree plantations as arrested succession due to human influence. Fig. 1 shows a scheme of this principle.

Another point of view is that cultivated land held up at the same level of successional sequence and having quite similar features at this level would develop into a remarkably different

Table II. Characteristics of arboreal components of seral stages in tropical American humid forests (selected from Budowski 1965).

	Pioneer	Early secondary	Late secondary	Climax
Age of communities observed, years	1—3	5—15	20—50	more than 100
Height, meters	5—8	12—20	20—30, some reaching 50	30—45, some up to 60
Number of woody species	few, 1—5	few, 1—10	30—60	up to 100 or a little more
Floristic composition of dominants	*Euphorbiaceae*, *Cecropia*, *Ochroma*, *Trema*	*Ochroma*, *Cecropia*, *Trema*, *Heliocarpus* most frequent	mixture, many *Meliaceae* *Bombacaceae*, *Tiliaceae*	mixture, except on edaphic association
Natural distribution of dominants	very wide	very wide	wide, includes drier regions	usually restricted, endemics frequent
Number of strata	1, very dense	2, well differentiated	3, increasingly difficult to discern with age	4—5, difficult to discern
Upper canopy	homogeneous, dense	verticillate branching, thin horizontal crowns	heterogeneous, includes very wide crowns	many variable shapes of crowns
Lower stratum	dense, tangled	dense, large herbaceous species frequent	relatively scarce, includes tolerant species	scarce, with tolerant species
Shrubs	many, but few species	relatively abundant but few species	few	few in number but many species
Grasses	abundant	abundant or scarce	scarce	scarce
Growth	very fast	very fast	dominants fast, others slow	slow or very slow
Life span, dominants	very short, less than 10 years	short, 10—25 years	usually 40—100 years, some more	very long, 100—1000 y., some probably more

Table III. Yield potential for similar *Arrhenatherion* (grass)-stands in different habitats (adapted from Aichinger in Lieth 1962). Present hay yields of *Arrhenatherion* associations in t/ha.

Former vegetation	*t/ha*
peat bog	6
alluvial forest with stagnating soil water	6
alluvial forest near running water	8
upland forest, moist lower slope	8
upland dry forest, acid soil	5
upland dry forest, calcareous soil	4

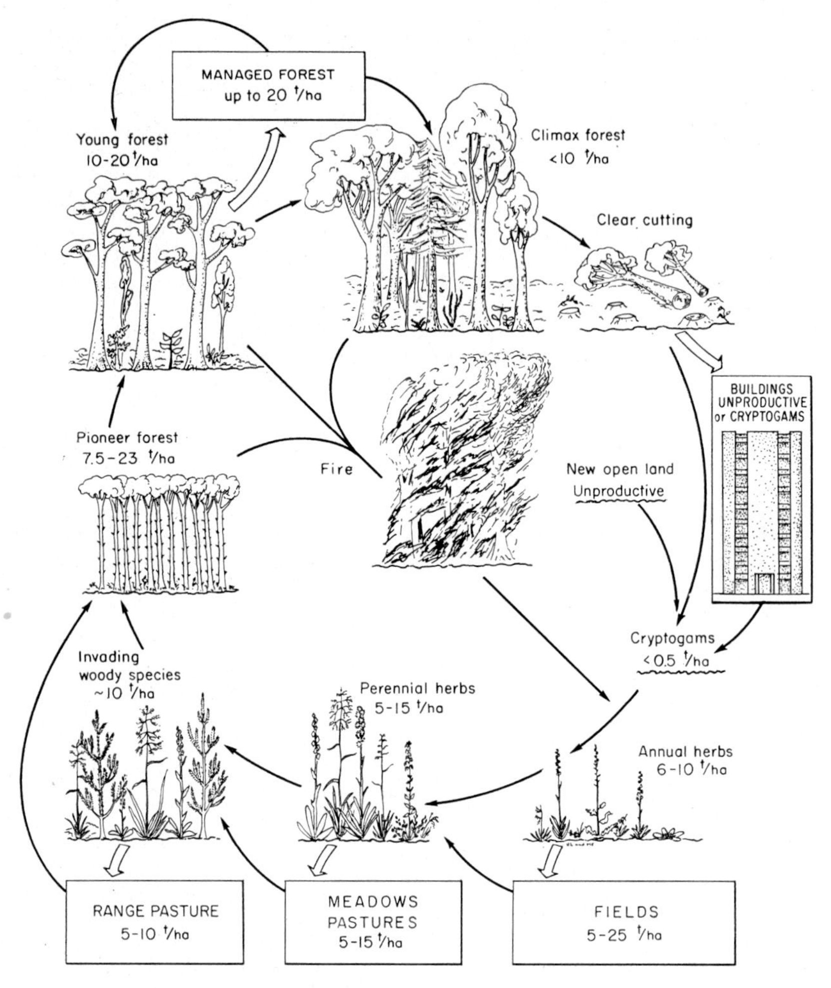

Fig. 1. Schematic diagram of the successional sequence in a deciduous forest area and the stages where human management arrested the successional sequence to create distinct management systems. Single lines indicate input and naturally occurring flow of succession. Double lined arrows lead to arrested stages, shown as labeled boxes. The numbers at each stage indicate the annual production of dry matter in metric tons/hectare. 1 t/ha = 100 g/m².

type of vegetation if left to the natural environmental forces. This situation was demonstrated by AICHINGER (1962). He calls such a unit a pleogenetic vegetation unit and points out that the productivity of a very similar set of species differs widely in its productivity (Table III).

TABLE IV. Productivity values from various stages of secondary succession. The figures represent dry matter produced within a 12 month period.

Successional stage	Biome type and geographical location species (N.A. = North America, E. = Europe)	Productivity (g/m²) a = above ground only t = total	Author
early herbaceous 1—3 years	old fields in deciduous forest, Georgia/N.A.	331—494(a)	ODUM (1960)
	open lake bottom, primarily *Lespedeza*, Tennessee/N.A.	471—1027(a)	DE SELM & SHANKS (1961)
late herbaceous 3—10 years	grassland in deciduous forest area 3—5 y, Germany/E.	685—1031(a) yearly averages over 3 years different habitats	BOEKER (1966)
	Urtica dioica in deciduous forest area on solid waste field, Germany/E.	909(a)	DAPPER (1966)
	Lespedeza on old lake bottom in deciduous forest area, Tennessee/N.A.	1041(a)	DE SELM & SHANKS (1961)
	old field in deciduous forest area, Michigan/N.A.	1280—1409 g(t)	GOLLEY (1960)
	Carex and *Juncus* growth in deciduous forest area, old lake bottom, Tennessee/N.A.	691—400(a) 332—378(a)	DE SELM & SHANKS (1961)
	Pteridium thicket in coniferous forest area, Black Forest, Germany/E.	577(a)	MEDINA & LIETH (1963)
early woody 5—15 years	young beech plantation in deciduous forest area, 9 y old, Denmark/E.	775(t)	MÖLLER, MÜLLER & NIELSEN (1954)
	forest fallow in Yangambi/Congo/Africa, 5—8 y in tropical, rain forest area	∾2000(t)	BARTHOLOMEW, MEYER & LAUDELOUT (1953)
	Salix thicket in deciduous forest area, old lake bottom, Tennessee/N.A.	1324—2150(a)	DE SELM & SHANKS (1961)
	Populus plantation in deciduous forest, 8—9 y old, Germany/E.	2300(t)	LIETH, OSSWALD, MARTENS (1965)
pioneer forest (including plantations) 10—40 years	*Pinus taeda* in deciduous forest, North Carolina/N.A., 16 y old plantation	1966(t)	WELLS & LIETH (unpublished)
	young beech plantation in deciduous forest area, Denmark/E., 25 y old plantation	1450(t)	MÖLLER, MÜLLER & NIELSEN (1954)

Usually there are a number of features that change during the successional development. The main trends of interest to ecologists are listed in Table I and II, compiled from ODUM (1969) and BUDOWSKI (1965).

Productivity values from primary successional stages are rarely measured. They measure only a few grams per m² and year if developing on rock or lava or other uninhabitable substrates. If the substrate is fertile, production starts with the arrival of the first seeds, as the polders in Holland and the orchid fields on volcanic ash in Hawaii show.

18.3 The Primary Productivity

Ecological productivity studies are made for a variety of reasons. With regard to succession they show either the increase of living biomass or total organic matter over a certain period of time, or they concentrate on a comparison of primary productivity in the consecutive successional stages. In more recent times full studies of energy dynamics, trophic level studies or complete systems analyses are being attempted.

The increase of biomass (standing crop) on any given location on earth is largely dependent on the environmental forces. In places where tundra or grassland forms the climax vegetation, the standing crop will never reach more than 5—25 t/ha. This level may be reached within 3—15 years. In places where forest formations form the climax, values are much higher. RODIN & BAZILEVICH (1966, 1967) give values of 400 t/ha in a little over 200 years for a tall oak forest in the temperate zone. Three-fourths of this value is reached within 100 years. Tropical forests can accumulate 120 t/ha within 10 years. One of the main characteristic differences between perennial herbaceous and perennial forest vegetation is the fact that herbaceous vegetation maintains about 50 % of the total maximum biomass measured each year in the underground root system. Forest, in contrast, rarely goes as high as 25 % investment in root biomass.

The total primary productivity (net production per year) is difficult to measure in perennial vegetation. It is only through the efforts of the International Biological Program that these values are now being gathered at many places and will be available in another decade. For a number of places figures are available for various stages in successional series. Table IV shows a selection of values, mostly from the temperate zone. A number of additional figures have been presented in a preliminary form, and will soon appear in literature. The recent UNESCO symposium in Brussels on forest ecosystems will contain figures from various successional stages and many different forest types.

The ecosystems analyses incorporate successional productivity studies as one parameter for the total model. Work in this direction has only recently been started. Partial elaborations are available in the books of SUKACHEV & DYLIS (1964), DUVIGNEAUD (1967) and ODUM (1970).

19 BIOMASS ACCUMULATION IN SUCCESSIONS

J. Major

19 BIOMASS ACCUMULATION IN SUCCESSION

Some generalizations on plant community biomasses and their accumulation during succession are offered by WOODWELL & WHITTAKER (1968). Their overall curve agrees more or less with Fig. 1. (of article 2). They postulate oscillations following a maximum although they define the climax as a steady state. In their discussion of a seral pine forest (WHITTAKER & WOODWELL 1968, 1969), they calculate its respiratory losses are 83 % (correcting their net ecosystem production figures) of annual gross productivity when 100 % would be a steady state. This criterion cannot be used alone, however, since the increasing rate of litter losses (tree mortality, branch and root shedding, etc. cf. RODIN and BAZILEVICH 1965, 1967) is also important. A 17 % growth rate on the curve of Fig. 1 would be a very high growth rate indeed.

In interior Alaska VAN CLEVE, VIERECK & SCHLEUTNER (1970) have studied biomass and N accumulation in sandbar succession along the Tanana River near Fairbanks (64°45′ N, 148 °W, 133 m elevation). The area has a very continental climate with annual mean temperature —3.6 °C, amplitude of monthly means 38.8 °C, precipitation 305 mm/yr, potential evapotranspiration 440 mm/yr, actual evapotranspiration 280 mm/yr, water surplus perhaps 10 mm/yr, deficit 150 mm/yr. The vegetation consisted mostly of *Alnus incana* ssp. *tenuifolia* with willows (*Salix brachycarpa* ssp. *niphoclada* giving way to *S. alaxensis*). Over 20 years, dry weight of biomass accumulated at a linear rate of 350 g/m² . yr. An additional 100 g/m² . yr accumulated as dead, standing organic matter.

Shifting to an entirely different kind of ecosystem, BARTHOLOMEW, MEYER & LAUDELOT (1953) present biomass data for forest fallow vegetation at Yangambi (0°46′N, 24°25′E, 470 m elevation) in the Congo. Here annual mean temperature is 24.4 °C, amplitude of monthly means 2 °C, precipitation 1870 mm/yr, potential evapotranspiration 1310 mm/yr, actual evapotranspiration also 1310 mm/yr, deficit zero, surplus 560 mm/yr (THORNTHWAITE & MATHER 1962). Over an 18-year period the new forest of *Musanga cecropioides* R. Br. accumulated biomass at first at the maximum rate of 2,660 g/m² . yr with an additional 240 g/m² . yr dead material. The rates appear to decrease after 5—8 years. However, if a 50-year old secondary forest in Ghana on a soil with much more soil nitrogen, exchangeable K, and especially exchangeable Ca and Mg (GREENLAND & KOWAL 1960) can be placed in the same chronosequence, the scatter of the original points looks more like random error than a smooth curve. The rate of increase of biomass then drops to 420 g/m^2 . yr after 10 years with an additional 100 g/m^2 . yr dead material.

Considerable data on increase of above-ground biomass in forest stands with time for different tree species or broadly different ecosystems are collected and diagrammed by RODIN & BAZILEVICH (1967: 210). Since many biomass data do not include below-ground parts, rates of accumulation of above-ground biomass from this diagram are of interest. A caution is that the data plotted for one species do not form a chronofunction, but the curves are composites of data from several different kinds of sites brought arbitrarily into one time series. Taking the African tropical rain forest data as being more or less ecologically homogeneous, however, they show a rate of above-ground biomass accumulation of about 1900 g/m^2 . yr for the first 10 years, decreasing to 300 g/m^2 . yr toward 50 years. Since the root proportion of the biomass in this kind of vegetation is roughly 20 %, the check with the rates given above is not exact. Discrepancies are due to the grossness of the diagram in the latter case. Other rates of accumulation of above-ground biomass in RODIN & BAZILEVICH's tree stands are 290 g/m^2 . yr for spruce (*Picea abies*) and 215 g/m^2 . yr for *Pinus sylvestris*. The oak forests in this diagram are definitely from two different locations, each of which we shall treat as a chronofunction below.

Instead of using only RODIN & BAZILEVICH's (1967) summary, we can find by following up their references several stands of different ages of a given tree species segregated into regional series which may be considered chronofunctions or successional series. REMEZOV, BYKOVA & SMIRNOVA (1959) give such data, but they have in many cases neglected the understory vegetation. Where RODIN & BAZILEVICH use REMEZOV et al.'s data, they add in calculated amounts to make biomass figures, and I have preferred their figures.

In all these studies of chronofunctions of forest stands the development of monospecific, even-aged stands of trees could be regarded not as a plant succession — a replacement of one plant community by another — but as simply growth stages of a kind of forest. The distinction between plant succession and forest growth is not absolute, however, and in fact most of the forest chronofunction data to be discussed below do indicate some replacement of the original species by other species within the total biomass.

REMEZOV et al. (1959: 104, RODIN & BAZILEVICH 1967: 44ff) give biomass data for 51, 83, and 115 year old mixed spruce forests on well-drained soils of site class I near Veliki Luki in European

Russia (56°21′N, 30°31′E, 105 m elevation). Here annual mean temperature is 4.8 °C, annual range of monthly means 24 °C, annual precipitation 543 mm, potential evapotranspiration 547 mm/yr, actual evapotranspiration 466 mm/yr, water deficit 81 mm/yr, and surplus 77 mm/yr. Rate of biomass accumulation is 115 g/m² . yr for the period of record but 335 g/m² . yr overall. Spruce-green or feather moss forest occurs in the same region but on more level terrain, on site II (REMEZOV et al. 1959: 69). Its rate of biomass accumulation after an initial lag to 20 years is 486 g/m² . yr, decreasing after 70 years, with an overall (93 yrs) rate of 338 g/m² . yr.

Pinus sylvestris L. in the Smidovich reserve in the Mordvinian ASSR (54°55′N, 43°15′E, $<$ 200 m elevation) occurs in a feather reaches 438 g/m² . yr, then declines after 50 years. Overall rate is 304 g/m² . yr.

Continuing to operate under the idea that different areas of different ages since t_0 can be legitimately strung together in a chronosequence, data from forest plantations can be used. OVINGTON'S data on tree (1957 : 305) plus ground flora (1959b : 230) in *Pinus sylvestris* plantations in England show a quite constant rate of increase of above- and below-ground biomass of 560 g/m² . yr from about 15 to 35 years, with progressively lower rates before that toward t_0. For natural *Betula pendula* and *B. pubescens* mixed stands in England on fen soils OVINGTON and MADGWICK (1959a : 279) found above-ground rates of increase of standing tree crop averaging 435 g/m² . yr from 20 to 55 years. Roots $>$ 0.5 cm (diameter) add 93 g/m² . yr. Both curves increase in slope somewhat with time. In Mississippi, USA, with a climate at West Point (33°25′ N, 88° 34′W, 72 m elevation) having precipitation of 1269 mm/yr, potential evapotranspiration 948 mm/yr, actual evapotranspiration 788 mm/yr, water deficit 160 mm/yr, and water surplus 598 mm/yr SWITZER, NELSON and SMITH (1966) found weight of above-ground dry matter of sample trees assembled into pure, even-aged stands of *Pinus taeda* followed logistic curves over 60 years with rates of increase at the inflections of 900 g/m² . yr on lowland sites of high quality (site index 32.1 m at 50 years) and 502 g/m² . yr on ridgetop sites of low quality (SI = 19.9 m). The authors give average rates over the first and second 30 years of 560 and 282 g/m² . yr on good sites and 303 and 259 g/m² . yr on poor. *Pinus radiata* planted as an exotic in New South Wales, Australia, over the first 12 years gave above-ground biomass increase of 1540 g/m². yr after a low rate of ca. 200 g/m². yr until 5 years (FORREST & OVINGTON, 1970). WIEGERT and MONK (1972) show a rate of increase of energy contained in above-ground standing crop of $dV/dt = 0.82\ V$ where V =

kcalories/m. and t = time in years from 7 to 13 for *Pinus palustris* plantations in South Carolina, USA. On the other hand, they present data from OVINGTON (1959a) which show a constant rate of increase of above-ground caloric content of *Pinus sylvestris* in England over 7—14 years, the rate being 1420 cal/m² . yr.

These are only some examples for biomass accumulation during the development of forest plantations published in recent times. An highly ample additional material of data can be found in the literature of forest science (e.g. WIEDEMANN 1951, SPURR 1952, ASSMANN 1961, PRODAN 1968; in these books collections of farther references).

For biomass accumulation in an English (Moor House) blanket bog with *Calluna vulgaris*, *Eriophorum vaginatum* and *Sphagnum* GORE & OLSON (1967) have modeled production and decomposition to produce accumulation curves with a maximum slope of 90 g/m² . yr, whereas standing dead material accumulated at the rate of 53 g/m² . yr, exclusive of the *Sphagnum*. Moor House in the Northern Pennines at about 550 m elevation with average temperature 6.7 °C, mean summer temperature 13 °C and winter —2 °C, annual precipitation 1950 mm/yr (STAMP 1969: 159), and a continuous excess of precipitation over potential evapotranspiration.

In southeastern Australia successions following fires in heath vegetation have been described by SPECHT, RAYSON & JACKMAN (1958) and GROVES & SPECHT (1965). The climate on the dry end is mediterranean with annual mean temperature of 15.0 °C, range of monthly means 11.7 °C, precipitation 457 mm/yr, potential evapotranspiration 771 mm/yr, actual evapotranspiration 457 mm/yr, water deficit 314, water surplus 0 (SPECHT & RAYSON 1957). Soils are nutrient-poor deep sands or sandy ground water podsols. The accumulation of biomass is influenced by the differential recovery rates of the perennial shrubs making up the vegetation. *Banksia ornata*, the ultimate dominant in some stands, grows rapidly only after some 10 years when *Casuarina pusilla* begins to die out. After 25 years from t_0 above-ground biomass had accumulated at an overall rate of 110—120 g/m² . yr on sand heath. On the ground water podsols in a wetter climate rates were higher, 160 g/m² . yr, but only to 5 years or less when they leveled off. Roots were about 5x tops in the sand, 4x in the ground water podsol.

EGUNJOBI (1969) studied rates of biomass and N-accumulation in secondary successional recovery of gorse, *Ulex europaeus*, after burning near Wellington, New Zealand. The climate at the experimental site (Taita) has precipitation of 1450 mm/yr, and if potential evapotranspiration is the same as at Wellington it would be 667 mm/yr, which is also actual evapotranspiration. Water deficit is zero and surplus 783 mm/yr. Biomass data were only for 1, 4, 7 and 10 years after t^1. There appears to be an inflection at 1—4 years with a rate of biomass accumulation of 2600 g/m².yr. Then the curve levels off to 470 g/m²yr.

An aspect of biomass accumulation is litter accumulation. JENNY, GESSEL & BINGHAM (1949) have treated this subject in detail. By using their determinations of production and decomposition rates of forest floor material we can calculate initial rates of accumulation using the equation of Fig. 1. JENNY et al. give times necessary to attain 95 % of equilibrium. At t_0 rates of accumulation of organic litter are low in a high altitude, sparse *Quercus kellogii* forest in the Sierra Nevada of California (78 g/m² . yr). The rate is a maximum (140) at a somewhat lower altitude and decreases (to 108) below in a drier climate. *Pinus ponderosa* at the lower altitude has a somewhat higher rate of accumulation than the oak (139), and the rate increases with altitude to 298 although there is considerable variability. The average rate of increase of litter at t_0 for the oak is 110 g/m² . yr and 200 for the pine. At still higher altitudes in a subalpine belt *Pinus contorta* spp. *murrayana* has a low rate of moss type with *Vaccinium vitis-idaea* L. on site I. The climatic data for nearby Yelatma (54°58′N, 41°45′E, 140 m) show average yearly temperature of 3.7 °C, range of monthly means 21 °C, precipitation 546 mm/yr, potential evapotranspiration 557 mm/yr, actual evapotranspiration 425 mm/yr, water deficit 132 mm/yr, and surplus 121 mm/yr. Rate of biomass accumulation is almost linear at 394 g/m² . yr which is also the overall rate (REMEZOV et al. 1959: 29, RODIN & BAZILEVICH 1967: 44ff). In pine with an understory of basswood (*Tilia cordata*) data for only 2 stands (36 and 57 years old) are given by REMEZOV et al. (1959: 51). Rates of biomass accumulation are 490 g/m² . yr at first, decreasing to 344, and 437 overall.

Oak (*Quercus robur*) in the Tellerman experimental forest in the forest-steppe (MINA 1955, RODIN & BAZILEVICH 1967: 117), located at about 51°10′N, 42 °E, elevation < 200 m on site I, has a rate of biomass accumulation of 360 g/m² . yr around age 50 and decreases thereafter so the overall rate to 220 years is 229 g/m² . yr. The climate here may be best represented by that nearby (to the south) at Uryupinsk (50°48′N, 42°00′E, elevation 86 m). The Tellerman forest is about 200 km east of Voronezh, just to the north of which REMEZOV et al. (1959: 175, 1964) collected other oak forest data. Uryupinsk has a cooler, more continental, drier climate than Voronezh. Mean annual temperature at Uryupinsk is 1.6 °C (Voronezh 5.6 °C), ranges of monthly means are 39 and 31 °C, precipitation 393 and 480 mm/yr, potential evapotranspiration

629 and 603 mm/yr, actual evapotranspiration 377 and 396 mm/yr, water deficits 252 and 207 mm/yr, and surpluses 16 and 84 mm/yr. According to RODIN & BAZILEVICH (1967: 117) the data on biomasses of the trees in the oak stands 40 km north of Voronezh on class I sites should be increased by 3,000 g/m² in the young stands, probably more in the older which have an undergrowth layer of higher coverage, to give actual biomasses. The rate of actual biomass accumulation, then, is 533 g/m² . yr, decreasing after 50 years and with an overall rate of probably 290 g/m² . yr over 130 years.

In the same region aspen (*Populus tremula*) stands 10, 25, and 50 years old on site I (REMEZOV et al. 1959: 210, RODIN & BAZILEVICH 1967: 116) gave a maximum rate of biomass accumulation of 1090 g/m² . yr after an initial lag period, followed by a declining rate after 25 years. The overall rate to 50 years was 610 g/m² . yr.

Tilia cordata stands on site II in the Mordvinian reserve (see above) are represented by 13, 40, and 74 year old stands (REMEZOV et al. 1959: 150, RODIN & BAZILEVICH 1967: 117) following cutting of mixed spruce. After an initial lag the rate of biomass accumulation 100. At mid-altitudes the very productive mixed coniferous forest has 420. These rates contrast with those from Colombia in South America where a lowland station (Calima at 30 m elevation) has a rate of 270 whereas a station at 1630 m has a rate of 570 g/m² . yr.

Times to equilibrium are only 3 years in the lowland tropics, 6 in the upland, 24—52 for black oak, 100 for mixed conifers, 100—200 for *Pinus ponderosa*, and 330 for lodgepole pine.

Using the same approach, in the annual grassland on the western slopes of Mt. Hamilton in the central Coast Range of California (37°20′N, 121°39′W, 1310 m) data on (possibly equilibrium) nitrogen content of soils and annual addition to soil N by the roots of the almost exclusively therophytic vegetation allow calculations of initial rates of accumulation of soil N and times to equilibrium (MAJOR 1953). Soil content of N to 50 cm depth were 300—800 gm². Annual additions in roots were 3—6 g/m² which are also essentially the rates of accumulation of soil nitrogen. Times to 95 % of equilibrium were 100—200 years. The climate on this altitudinal transect from 500 to 1300 m changed as follows: Mean annual temperatures from 15 at low to 11.5 °C at high altitudes, range of monthly means 12 to 16.6 °C, precipitation 400 to 744 mm/yr, potential evapotranspiration 730 to 672 mm/yr, actual evapotranspiration 360 to 315 mm/yr, water deficit 370 to 357 mm/yr, surplus 40 to 429 mm/yr.

In southeastern Alaska (see below under nitrogen accumulation) there have been several determinations of litter or forest floor accumulation. At Glacier Bay (CROCKER & MAJOR 1955: 436, 445) over 50 years in alder (*Alnus crispa* ssp. *sinuata*) thickets after a 10-year delay an almost linear rate of litter accumulation of 140 g/m^2 . yr held to 40 years when it decreased. Over a longer time under successive kinds of vegetation the rate was about 96 g/m^2 . yr after an initial lag, followed by a decrease after 80 years. The overall rate was 54 g/m^2 . yr. UGOLINI (1966: 69) found a maximum rate of 150 g/m^2 . yr at about 40 years with the curve flattening by 150 years. Near Juneau CROCKER & DICKSON (1957: 180) found more or less linear early parts of the curves (180 g/m^2 . yr for the Mendenhall glacier and 90 for the Herbert) which flattened out at 120 and 150 years respectively. The rates to these equilibria were 134 and 67 g/m^2 . yr.

Still farther south on Prince of Wales Island (55°25'N, 133°W) near Ketchikan forest develops on landslides from pioneer *Alnus rubra* to *Picea sitchensis* and after 200—300 years to *Tsuga heterophylla* (GREGORY 1960). The forest floor increases in depth at about 0.4 mm/yr or 58 g/m^2 . yr.

In the colder and more continental, as compared to southeastern Alaska, subalpine belt of the Swiss National Park the study of river terrace plant succession over dolomitic gravel (BRAUN-BLANQUET 1964: 669—673, BRAUN-BLANQUET, PALLMANN & BACH 1954: 43, 153—164, PALLMANN 1942: 166) (see below under nitrogen accumulation) shows an increase in depth of A0+A1 horizons of about 0.35 mm/yr after an initial lag of a few decades. Organic matter in these humic soil horizons (the regional soil is an iron podsol) accumulates at the rate of 66 g/m^2 . yr.

In the Piedmont of North Carolina BILLINGS (1941) found that the linear rate of increase of soil organic matter in the A1 horizon under *Pinus echinata* was 0.02 %/yr to 110 years. At t_0 the soil contained 0.75 % organic matter.

The forests discussed above are not in a steady state. Most are too young, but even the old ones have true increment of biomass of 40 g/m^2 . yr for the 220 year-old oak forest to 15 g/m^2 . yr for the older *Tilia cordata* stand. RODIN & BAZILEVICH (1967: 210 ff) even calculate a true increment of biomass of 670 g/m^2 . yr for an average tropical rain forest, 125—200 years old. Their data on old coniferous forests all show true increments of biomass — 200 g/m^2 .yr in 200 year-old spruce in the Archangelsk Province down to 60 g/m^2 . yr in a *Pinus sylvestris-Sphagnum* bog (op. cit. p. 44 ff). For central European beech (*Fagus sylvatica*) they list a true increment of biomass of 400 g/m^2 . yr (op. cit. p. 116) at 120 years.

20 NITROGEN ACCUMULATION IN SUCCESSIONS

J. MAJOR

Nitrogen in ecosystems has been determined as a function of time of plant succession in the entire ecosystem, in its biomass, and in its soil alone.

JENNY's replicated study (1965) of N accumulation in vegetation and soil on land uncovered by the retreating Rhone glacier found $dN/dt = 0.96\ g/m^2 \cdot yr$ at t_0 and $dN/dt = 0.69\ g/m^2 \cdot yr$ at 300 years. Following the curve hypothesized by JENNY (the solid line of Fig. 1, p. 14), the asymptotic value of 963 g $N/m^2 \cdot yr$ in this ecosystem would be 95 % reached by 3,000 years, and the overall rate of N accumulation in the ecosystem would have been 0.30 g $N/m^2 \cdot yr$. The area, at 1800 m elevation, is in the subalpine belt (LÜDI 1958: 394) and has an average yearly temperature of 1.6 °C, precipitation 1700 mm/yr, potential (and actual) evapotranspiration about 400 mm/yr, water deficit zero, and surplus for leaching 1300 mm/yr. JENNY (1965) specifies the levels of the other state factors determining this ecosystem.

The splendid study of river terrace plant succession over dolomitic gravels in the Swiss National Park (BRAUN-BLANQUET 1964: 669—673, BRAUN-BLANQUET, PALLMANN & BACH 1954: 43, 153—164, PALLMANN 1942: 166) which has been discussed twice above shows an initial lag in accumulation of soil nitrogen in the A0+A1 horizons to a few decades when annual flooding ceases and then a linear increase to about 1000 years of 1.3 g $N/m^2 \cdot yr$. The climate at Buffalora (46°39'N, 10°15'E, 1977 m elevation near the Ofenpass has a mean annual temperature of —0.1 °C, range of monthly means 20.4 °C, precipitation 923 mm/yr, potential and actual evapotranspiration 400 mm/yr, no water deficit, and surplus 523 mm/yr.

Other studies of accumulation of N in the soil portion of ecosystems have been made in maritime, relatively warm, coastal southeastern Alaska. This is the area studied by COOPER (1923, 1931, 1939), LAWRENCE (1958), and others (see below). Climatic data for the region are given in Table I and the climate is discussed by LOEWE (1966). At Glacier Bay (58°30'N, 136 °W) (CROCKER & MAJOR 1955) inland to Muir Inlet annual rates of N accumulation in soil under *Alnus crispa* ssp. *sinuata* thickets are about 8.0 into the soil as a whole after an initial lag period of about 18 years, decreasing after 40 years. The mineral soil (to 61 cm depth) accumulates N at a linear rate of 2.6 $g/m^2 \cdot yr$. The overall rate over 40 years into

Table I. Climates in southeastern Alaska.

Station Latitude (N), Longitude (W)	Topographic position	Elevation	Mean annual temperature	Amplitude of monthly mean temperature	Precipitation	Potential evapotranspiration	Actual evapotranspiration	Water deficit	Water surplus
		m	°C	°C	mm/yr	mm/yr	mm/yr	mm/yr	mm/yr
Cape Spencer 58°12′N, 136°38′W	Facing the Pacific Ocean	25	5.8	11.0	2924	533	533	0	2391
Gustavus CAA 58°25′N, 135°42′W	Mouth of Glacier Bay	7	5.1	15.2	1419	512	512	0	906
Haines 59°13′N, 135°26′W	Head of Lynn Canal	78	4.7	18.8	1531	511	423	88	1108
Juneau 58°18′N, 134°24′ W	Mainland, protected by islands from Pacific	22	5.7	16.1	2134	533	533	0	1501
Juneau Airport 58°22′N, 134°35′W	More exposed than Juneau	5	4.6	17.4	1404	501	501	0	903

the soil as a whole is 6.8 g/m^2 . yr. Over a longer time span (to 200 years) as the short-lived alder is replaced by poplar and spruce, the rate of N accumulation peaks at about 80 years and declines thereafter at the rate of 1.0 g/m^2 . yr). The maximum rate of increase occurs just prior to 50 years (5.4 g N/m^2 . yr). The mineral soil alone has a maximum rate of 1.65, and after about 125 years decreases at 0.2 g N/m^2 . yr. The overall rate to the peak is about 3.5 g N/m^2 . yr (280 g N/m^2 in the soil after 80 years).

Ugolini (1966, 1968) later studied chronofunctions of soil properties in this same area. He concluded the soils were formed under the impact of both a changing biota (flora) and autogenic plant succession (time). However, the disseminule rain over the entire lowland area can be taken as at least qualitatively constant, and this is the only even theoretically operational definition of the flora as an independent ecosystem factor that we have (JENNY 1941, 1958, 1965, CROCKER 1952, MAJOR 1951). As noted above, we consider the vegetation properties, including soil properties dependent on vegetation and also the species composition of the vegetation which has been selected from the available flora by the particular levels of independent state factors of the ecosystem, to be dependent. We take UGOLINI's data as chronofunctions. His rate of N accumulation in the mineral soil to 25 cm depth was also linear at 0.60 g N/m^2. yr. It showed no maximum to a 250-year old podsol. The forest floor added 3.3 g N/m^2 . yr after an initial lag, but a maximum was reached after 50 years or so and the rate then became negative. The 250-year old podsol showed a N accumulation rate overall of 0.9 g/m^2 . yr.

Some 70 km to the southeast of Glacier Bay on the mainland near Juneau CROCKER & DICKSON (1957) studied the terminal areas of the Herbert (58°33′N, 134°40′W) and Mendenhall (58°24′ N, 134°35′W) glaciers. Maximum rates of N accumulation in the mineral soil (to 61 cm depth) were about 1.5 g/m^2 . yr for the Herbert and 1.7 for the Mendenhall. After 100 years the rates decreased. The N accumulation rate in the total soil was a maximum shortly after an initial lag and was 3.0 g/m^2 . yr for the Herbert and 4.3 for the Mendenhall. Overall rates to 200 years for the soil part of the ecosystem were 1.5 g N/m^2 . yr for the Herbert and 1.8 for the Mendenhall.

STORK's data for the Stor glacier moraines in the Kebnekajse area of northern Sweden (1963) are unfortunately for our purposes not on a comparable areal basis. Her data show a more or less linear rate of increase, after a beginning maximum rate, of 0.2 mg N/100 g of the $<$2 mm fraction of the soil per year. On a comparable basis VIERECK's (1966) data for the Muldrow glacier moraines in central Alaska show an initial lag, then a maximum rate of 1.7 mg N/100 g soil . yr from 50—100 years, and then a greatly reduced rate of 0.28 mg N/100 g soil . yr.

In the study of sand bar establishment of alder (*Alnus incana* ssp. *tenuifolia*) along the Tanana River of central Alaska which was discussed above under biomass, VAN CLEVE, VIERECK & SCHLEUTNER (1970) obtained complete ecosystem data on N contents as well as biomass data. Nitrogen accumulates in the total ecosystem at a decreasing rate, after 5 years becoming a more or less linear rate. From t_0 to 5 years N accumulates in the total ecosystem at a rate ot 36.2 g/m^2 . yr, in the vegetation alone 4.8 with 3.6 living and 1.2

dead litter, and in the mineral soil 31.4 g/m^2 . yr. Over the next 15 years the rates are 9.4 for the ecosystem, vegetation alone 2.6 with 2.1 living and 0.5 dead (standing since litter does not increase), and mineral soil 6.8. Overall rates over the 20 years are 18.0 g/m^2 . yr for the ecosystem, vegetation alone 3.5 with 2.6 living and 0.9 dead, and mineral soil 14.5 g/m^2 . yr.

Another rate of accumulation of N under alder was studied by VOIGT & STEUCEK (1969) in northeastern Connecticut. Storrs (41°48'N, 72°15'W, 183 m elevation) has a mean annual temperature of 8.9 °C, range of mean monthly temperatures 24.3 °C, precipitation 1134 mm/yr, potential evapotranspiration 619 mm/yr, actual evapotranspiration 603 mm/yr, water deficit 16 mm/yr, and surplus 531. The alder (*Alnus rugosa*) vegetation developed on the bed of a drained pond over 16 years and, after an average initial lag of 8 years, accumulated N at the rate of 8.6 g/m^2 . yr in the ecosystem, 0.9 in the living vegetation biomass, and 7.7 in the soil, including 0.7 in the litter. These rates could be reduced by 1/3 if the lag is neglected. Under the same species of alder near Montreal, Quebec DALY (1960) found the rate of increase of soil N was 17 g/m^2 . yr to a 61 cm depth.

A differential rate of increase of soil N under *Alnus rubra—Pseudotsuga menziesii* as compared to pure Douglas fir showed 3.9 g/m^2 . yr additional to 91 cm soil depth over 26 years (TARRANT & MILLER 1963).

In the tropics initial rates of increase of N in vegetation are about 8.6 g/m^2 . yr, dropping to 3.1 after 5—10 years, assuming that GREENLAND & KOWAL's Ghana data (1960) do form one chronosequence with BARTHOLOMEW et al.'s (1953). At any rate, the 50-year old Ghana forest has an overall rate of biomass N accumulation of 3.6 g/m^2 . yr with an additional 0.5 in litter.

In a temperate region forest succession OLSON's (1958, JENNY 1965) sand dune plant succession on the south end of Lake Michigan had a rate of N accumulation in the upper dm of the soil of 0.45 g/m^2 . yr at t_0, decreasing to essentially zero in 1000 years. The asymptotic level of accumulation was 151 g N/m^2 so over the 1000 years the rate of accumulation was 0.15 g N/m^2 . yr. This area near Chicago, Illinois (41°47'N, 87°45'W, 186 m elevation) has a yearly mean temperature of 10.0 °C, amplitude of monthly means 26.8 °C, precipitation 838 mm/yr, potential evapotranspiration 675 mm/yr, actual evapotranspiration 598 mm/yr, water deficit 77 mm/yr, and surplus 240 mm/yr.

The chronofunction of soils and vegetation on mud flows at the south base of Mt. Shasta in northern California (DICKSON & CROCKER 1953—54) has subsequently been studied by JENNY

1965: 27—28) and GLAUSER (1967). The area is at an elevation of 1130 m, forested with *Pinus ponderosa—Purshia tridentata* and at nearby (4 km distant) McCloud (41°16'N, 122°07'W, 990 m) mean annual temperature is 9.7 °C, range of monthly means is 17.4 °C, precipitation 1190 mm/yr, potential evapotranspiration 626 mm/yr, actual evapotranspiration 403 mm/yr, water deficit 223 mm/yr, and surplus 787. Fresh mudflow areas are colonized by pine or bitterbrush. The rate of accumulation of N to 22 years under the N-fixing bitterbrush levels off at 4.0 g/m^2 . yr for the total ecosystem, peaks at 3.0 g N/m^2 . yr at 12 years for the organic forest floor, and peaks at 2.0 for the 0—61 cm mineral soil profile at 7 years. Under the pine N accumulates at an increasing rate to 5.75 g N/m^2 . yr at 27 years in the total ecosystem, with the forest floor plus mineral soil (0—61 cm) rate peaking at 18 years at 1.25 g/m^2 . yr and the mineral soil alone peaking at 13 years at 1.0 g/m^2 . yr. The overall ecosystem rate under pine to 27 years is 2.26 g N/m^2 . yr. In the wider ecosystem, consisting of the woody plants replacing one another successively in space, the rates of N accumulation in the total ecosystem drop only slightly over the 28 to 300-year period (Table II), and this decrease in rate takes place in the

TABLE II. Overall rates of N accumulation (gN/m^2 . yr) under 22-year old *Purshia tridentata* and *Pinus ponderosa* and in wider areas dominated by these two species together for 28, 42, and 300 years.

	22-years		Both species		
	Purshia	*Pinus*	28-	42-	300-years
Total ecosystem	3.4	1.7	1.1	1.0	1.0
Vegetation	1.2	0.7	0.4	0.3	0.2
Total soil	2.2	1.0	0.7	0.7	0.8
Mineral soil (0—61 cm)	1.3	0.8	0.4	0.6	0.7

vegetation, not in the mineral soil which has a slowly increasing rate.

In the southeastern Australian *Banksia ornata* ecosystem discussed under Biomass above, SPECHT & RAYSON (1957) give an N accumulation rate of 6.7 g/m^2 . yr.

Considerable date on nitrogen accumulation in the Russian forests of different ages already discussed under biomass are summarized by RODIN & BAZILEVICH (1967). The original data (e.g. REMEZOV et al. 1959, MINA 1955) usually include only the tree part of the ecosystem. It is a good generalization that understory vegetation contains an amount of nitrogen greater in proportion than its contribution to the biomass (Scott 1955, Remezov et al. 1964, Pyavchenko 1960). Thus, the nitrogen data given by Rodin & Bazilevich for these forests are increased above the originally published data. Where Rodin & Bazilevich did not re-work the data for a given stand of a particular age, my plotting of the chronofunctions had gaps.

The mixed spruce forests near Veliki Luki show a decrease in N content in the trees' biomass with increasing age from 51 to 115 years (Remezov et al. 1959: 121). The 83-year old stand has 94.7 g N/m² according to Rodin & Bazilevich (1967: 44ff) for an overall rate of nitrogen accumulation of 1.0 g/m². The spruce-feather moss forests in the same region have 12.0 g N/m² more in the biomass of a 72-year old forest according to Rodin & Bazilevich (loc. cit.) than is given for this same stand by Remezov et al. (1959: 79). The overall rate of N accumulation in the biomass of this stand is 0.78 g/m² . yr. However, the N accumulation in the tree biomass of the stands increases almost linearly with age from 27 to 60 or 72 years, then flattens. This linear increase is a low 0.38 to 0.26 g N/m² . yr. Obviously in these forests initial rates of N accumulation must be $\gg$ 0.8 g/m² . yr.

Pinus sylvestris with *Vaccinium vitis-idaea* and feather mosses in the Mordvinian Reserve (Remezov et al. 1959: 38) has only 1.9 g/m² more N in the total biomass at age 71 years than in the tree part of the biomass alone according to Rodin & Bazilevich (1967: 44ff). The rate of N accumulation in the tree stand is constant, just as was biomass accumulation, at 0.81 g N/m² . yr. The intercept of this line, however, is above the origin. Thus the initial rate must be greater than the linear rate, and in fact the initial rate must be much greater than the overall rate for the total biomass of the 71-year old stand for which we have data (0.94 g N/m² . yr). In the pine with *Tilia cordata* understory the two stands of ages 36 and 57 years had tree biomass nitrogen contents of 41.5 and 62.7 g/m², giving rates of N accumulation of 1.2 g N/m² . yr at first, decreasing to 1.0, and 1.1 overall.

Two stands (43- and 220-years old) of *Quercus robur* in the Tellerman Forest (Mina 1955) have N accumulations listed by Rodin & Bazilevich (1967: 117). The rate to 43 years is 1.46 g N/m² . yr, decreasing to 0.36, with an overall rate of 0.57 g N/m² . yr. In the oak stands north of Voronezh (Remezov et al. 1959: 121, Remezov et al. 1964, Rodin & Bazilevich 1967: 117) the data for 12- and 48-year old stands show a rate of early N accumulation of 2.3 g N/m² . yr, declining to 1.9, with an overall rate of 2.0 g N/m² . yr.

Aspen stands in this same region (Remezov et al. 1959: 218,

RODIN & BAZILEVICH (1967: 116) show a lag to 10 years, then a rapid rate of N accumulation decreasing to 50 years. The overall rate of N accumulation to 25 years is 2.7 g N/m^2 . yr, and the rate between 10 and 25 years may be as high as 3.6 g/m^2 . yr. The rate declines toward 50 years to 1.66 g N/m^2 . yr, with an overall rate to 50 years of 2.2 g N/m^2 . yr.

The *Tilia cordata* stands in the Mordvinian Reserve (REMEZOV et al. 1959: 157, RODIN & BAZILEVICH 1967: 117) have only slightly decreased rates of N accumulation from 13 to 40 to 74 years. The rates for the 3 segments of this curve (assuming as usual that it started at a zero level of N) are 2.4 g N/m^2 . yr, 1.7, and 0.9, with an overall rate of 1.5 g N/m^2 . yr.

The natural birch stands studied by OVINGTON & MADGWICK (1959b) in England accumulated N in the trees above-ground parallel to organic matter at a rate of 0.72 g/m^2.yr between 20 and 55 years.

The Mississippi *Pinus taeda* of SWITZER et al. (1966) accumulated N parallel to dry matter above-ground at rates of 1.19 and 0.93 g/m^2.yr on the good and poor sites at the inflections of the logistic accumulation curves. Data in a later paper (SWITZER, NELSON & SMITH 1968 : 6) show N in above-ground biomass accumulating on good sites at a decreasing rate, sharply after 30 years and at 0.77 g/m^2 . yr until then. An apparent asymptote of 25.0 gN/n^2 is reached by 50 years.

In the prairie grassland region of eastern Nebraska over a period of 75 years the rate of N accumulation in the top 14 cm of a soil developed from calcareous glacial till was 2.15 g/m^2 . yr in 55.8 g of organic matter/m^2 . yr (ANDREW & RHOADES 1947). The vegetation which had developed on the site from t_0 was dominated by *Andropogon gerardi* and *A. scoparius*. The climate (Lincoln at 40°49′N, 96°42′W, 351 m elevation) has a mean annual temperature of 11.5 °C, range of monthly means 30.1 °C, precipitation 701 mm/yr, potential evapotranspiration 742 mm/yr, actual evapotranspiration 650 mm/yr, water deficit 92 mm/yr, and surplus 51.

In sand dunes of coastal New Zealand SYERS, ADAMS & WALKER (1970) have described a decreasing rate of N accumulation over 10,000 years. The rate is 2.0 g/m^2 . yr from 50 to 500 years, 0.46 to 3000 years, and 0.036 thereafter. The climate (nearby Wanganui at 39°55′S, 174°40′E) has an annual mean precipitation of 921 mm, potential evapotranspiration 702 mm/yr, actual evapotranspiration 682 mm/yr, water deficit 20 mm/yr, surplus 239.

EGUNJOBI's 1—10 year gorse succession (cf. biomass accumulation) showed an N-accumulation rate at the inflection of just 0.01x the dry matter, namely 26.0 g/m^2 . yr, but a pronounced decrease from 4—10 years with a rate of accumulation of 1.5 g/m^2. yr.

21 ACCUMULATION OF ASH ELEMENTS AND pH CHANGES

J. Major

21 ACCUMULATION OF ASH ELEMENTS AND pH CHANGES

The data on rates of accumulation of ash elements in ecosystems, or parts of ecosystems, is even more scanty than that for biomass or nitrogen. However, the data in Table I from young *Quercus robur* forests north of Voronezh (REMEZOV et al. 1964, 1959, RODIN & BAZILEVICH (1967: 112) will provide an example of orientation. The tropical Yangambi data (BARTHOLOMEW et al. 1953) show linear rates of Ca+Mg accumulation of about 6.1 $g/m^2 . yr$ with an additional 2.0 in dead material. The rate for accumulation of K is about 9.6 $g/m^2 . yr$ with only 0.2 more in dead material. These rates of accumulation have a slight initial lag and drop off precipitously after 8 years.

TABLE I. Rates ($g/m^2 . yr$) of accumulation of ash elements other than N in 12- and 48-year old oak forests north of Voronezh. (From REMEZOV, SAMOYLOVA, SVIRIDOVA & BOGASHOVA 1964: 223.)

Time span (years)	Si	Ca	K	Mg	P	Al	Fe	Mn	S
0—12	0.60	4.90	2.06	0.50	0.53	0.25	0.07	0.06	0.42
12—48	0.04	2.28	0.79	0.14	0.12	0.67	0.02	0.03	0.13
0—48	0.18	2.94	1.11	0.23	0.22	0.11	0.03	0.04	0.20

More complete data have been supplied by VAN CLEVE & VIERECK (1972) for the 20-year alder ecosystem succession in central Alaska. If linear rates of ash element accumulation are picked off they are 2.5 g $Ca/m^2 . yr$, 1.1 g $K/m^2.yr$, 0.35 g $Mg/m^2.yr$, 0.22 g $P/m^2.yr$, and all the lines could go through the origin at t_0. The natural birch stands of OVINGTON & MADGWICK (1959b) are also irregular, but after 20 years and until 55 years accumulation rates are 1.5 g $Ca/m^2.yr$, 0.89 g $K/m^2.yr$, and from 10—55 years 0.011 g $Mg/m^2.yr$ and 0.0067 g $P/m^2.yr$. SWITZER, NELSON & SMITH (1968 : 6) give curves for mineral accumulation in above-ground *Pinus taeda* to 60 years. Rates are approximately linear until 30 years for K and P, until 40 years for Ca and S, 50 for Mg. Then the slopes decrease. The linear portions are 0.45 g $K/m^2.yr$, 0.39 g $Ca/m^2.yr$, 0.13 g $Mg/m^2.yr$, and 0.067 g S or $P/m^2.yr$. The K and Ca are approaching asymptotes of about 20 and 17 g/m^2 at 60 years, and the Mg, S, and P have reached asymptotic values of 7, 5.5, and 5 g/m² by that time.

$CaCo_3$ content of soils decreases with time in a leaching environment. The process is solution, but it is aided by the presence of vegetation. Although under 641 mm ppt/yr on the east coast of England a more or less linear rate of decrease from 0.42 % $CaCO_3$ of 0.185×10^{-2} %/yr held to about 235 years, on the west coast under 813 mm ppt/yr the decrease from 6.5 % $CaCO_3$ was very rapid at first, with the curve flattening at about 0.2 % by 200 years (SALISBURY 1925, 1952: 300, JENNY 1941: 42). BURGES & DROVER (1953) in a humid environment in New South Wales, Australia, under 1233 mm ppt/yr found a rate of decrease of $CaCO_3$ in the upper 10 cm of beach sand of 0.035 %/yr to almost 100 years, decreasing to 0.003 %/yr to 200 years when all the $CaCO_3$ was gone. In eastern Nebraska under bluestem prairie grasses the rate of $CaCO_3$ leaching from a glacial till was 123 g/m² . yr to 14 cm depth over 75 years (ANDREW & ROADES 1947).

Soil acidity would seem to be under more direct influence of vegetation. In dunes in England pH dropped about 0.9 unit over 235 years on the east coast but 2.7 units over 280 years on the west (Salisbury). In the Australian example pH decreased 3 units to pH 6 at 200 years. At Glacier Bay in southeastern Alaska (CROCKER & MAJOR 1955: 435) the decrease was 3.1 units down to pH 5.0 in about 40 years under *Alnus crispa* ssp. *sinuata* thickets. The maximum rate of decrease was about 0.13 pH/yr under alder whereas on bare surfaces pH decreased only 0.005 pH/yr and under tall *Salix barclayi* only 0.025 pH/yr. Even under *Dryas drummondii*, which also fixes atmospheric N, to 20 years the drop of pH was only 0.02 pH/yr. $CaCO_3$ (originally 4 %) vanished from the profile under alder in about 20 years. UGOLINI's data for the same area (1966: 70) show a smoothly decreasing rate of pH decrease amounting to 3.7 units over 250 years. Near Juneau CROCKER & DICKSON (1957) found a decrease in the 0—8 cm mineral soil of 0.10 pH/yr over the first 30 years at the Mendenhall glacier but only 0.0065 pH/yr at the Herbert from about 20 to 200 years. STORK (1963) found a pH decrease of 0.09 pH/yr for the first 50 years before the Stor glacier in northern Sweden. In the subalpine belt of the Swiss National Park (BRAUN-BLANQUET 1964: 669—673, BRAUN-BLANQUET, PALLMAN & BACH 1954: 43, 153—164) plant succession over dolomitic gravel produced at a rate of pH decrease of about 0.0055 pH/yr to about 500 years. Before the Muldrow glacier in central Alaska at the upper limit of the forest tundra VIERECK (1966: 187) found a decrease of pH of 2.3 units over 250 years.

22 VEGETATIONAL CHANGES ON AGING LANDFORMS IN THE TROPICS AND SUBTROPICS

J. S. Beard

22 VEGETATIONAL CHANGES ON AGING LANDFORMS IN THE TROPICS AND SUBTROPICS

The insistence of plant geographers in giving undue importance to the influence of climate has been paralleled among ecologists ever since CLEMENTS proposed the concept of the climax, and among pedologists since the Russians developed the zonal classification of soils. There is a discussion in BEARD (1945) of the difficulties surrounding the monoclimax theory in the tropics, while BEADLE (1951) has spoken against it in an Australian context. The catenary sequences of soils and vegetation on the West Australian plateau immediately pose questions of status and development, as the example given for the Boorabbin area by BEARD (1969) well shows. Which of the five vegetation-types concerned would be considered as the climatic climax, and what is the status of the others?

According to the definition of TANSLEY & CHIPP (1926), i.e. that "the highest type of plant community that can exist in a given climate is known as the climatic climax", the latter is evidently the salmon gum woodland. Since however this type of woodland occurs on enriched alluvial soil, it clearly does more than reflect the simple potentialities of the climate, and one feels that the conclusion is unsatisfactory. To go back to the original intention of CLEMENTS, one would no doubt have to propose that the mixed eucalypt woodland is the climatic climax as it is developed on a soil produced *in situ* on the underlying rock without either abnormally favourable or unfavourable characteristics. It could certainly be described as the prevailing climax in the sense of WHITTAKER (1953). The status of the other communities in the catena is then very much open to question. If a developmental relation is sought, they will not appear in the usual subordinate position where their soils might be considered as capable of developing in time into that which supports the climatic climax. Quite the reverse seems to be the case. If anything, the various soil types have developed *from* that of the climatic climax, acquiring less or more favourable characteristic as they did so. They are in fact a postmature phenomenon.

The existence of this state of affairs has been noted by BEARD on previous occasions (1945, 1953). In the concluding chapter of "The Natural Vegetation of Trinidad" (1945) it was suggested that communities capable of evolution towards the climatic climax — the swamp sere being quoted as a typical example — should be styled "edaphic pre-climaxes". There could be more than one climatic climax in a region according to inherent soil properties, but it would always develop on relatively young topography and soil under optimum drainage. It was envisaged that aging of soil and landform in a tropical region of high rainfall would produce a somewhat flat landscape of poor drainage. At the same time the soils internally would become leached and profile-differentiated, intensifying the adverse drainage factor. A series of "edaphic post-climax" types was recognised, reflecting this degradation of the optimum and culminating in the neotropical savanna as the ultimate edaphically controlled community of senile soils.

The question was further examined in a study of savanna vegetation as a whole in northern tropical America (BEARD 1953), with this conclusion: "Savanna is the natural vegetation of the highly mature soils of senile land-forms (or in some cases of very young soils in juvenile sites) which are subject to unfavourable drainage conditions and have intermittent perched water tables, with alternating periods of waterlogging and desiccation. Frequent fires occur, but are not a necessity for the maintenance of the savanna, which is an edaphic climax."

To clarify this issue, the above applies strictly to the type of savanna found in northern tropical America, a relatively well-watered region. In the drier climate of the Brasilian plateau, possibly, and in Central Africa certainly, there are savannas of different physiognomy which appear to be woodlands degraded by repeated fires. Even here, however, there are local occurrences of edaphically controlled grasslands on seasonally water-logged soil, e.g. the "dambos" of the Rhodesias and Angola. In northern tropical America there is an obvious and striking relationship between the distinctive savanna vegetation and old land surfaces where there is drainage impedance. Beard cited the work of BENNETT & ALLISON (1928), CHARTER (1941) and MARBUT (1932) to establish the soil-forming processes involved. It is worth quoting MARBUT in full:

"Throughout the Amazon valley the soil consists of (1) top-soil, (2) iron-oxide layer, porous and slag-like, (3) mottled layer, (4) grey layer, (5) unconsolidated clay and sand. The mottled zone, with or without the induration of its upper part to an iron oxide crust, is found invariably beneath the surfaces or where the relief shows that dissection is very recent. It is not found on mountain slopes or in thoroughly dissected regions where the dissection is old."

This then is the pedological result of peneplanation under a high rainfall. The rain forest which so largely covers the Amazon basin can tolerate the process up to a certain point. A nutrient cycle is established in virgin forest which enables the forest to maintain itself on a leached and infertile substratum.

Adaptations of structure and floristic composition afford a forest type tolerant of poor drainage. A breaking-point, however, may eventually be reached, either through a desiccation of the climate to a regime of greater seasonality or by attainment of an extreme of adverse drainage. Both of these processes can be illustrated in tropical America. In the former case fire becomes a possibility, and ruptures the nutrient cycle. Dehydration hardens the mottled zone of the soil to laterite, seasonality of water relations becomes more pronounced, and the forest disappears.

South America today may give us a very good idea of conditions prevailing in Western Australia during the Miocene, and help to explain the changes in soils and landforms which have taken place. The analogy is a close one and the only real difference is that the Mediterranean climate has produced a sclerophyll scrub as the end-product in vegetation instead of grassland. Otherwise we can see the whole process faithfully reproduced: peneplanation and laterisation under a high rainfall, desiccation, fire, the disappearance first of forest then of woodland, finally the establishment of a depauperate vegetation specialised and adapted to highly unfavourable physical and chemical soil properties on an ancient landscape.

It is necessary that it should be more widely realised by ecologists and plant geographers that weathering and developmental changes in soils are not always ameliorative. One does not have to have tropical experience to appreciate this since the temperate-zone process of podsolisation is a case in point, and many of the heath communities of the northern hemisphere must be degradation products comparable to the neotropical savanna. The sclerophyll scrub of Western Australia is often termed heath by Australian ecologists and in this sense could be substituted for savanna in BEARD's definition of 1953: "Heath" is the natural vegetation of the highly mature soils of senile landforms etc. In the West Australian climate it is not clear to what extent internal drainage of the soil is still significant. Impedance is present, but waterlogging on most sites probably does not occur very often. Under the prevailing low rainfall water holding capacity may assume greater importance, and the precise role of low nutrient status is not understood. More investigation is required, though it is quite clear that in these respects also there have been retrogressive changes.

It is well understood in geomorphology that landforms may be young, mature or senile. Soils are known to age in a similar way. There are therefore young, mature and senile plant communities. CLEMENTS was only able to grasp the facts of youth and maturity, and it was a dictum of his that "the climax can be destroyed but cannot retrogress". The examples given above should suffice to prove that the climax can very frequently retrogress. Under the circumstances it seems necessary to seek some better definitions and terminology for communities of different status: for the climatic climax (or as CLEMENTS would have it, "the climax"), for those edaphically controlled communities showing a developmental evolution towards the climax and for those with a devolutionary history from the climax. BEARD suggested (1945) that the latter two groups be styled edaphic pre-climaxes and edaphic post-climaxes respectively, a possibly unfortunate choice of terms, since they are neither pre- nor post-climax in the sense in which CLEMENTS himself proposed the terms. "Infra-climax" might be a better choice for the former, and "serclimax" for the latter (*ser-* in the sense of *after*). The true climax itself must be one which best expresses the potentialities of local topography and soils. One may perhaps suggest a definition as follows:

The climatic climax is the natural vegetation of young and mature land-forms on regionally prevalent soil-types which are free from special local features either favourable or unfavourable to plant growth.

G. EXAMPLES OF FLUCTUATIONS

23 FLUCTUATIONS IN CONIFEROUS TAIGA COMMUNITIES

A. A. Korchagin & V. G. Karpov

23 FLUCTUATIONS IN CONIFEROUS TAIGA COMMUNITIES

The principal climax types of the Eurasian taiga are coniferous forests consisting of evergreen coniferous trees, such as the species of spruce (*Picea abies*, *P*, *obovata*, *P. ajanensis* and others), fir (*Abies sibirica*, *A. nephrolepis* and others), pine (*Pinus sylvestris*, *P. sibirica*, *P. koraiensis*) and summergreen coniferous trees, the larches (*Larix sibirica*, *L. dahurica* and others).

In the middle subzone of the coniferous taiga zone primary coniferous forests have no admixture of leafy trees that appear only in burns and cutting-areas, forming there temporary secondary small-leaved forests of birches (*Betula pendula*, *B. pubescens* and others), aspen (*Populus tremula*) or alders (*Alnus incana* and others). Only in the extreme north in the northern taiga subzone and high in the mountains the admixture of birch in coniferous forests is primary. In the south, in the Southern taiga subzone the admixture of broadleaved trees (*Quercus robur*, *Acer platanoides*, *Tilia cordata* and other trees) is permanent.

In typical climax taiga-type forests the lowest layer consists mainly of dwarf shrubs (*Vaccinium vitis idaea*, *V. myrtillus*, *V. uliginosum*, *Empetrum nigrum* etc.) with a slight admixture of forest herbs (*Oxalis acetosella*, *Majanthemum bifolium*, *Trientalis europaea*, *Equisetum sylvaticum* etc.) growing over the dense moss layer, formed by *Hylocomium proliferum*, *Pleurozium schreberi*, *Dicranum scoparium*, *Rhytidiadelphus triquetrus*, *Polytrichum commune* and sometimes by lichens, such as the species of *Cladonia*, *Peltigera* etc. It is only in the southern taiga subzone that nemoral herbs (species of *Pulmonaria*, *Aegopodium*, *Convallaria*, *Asarum*, *Hepatica* etc.) prevail in the lowest layer. (Korchagin 1966, Sochava 1953, 1956.)

A slight expression of seasonal phenological variation of both evergreen coniferous trees and the dwarf shrubs is characteristic of typical taiga forests. Still less pronounced in the coniferous taiga are variations from year to year (fluctuations), associated with different yearly variations in meteorological conditions (precipitation, temperature etc.), although the range of the latter, as well as that of the variations of soil factors (moisture, aeration, mineral regime etc.), is rather wide.

The communities of coniferous forests are rather stable systems, well adapted to considerable variations, from year to year, of meteorological conditions and regimes of the soil. Apparently the natural selection of life forms and species, the development of the phytocoenotic structure of communities proceeded primarily in the direction of the adaptation to critical levels of environmental factors, but not to the optimal conditions which accounts for the high degree of stability of the floristic composition and of the synusial structure in different years differing widely from one another in the temperature regimes, atmospheric precipitation and soil conditions.

The fluctuations in the proportions of different species in the tree stand, in the undergrowth, in the herbage-dwarf-shrub and in the moss layers of coniferous forests are pronounced but very slightly and practically imperceptible by sight. The hylocomious boreal coniferous forests (Pineta, Piceeta, et Lariceta) are more stable and characterized by a slighter response to the differences in the environmental conditions between different years as compared to the nemoral-type forests of the coniferous-broadleaved forests zone, more complex both floristically and ecologically.

A mass death of thermophilous trees, shrubs and herbs of the nemoral complex is observed occasionally in the communities of coniferous-broadleaved forests in the years with critically low temperatures of winter and spring. This results in considerable changes of the species composition and the structure of these communities, the re-establishment of the initial state, requiring long periods of time (SUKACHEV 1931, TIMOFEEV 1936, ILJINSKY 1939). Still in the communities of coniferous-broadleaved forests, even slight variations in the meteorological conditions result in perceptible variations of the number of shoots per unit area, of the vigour of separate ecological and phytocoenotical groups of plants and of their course through separate phenophases of the seasonal development (KOZHEVNIKOV 1937).

However, many dynamic phenomena and processes in taiga-type coniferous forests are very sensitive to the variations, from year to year, of meteorological conditions and of the regimes of the soil. If these variations do not affect significantly either the proportions of different species or the structure of the tree stand, the undergrowth, the herbage-dwarf-shrub and the moss layers, they do affect very conspicuously such characteristics, as the growth rate of trees and of the plants of the lower layers, the degree of foliage development and the variations in the working surface of leaves, the density of the canopy, the projective coverage of the soil surface, the dates and the duration of separate phenological phases of the seasonal development of different components of the coenoses and also the phenomena of fructification and reproduction of plants by seeds. All these characteristics exhibit a quantitative dependence on the variations, from year to year, of meteorological factors and of the soil conditions. Sometimes this dependence is very complicated and bears the character of responses of a community to the combinations of factors not only of the present year, but also of previous years.

The following main types of phenomena of fluctuations can be mentioned: (a) the fluctuations of the growth rate of trees and of the plants of lower layers; (b) the fluctuations in the rate of fructification and in the abundance of plants (particularly of trees) of new generations; (c) the fluctuations of the seasonal developmental cycles of plants.

A special category of reasons for fluctuations of the communities of the coniferous taiga are the changes in the structural characteristics of communities caused by the periodical managemental human activity (cleaning, cutting, grazing, etc.), of the damage caused by insect pests, large vertebrates and pathogenic microorganisms, particularly by lower fungi.

As it is shown by the recent investigations, the fluctuations can be traced quite distinctly also in the underground parts of the communities where they affect the periodic withering of absorbing and growing root tips of trees during the period of a considerable excess in the soil moisture and their regeneration in certain parts of the soil profile in the years of sufficient aeration (ORLOV 1966).

(a) The fluctuations of the growth-rate of trees and of the plants of the lower layers are inherent in all the communities of the coniferous taiga. However in different types of forest and in different phyto-geographical zones the range of the fluctuations of the increment of trees is different. As it is shown by numerous investigations, the fluctuations of the increment of trees can be very wide (the deviations from the longterm means attaining 50—150 %). The correlation between the annual variations of the increment on the one hand and the amount of precipitation and the air temperature on the other hand is quite definite (DOUGLAS 1936, HUBER 1954, GORTINSKY 1969, GORTINSKY & TARASSOV 1969). It is also established that the fluctuations of the increment of spruce forests, belonging to the same group of forest types, but situated in different geographic localities are similar in character. On the contrary, in the same years different groups of types of spruce forests can differ significantly in the size and congruence of the annual increment of tree stands (GORTINSKY 1969). The information on the fluctuations of the growth of shrubs, herbs and mosses of coniferous forests, though very scanty, nevertheless shows sufficiently distinctly the dependence of the noticeable annual increase in the height of plants of lower layers, of the annual increase in the length of their shoots, of the number of leaves and of the general vigour of these plants (PERTULLA 1941, TAMM 1950, KORCHAGIN 1960a, 1960b, 1960c, MALYSHEVA 1969, IPATOV 1969) and also of the linear annual increment of forest mosses and lichens. These fluctuations are closely associated with the cycles of the age development and with the annual dynamics of the most important environmental factors (annual variations of the atmospheric humidity, soil moisture, temperature of the air layer adjacent to the soil surface etc.).

(b) Fluctuations of the seed yield and of the abundance of new generations are one of the most characteristic traits in the dynamics of the coniferous taiga communities.

Thus, the seed yield of spruce stands of the middle and south taiga zone in the productive years attains 7.5—10 million seeds per hectar, while in the years of poor productivity it falls to a few hundreds of thousands per hectar. In case of the temperature and moisture of the litter optimal for the germination of spruce seeds the number of emerged seedlings can attain 1.5—2.0 million per hectar. The immense majority of the emerged seedlings die during the first vegetative period and their abundance falls abruptly the next year (MOROZOV 1928, YLI-VAKKURI 1961, KARPOV 1969). The intensity of elimination of emerged seedlings depends on the species composition and the structure of the community, as well as on the seasonal and annual dynamics of meteorological factors. It is much higher in the spruce stands with a well developed undergrowth, dwarf shrubs and herbs and is significantly lower in spruce stands of the hylocomiosa group having a simpler structure. In the years of poor seed productivity, as well as in cases of unfavourable combinations of environmental factors controlling the germination of seeds (mainly the regimes of temperature and moisture of the litter), the number of emerged spruce seedlings is extremely small (not over 500—1000 per hectar).

The periodic fluctuations of the abundance of emerged spruce seedlings and the peculiar "waves" of their mass appearance and death are very typical for the life of a spruce forest. This periodicity of such fluctuations is based on complicated relationships between the seed yield of stands, the annual variations of the main meteorological factors, the physiological processes in trees and the ecological phytocoenotic conditions of germination of seeds and of the development of new generations. Up to the present time these relationships were analysed obviously insufficiently, although an ample evidence has been accumulated in the literature on the fluctuations of the seed yields of stands and the abundance of emerged seedlings and saplings of spruce as related to the dynamics, from year to year, of meteorological factors and the regimes of seed germination in taiga communities (NEKRASOVA 1958, KORCHAGIN 1960, YLI-VAKKURI 1961, GORTINSKY 1964, KARPOV 1969). Dwarf shrubs and herbs of coniferous forests are mainly reproduced vegetatively (KUJALA 1926, SUKACHEV 1934). Therefore the fluctuations of the abundance of emerged seedlings in the lowest layers are expressed but slightly and are perceptible only in a small number of species (e.g. *Oxalis acetosella, Luzula pilosa)* (PERTTULA 1941).

(c) Fluctuations of the cycles of seasonal development of the coniferous taiga communities are sufficiently conspicuous in certain years and their correlation with the variations, from year to year, of the meteorological conditions is quite distinct. However but a slight saturation of the floristic composition of boreal-type coniferous forests with sufficiently "aspective" species produces an impression of slight fluctuations of the cycles of the seasonal development of communities in these forests. On the contrary, coniferous forests well saturated with nemoral species are characterized by more distinct fluctuations of their phenological state (KOZHEVNIKOV 1937). However, it was shown by detailed investigations, that in the communities of typical boreal forests the rhythms of their seasonal development also vary depending on the meteorological conditions. In the years with a cold and protracted spring in many species (including tree species) the beginning of the vegetative period is shifted to later dates, the duration of the colour aspects is changed, the curves of flowering of the species of a community are less steep, sometimes scalariform in character. Some species in certain years are represented by but a small number of flowering specimens, while other species blossom abundantly in the same years. However, the data on the fluctuations of the phenological state of coniferous forests as yet can give no more than a general notion, being insufficient for any quantitative evaluation of the factors causing the changes of rhythms of the seasonal development of communities in years with different meteorological conditions.

24 FLUCTUATIONS IN NORTH AMERICAN GRASSLAND VEGETATION

R. T. Coupland

24 FLUCTUATIONS IN NORTH AMERICAN GRASSLAND VEGETATION

The species composition of natural grassland fluctuates much more in response to fluctuations in environment than does that of the woody component of shrublands and forests. This is because the herbaceous shoots of grasslands have a much shorter life-span than do woody stems, but also because the degree of fluctuation in weather increases along the forest-shrubland-grassland gradient, especially in respect to precipitation.

Life span of herbaceous plants also (as well as shoots) has an effect on the response of vegetation within grassland to environmental changes. The grassland communities that are most responsive to changes in weather are presumably those dominated by annuals, which depend on germination to provide new plants at the beginning of each growing season. In grasslands dominated by perennials, life form of dominant grasses determines the degree of change that will take place in the density of plants or shoots. Bunch grasses respond slowly to changes in environment because their populations tend to be balanced by the necessity for extremely adverse conditions to prevail before the plants die and because the production of new plants from seed is precarious. Rhizomatous species are more responsive, since the number of shoots in any growing season is apparently dependent on factors favorable or unfavorable to the initiation of shoot development. With these species death of plants does not seem to be a factor, since part of the rhizome system survives even in extreme environmental adversity.

Subdominants are more variable than dominants in respect to their contributions to the herbaceous plant cover under changing environments. Many species are adapted to produce stems only when conditions are favorable to them, so that interseasonal (between growing seasons) fluctuation is sometimes extreme. Some species have been observed to remain dormant with no shoot production for several years in succession without expiring.

The longevity of individual shoots is, of course, an important factor in determining whether or not their density is affected by an environmental extreme of short duration. Recent intensive studies in herbage dynamics, stimulated by the need for estimating primary productivity as part of the International Biological Programme,have indicated that shoots of grasses frequently stay green for only a few weeks, so that it can be expected that the species composition of grass swards will change intraseasonally (within the same growing season) as well as interseasonally. In regions where the non-growing season is characterized by extremely low temperatures (e.g. — 40 °C in the northern Mixed Prairie) or extreme drought (as in the Desert Plains Grassland), presumably all shoots are killed annually back to (or near) the soil surface. It is possible, although improbable, that there are natural grasslands that exist in a climate in which survival of herbaceous shoots does occur from one growing season to the next after they have started to elongate. Perhaps the tussock grassland of New Zealand is an example. However, consideration should be given as to whether this is a natural grassland situation. Perhaps with a degree of environmental uniformity that fosters continued growth of herbaceous shoots for periods greater than one year, shrubs or trees would abound in the absence of man's influence.

Although interseasonal fluctuations in the relative abundance of species (as measured by numbers of plants or shoots or by amount of soil surface occupied) in grassland increase with the degree of fluctuation in environment, it seems probable that the homogeneity of the plant cover as a whole is increased by environmental fluctuation. Species are excluded from the community that could occupy favorable local niches, but are not adapted to surviving environmental extremes. Texture of soil and regularity of the soil surface is rarely so uniform that horizontal differences do not occur in moisture distribution. In the semi-arid conditions of grassland this creates a patterning of vegetation along a moisture gradient. The complexity of the vegetation along this gradient increases with the supply of soil moisture. However, towards the moist end of this gradient, species capable of flourishing in moist periods cannot survive the driest portion of the weather cycle. This results in exclusion of many species from grassland communities for which conditions for survival are adequate most of the time. Perhaps the most conspicuous example of this phenomenon is along the forest-grassland margin where trees are prevented from surviving within the grassland because of very short, dry periods that occur infrequently, but sufficiently often to regularly kill slow-growing woody plants. One might argue that a similar situation is reflected all along the moisture gradient, so that the reaction of the plant cover at the drier end of the gradient will compensate for that at the moister end. However, no exclusion of species from the community results near the drier end, since drought-resistant species can endure the moist phase of the weather cycle as well as the dry phase.

The response of the vegetation as a whole to fluctuations in environment is a summation of the response of individual species. In the natural perennial grasslands of North America numerous reports have appeared concerning fluctuations in the proportion of the plant cover contributed by various species during periods of adversity. The environmental stresses on natural grassland that most commonly result in fluctuations in the floristic composition of the plant cover are variations in the supply of soil moisture and in the degree of grazing by domesticated animals. It is to be expected that droughts will occur from time to time in grassland regions where soil moisture is a controlling factor in determining the grassland nature of the landscape. WEAVER and his associates have made the most extensive studies of the reaction of perennial grasslands to drought by repeatedly sampling numerous stands of True Prairie and Mixed Prairie before, during and following the drought of 1934 to 1940 that occurred in Nebraska and adjacent states (WEAVER 1943, 1950, 1954, WEAVER & ALBERTSON 1936, 1939, 1940, 1940a, 1943, 1944, WEAVER et al. 1935, 1940). They have summarized these studies in two books (WEAVER 1954, WEAVER & ALBERTSON 1956). COUPLAND (1959) has recorded changes that took place in a period when supply of soil moisture was in excess of "normal" for a series of years. COUPLAND (1958) has reviewed studies of the effects of fluctuating weather on grasslands of the Great Plains. Numerous reports appear in the literature on range science of the effects of excessive grazing of natural grasslands by domesticated livestock (ALBERTSON et al. 1953, ALBERTSON & WEAVER 1944, BLACK & CLARK 1942, CLARKE et al. 1943, COUPLAND et al. 1960, SARVIS 1941, WHITMAN et al. 1943).

The first response to decreased supply of soil moisture is reduction in height and biomass of stems. Such a response results from exposures of only a few weeks and causes intraseasonal fluctuations in the grassland community, even in "normal" years. This is followed closely by a decrease in density of shoots. Intraseasonal fluctuations in the rate of production of shoot biomass have not concerned investigators of natural grasslands until recently. Instead, they have been satisfied with the estimation of differences in production between years by annual harvests of the peak standing crop. The best correlation between these harvests and an environmental factor has been shown to be with precipitation two to three months before harvest (SARVIS 1941, SMOLIAK 1956). Extremes in height of dominant grasses between years have been reported in the ratio of as great as 15 : 1 and of maximum standing crop as great as 9 : 1. Density decreases of 90 percent have been reported in extreme drought, but increases have also occurred as short grasses have replaced mid grasses in Mixed Prairie.

In the semi-arid environment of much of the North American grassland there is a tendency for years of below average precipitation to be grouped (Borchart 1950), so that drought periods are of sufficient length to cause a reaction by the plant community far beyond the short-term response of shoot size. In successive years precipitation may differ in a ratio of 3 : 1 and for several successive years the mean precipitation may be 20 percent or more above or below the long-term mean. The effect of these fluctuations in decreasing (or increasing) the supply of soil moisture is accentuated by the occurrence of above average temperatures (by several degrees) in dry growing seasons and below average temperatures in moist ones.

During prolonged drought there is a differential species response related to capability of evading, escaping, enduring or resisting drought. In the Mixed Prairie, for example, growth of physiologically more drought resistant dominants (e.g. the short grasses, *Bouteloua gracilis* and *Buchloe dactyloides*) is less adversely affected by increasing aridity than that of the dominant mid grasses. As a consequence, the mixed grass cover of "normal" periods attains a short grass character in periods of prolonged drought, and, on a regional basis, the Mixed Prairie tends to move eastward into the True Prairie. Early season subdominants (e.g. *Koeleria cristata* and *Carex* spp.) are less affected than the dominants because of their demand for moisture being restricted to that portion of the growing season when moisture supply is more dependable. Deep-rooted forbs are able to endure drought because of their lack of dependance on surface moisture. Others apparently have the ability to become dormant so that their underground parts may survive without shoots appearing above ground. The result is that only after prolonged, severe periods of deficiency of soil moisture do many plants succumb. Bunch grasses adjust their contributions to the vegetative cover by a decrease in diameter of the bunch and succumb only after long periods of severe drought. In rhizomatous grasses the adjustment is by the number of shoots that emerge from the rhizomes. During the early stages of drought there is a tendency for the roots of some species to penetrate more deeply in the soil so as to increase the effective volume of soil from which moisture may be extracted. As the plants become dependent on periodic surface moisture, the deeper roots die and the community becomes more shallowly rooted than in average years.

Another response to drought is absence of flowering in both grasses and forbs. When soil moisture again becomes abundant the plants reproduce profusely as a result of reduced competition. Intraseasonal environmental conditions are also important in relation to the flowering characteristics of various species. In the northern Mixed Prairie, for example, the cool season plants (*Agropyron* spp. and *Stipa* spp.) need sufficient moisture during the early spring if flowering is to be initiated. Abundant moisture later in the season will not compensate for that which is missed during the cool season.

The response of grassland to drought is modified by the texture of the soil in which the grasses are growing. Sandy soil tends to be more difficult than is loam or clay for shallow-rooted species during drought, because of the rapid loss of surface moisture by percolation and evaporation. The moisture reservoir is greater in sandy soils for deep-rooted species than is that of clay for short-rooted ones. This difference sometimes permits tall grasses to suffer less in sandy areas during drought than do mid grasses in soils of finer texture.

The relief from drought is just as spectacular as initiation of drought in its effect upon the vigour of the plants and the changes in floristic composition. Spectacular increases in height growth occur with the onset of rains, apparently as a result of decreased competition caused by reduced density. Changes in floristic composition from a xerophytic towards a mesophytic condition, while lagging behind the changes in plant vigour, are nevertheless extreme. Morphology may provide an advantage to one species over another during this period of readjustment, just as the other species had the advantage during the onset of drought. For example, the rhizomatous *Bouteloua gracilis* has an advantage during drought because of its greater physiological ability to resist dry conditions than the stoloniferous *Buchloe dactyloides*. Before the drought of the 1930's these species occupied similar proportions of the soil surface as codominants over large parts of western Kansas and adjoining states. During eight years of drought *Bouteloua* suffered less than *Buchloe* so that the ratio of the first to the second in basal area occupied increased to 2 : 1. However, with the onset of rains *Buchloe* was able to make such rapid gains relative to *Bouteloua* because of its stoloniferous habit, that in two years it reversed its position to being five times as dense as *Bouteloua*. Only after ten years of favorable weather was *Bouteloua* able to regain its former equal importance with *Buchloe*.

Reversed trends (to those that occur as a result of drought) have been observed after a succession of years of above average precipitation. Such changes have not received the same attention as the ones associated with declining soil moisture. However, changes have been recorded that took place in the northern Great Plains following a five-year period during which precipitation averaged 20 percent above the long-term mean. In a general way these changes were opposite to those that occurred in drought, being an increase in height and a decrease in drought-resistant species relative to mesophytic ones.

The changes that take place as a result of overgrazing are similar to those that have been recorded as a result of drought. Since the short grasses of the Great Plains are both more drought resistant and more resistant to grazing than are the mixed grasses, they increase relative to the mid grasses during periods of drought and overgrazing. The effect of drought is everywhere more severe on heavily grazed ranges than on moderately grazed ones, but even moderately grazed ranges become overgrazed ones during periods of drought, unless livestock numbers are rapidly adjusted. When prolonged drought is superimposed upon overgrazing, the mid grasses are destroyed and a short grass cover develops. As a consequence, much of the drier part of the western Great Plains in the rain shadow of the Rocky Mountains has come to be characterized as short grass plains. While it is well known that this is really mixed grass (mid and short grasses) prairie in the natural state, the grazing economy of the area permits the short grasses to predominate. The success of the mid grasses in the driest part of the Great Plains, therefore, depends on both factors, drought and grazing. Since, during drought, it is often necessary for ranchers to reduce the number of animals that their ranges support, the rate of recovery from drought is presumably accelerated by low stocking rates, just as deterioration was favored by grazing during the onset of drought.

In landscapes where short grasses are naturally dominant, such as in the Desert Plains Grassland of the southwestern United States, the response to drought is again to cause a trend towards the characteristics of the vegetation of drier regions, in this case semi desert. Overgrazing intensifies these effects. Since the invasion of desert shrubs into a broken grass cover is a difficult (or impossible) trend to reverse, repeated droughts under grazing pressure have caused desert shrub vegetation to occupy parts of these areas naturally capable of supporting grassland.

Litter does not accumulate in grassland that is in adjustment with climate. However, in the portion of the grassland near the forest margin, where trees are discouraged by short drought periods (and perhaps by fires in the past), the net effect of the more mesophytic environment is to favour the accumulation of litter. This sometimes causes reduced yield of shoots and favours rhizomatous mid grasses as compared to bunch grasses. This trend can only be reversed by occasional fires, which liberate tied-up nutrients and have a desiccating effect on the habitat. Like overgrazing, protection from fire is a man-imposed condition. The degree of fluctuation depends on the frequency of occurrence of the antidote.

The natural grasslands of North America are thus seen to be in a constantly changing flux around a "normal" condition. The periodicity of the fluctuations are sufficiently brief in relation to man's life span to characterize them as part of a natural equilibrium within the "climax".

25 FLUCTUATIONS IN THE SEMIDESERT AND DESERT VEGETATION OF THE TURANIAN PLAIN

B. A. Bykov

25 FLUCTUATIONS IN THE SEMIDESERT AND DESERT VEGETATION OF THE TURANIAN PLAIN

The complex vegetation is the most widespread in the semidesert zone. This paper deals with the fluctuations of the three-component complex: *Artemisia pauciflora*+*Kochia prostrata*-ass. *Tanacetum achilleifolium*+*Agropyron desertorum*-ass. *Agropyron pectiniforme*+*Festuca sulcata*-ass.

The desert-steppe complex very typical of the northern part of the Caspian Area is represented by a desert association of xerophilous dwarf subshrubs on solonetz soils (about 50 %), by a steppe association of mesoxerophilous and xerophilous grasses on dark chestnut soils of small depressions (about 24 %) and finally, by a desert-steppe association of a mesoxerophilous dwarf subshrub and a xerophilous grass on saline light-chestnut soils. These main associations are accompanied by certain other ones (GORDEYEVA & LARIN 1965).

The climate in this region is continental with a severe winter and a hot summer. The average precipitation is 280 mm per year. The state of the communities, their productivity and their various fluctuations depend on hydrothermical factors, particularly upon the quantity and distribution of the precipitation and upon the reserve of the available water in the soil.

The rhythms of the communities and populations are determined by these factors. Thus, in years with deficient moisture, the efflorescence in *Artemisia pauciflora* is delayed for up to a month, while in the most droughty years a complete failure of flowering and fructification is observed in this species. The period of its summer dormancy varies from 25 to 125 days. In ephemeroids, such as *Tulipa biflora*, *Ornithogalum fischerianum* and others the developmental rhythms are more stable. The abundance of ephemers varies widely from year to year. Thus, the average abundance of *Alyssum desertorum* for the period of 1955—1959 was 5 specimens per sq.m., but varied from 1 to 11.

Very characteristic are the fluctuations of the abundance of emerged seedlings of plants affected both by the meteorological conditions of the spring and by the intensity of the previous years' fructification. This caused the alternation of high germinative capacity of seeds of certain groups. Thus, in 1952, a great abundance of the emerged seedlings was observed in *Artemisia pauciflora*, *Kochia prostrata*, *Agropyron pectiniforme*; in 1956 the emerged seedlings of *Tanacetum achilleifolium, Festuca sulcata, Medicago romanica,* and in 1958 of *Astragalus virgatus, Trinia hispida* were very abundant, respectively.

The range of variation of the fluctuations of the seed yield in semidesert plants is very wide. The productivity of the overground plant biomass is also fluctuating very widely from ±65 % in *Tanacetum* communities to ±90 % in *Agropyron* communities.

A direct correlation between these fluctuations and the meteorological factors was shown by the investigations of FEDOSEYEV, (1964) who established the following dependence of the productivity of overground plant biomass on the humidity coefficient:

$$K = b + c / \Sigma d$$

b is the spring reserve of the available water in the 1-metre soil layer, c is the amount of precipitation from the spring until the time of the maximal yield and d is the air humidity deficit during the same period (fig. 1a).

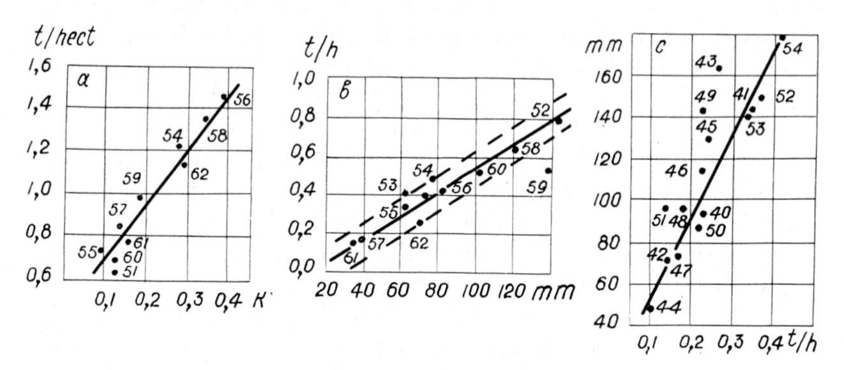

Fig. 1. The dependence of the productivity of the pastures: (a) of a complex semidesert on the coefficient of soil moistening; (b) of an Artemisieta desert on the reserve of the water in the soil; (c) of a Cariceta-Haloxyloneta desert on the amount of precipitation (during XI—V) (from FEDOSEYEV 1964).

In the desert zone, the most characteristic one among several types of vegetation is the semiligneous type (*sublignosa*) represented by the following subtypes: semiarboreous (*sublignoseta,* e.g. Haloxylonetum), semishrub (*subfruticeta,* e.g. Calligonetum) and dwarf semishrub (*subfruticuleta,* e.g. Artemisietum with the dominant belonging to the subgen. *Seriphidium*).

Fluctuations in the communities of *Haloxylon persicum* in the first subtype and *Artemisia terrae-albae* in the last one will be considered.

The association of *Artemisia terrae albae + Anabasis aphylla—Rheum tataricum* is very widespread in the northern part of the deserts of the Turanian Plain, in particular on the northern coast of the Aral Sea where it represents the climax vegetation (BYKOV 1968). The climate here is extremely continental with anal most snowless, severe winter (down to —42 °C) and a hot droughty summer (up to +44 °C) and with the precipitation of 120 mm per year.

This community is characterized by a distinct rhythm: the period of summer dormancy in xerophilous dwarf semishrubs (including the dominant species), the short vegetative period in ephemeroids and ephemers, the uninterrupted vegetative period in haloxerophytes (*Anabasis aphylla*) and in mesoxerophilous annual plants (*Ceratocarpus arenarius*).

Deviations from the normal hydrothermical regime result in considerable fluctuations of the developmental rhythms. In particularly droughty years even a dominant species does not flower and fructify sometimes for several years in succession; the period of summer dormancy is prolonged considerably during these years. However in rare years with a relatively abundant precipitation many plants are flowering a second time in summer, mostly even more abundantly than the first flowering in spring (e.g. in xerophilous shrubs *Atraphaxis spinosa*, *Convolvulus fruticosus*).

The fluctuations of the abundance of mature individuals of a species are to a certain extent correlated with the average longevity inherent in this species. Less long-lived dwarf semishrubs are characterized by a wider range of fluctuations. For instance, in 1967, the abundance of mature plants of *Kochia prostrata* was 67 specimens per 1 ar, while in 1969 it was only 32. The similar abrupt fall in the abundance of plant biomass was observed in *Anabasis aphylla* (from 11 to 5 specimens per 1 ar); this was obviously the result of a severe winter. The abundance of perennial ephemeroid plants particularly of *Rheum tataricum* remained almost unchanged, although the species with bulbils lying near the soil surface (e.g. *Colpodium humile*) exhibited considerable fluctuations (±30 %) of the abundance, while during the cold winter of 1968—1969 it decreased by more than one half. A still wider range of fluctuation was observed in ephemers, (e.g. *Leptaleum filifolium*) viz. ±70 % the average abundance being 668 specimens/ar.

The abundance of emerged seedlings fluctuates considerably both in dwarf semishrubs and in annual species (Fig. 2). The death of emerged seedlings is caused mainly by the drying and excessive heat of the soil, to a lesser extent by early-spring frosts.

The fluctuations of the primary productivity are investigated insufficiently. The total reserve of the plant biomass in a community is about 12 tons per 1 hectar (dry weight), the propor-

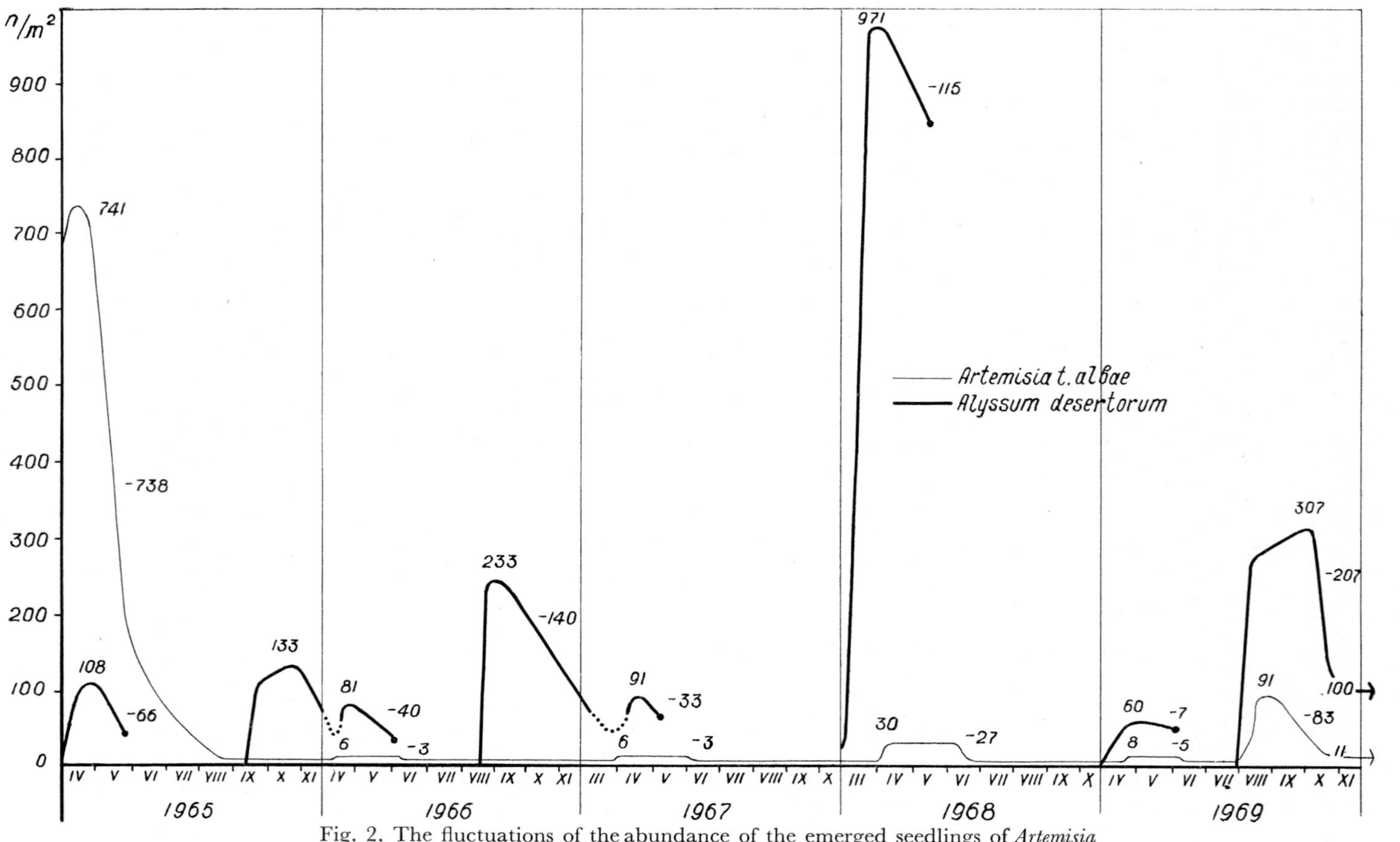

Fig. 2. The fluctuations of the abundance of the emerged seedlings of *Artemisia terrae-albae* and *Alyssum desertorum* during the period of 1965—1969. Or.

tion of the overground part being 29 % of this quantity (3.7 tons per hectar), the weight of one-year-old shoots being only 1 ton. The useful production comprises 0.4 tons per hectar of good fodder and about 0.5 tons per hectar of the roots of *Rheum*, a superb source of raw tanning material, containing 10 per cent tannids.

The productivity of the useful overground matter varies widely, from 0.2 to 0.6 tons per hectar (Fig. 3). A wider range of fluctuations, ±40 %, is known for *Artemisia* communities (Bykov 1955) and a still wider range, ±60 % to ±75 %, was observed in more southern *Artemisia* communities (Nechayeva 1956).

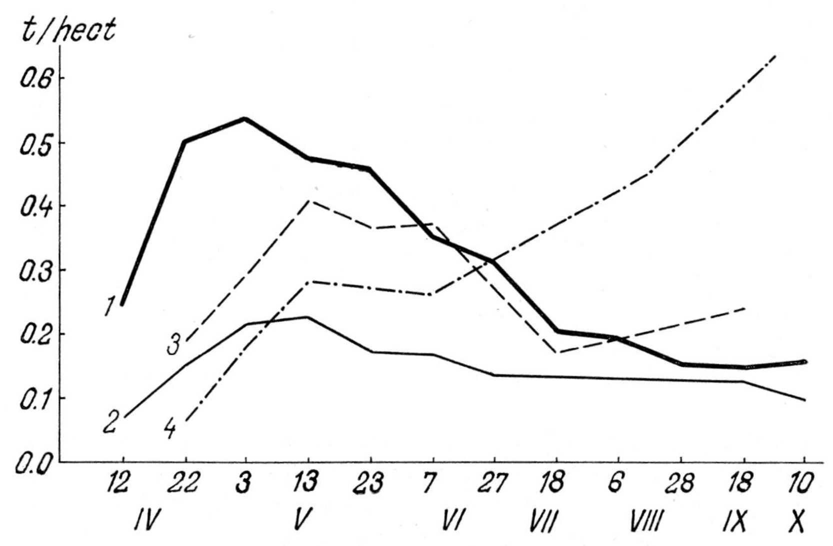

Fig. 3. The seasonal and annual fluctuations of the productivity of pasture: 1 — 1966; 2 — 1967; 3 — 1968; 4 — 1969. Or.

A notion of the seed productivity fluctuations can be formed, if the following circumstances are considered: The reserve of the available water in the soil is usually exhausted by plants already by the end of May; a great proportion of the total expenditure (35 %) is used for transpiration. In the years of unusually great abundance of ephemers the expenditure of the reserve of the available water is more rapid than usual; consequently *Artemisia terrae-albae* (flowering and fructifying in the autumn) remains in such years without a sufficient quantity of water and its fructification fails unless the available water reserve is replenished by the summer precipitation. Such ephemeroids as *Rheum tataricum*, even though possessing a great reserve of water in their fleshy roots, regularly rest in certain years, while certain other ephemeroids, such as *Megacarpaea megalocarpa*, apparently have no such years of rest. *Anabasis aphylla*, a vigorous phreatophyte draws the water from the loamy sand horizons, partly from gypsum horizons and consequently depends less on the air humidity than the other plants; this species fructifies abundantly almost every year.

Since most of the plant biomass of ephemeroid Artemisieta pastures is afforded by plants with not deep root systems (xerophilous dwarf semishrubs, ephemers and ephemeroids), there exists a direct correlation ($r = +0.822 \pm 0.066$) between the productivity of these pastures and the reserve of the soil water in the upper 1-metre soil layer (Fig. 1b), which is expressed by the following empirical equation:

$$U = 0.046\ h\ 0.0237,$$

U is the maximal yield in centners per hectar and h is the reserve of available water in mm in the upper 1 metre soil layer during the 10 days' period with a mean diurnal temperature of +10 °C (FEDOSEYEV, 1964).

The association of *Haloxylon persicum—Carex physodes*, despite its very wide distribution cannot be assigned to climax associations, because it exists on a very mobile substrate. Apparently the very slow and prolonged development of the ecosystem is directed to the climax dwarf semishrub type of vegetation with zonal grey-brown desert soils.

The best descriptions of *Haloxylon persicum* communities were made by RODIN (1948, 1968) and by KUROCHKINA (1966). This region (eastern part of the Kara-Kumy Desert) differs but slightly from the area to the north of the Aral Sea in the total amount of precipitation (115 mm per year), but its seasonal distribution is characterized here by a conspicuous autumn-winter maximum. The hot dry summer already begins in May and lasts until the middle of October. The unfavourable conditions of moistening are somewhat smoothed by the physical conditions afforded by sandy soils.

Meteorological conditions affect conspicuously the seasonal rhythms of phytocoenoses, which is expressed not only as the distinct alternation of synusia and as the summer dormancy in many plant species (accompanied by the branch fall in leafless semitrees and semishrubs), but also as a wide range of fluctuations both of the rhythms themselves and of the abundance and productivity (NECHAYEVA 1956, 1958). In *Haloxylon persicum*, dominating in the communities, the flowering usually begins about the middle of February; but in 1942 its onset was delayed for a month, while in 1943 it began a month earlier. In favourable years growth and development in many species was observed in winter.

The most constant abundance of mature individuals is observed in semishrubs, shrubs and semitrees, particularly in *Haloxylon*; it is most variable in ephemeroids, particularly in *Carex physodes*, a very valuable forage plant (from 260 to 425 shoots per sq.m. during the period from 1940 until 1945); the range of fluctuations is increased by grazing. The widest range of fluctuations of the abundance was observed in ephemers.

The reserve of seeds in the soil fluctuated considerably. Thus, the greatest abundance, 4.020 specimens per sq.m. (down to the depth of 5 cm), was recorded in 1941. During the next few years this number gradually decreased to 1.400 in 1945. The fluctuations of the abundance of emerged seedlings are determined by those of the seed productivity, the reserve of seeds in the soil and the hydrometeorological conditions.

While the total reserve of the vegetative matter equalled on an average 2.1 tons per hectar (91 % of this quantity being at the expense of *Haloxylon* shrubs and semishrubs) and its average annual increment was 0.4 t/ha, it varied from year to year during the period of 1960—1965 from 0.2 to 0.7 t/ha.

While the total reserve of the forage (chiefly *Carex physodes*) during the 15 years' period of observations (1940—1954) had the average value of 0.2 t/ha, its annual fluctuations during this period were from 0.09 to 0.42 t/ha, i.e. ± 72 %. Still greater fluctuations were observed under intense grazing. In this case just as in all the instances described above, a complete correlation exists between the productivity and the amount of precipitation during the period from November to May.

Thus, not only all the fluctuations in the phytocoenoses typical of the semideserts and deserts of the Turanian Plain, but the very existence of these phytocoenoses and particularly their productivity are controlled by the minimal levels of hydrometeorological factors viz. the water deficit, partly extreme temperatures. Besides, a certain role also belongs to a smaller or greater capacity or reducing the reserve of the available water in the soil by means of transpiration and accumulation in the plant biomass.

H. SYNCHRONOLOGICAL VEGETATION DYNAMICS

26 SYNCHRONOLOGIE WÄHREND DER EINZELNEN GEOLOGISCHEN ZEITRÄUME IN EUROPA

R. Tüxen

Contents

26 SYNCHRONOLOGIE WÄHREND DER EINZELNEN GEOLOGISCHEN ZEITRÄUME IN EUROPA

26.1 Einführung und Methoden

Wenn ein erfahrener Pflanzensoziologe eine Flora aus einem Nachbarlande durchsieht, kann er mit großer Sicherheit die dort zu erwartenden Assoziationen, Verbände, Ordnungen und Klassen voraussagen, weil er die Bindung vieler Arten an bestimmte syntaxonomische Einheiten kennt. Diese Erwartung wird aber um so häufiger getäuscht, je weiter das geprüfte Land von dem gut bekannten eigenen Arbeitsgebiet entfernt ist und je abweichender Klima und Vegetationsgeschichte sind. Dann zeigt sich, daß bestimmte Arten in verschiedene Gesellschaften eintreten können. In der gleichen Lage ist der Pflanzensoziologe, wenn er in der Zeit zurückschreitet und fossile Arten oder deren Kombinationen soziologisch beurteilen oder gar heute lebende Pflanzengesellschaften wieder erkennen soll. Ebenso gewagt wie aus dem Vorkommen einer oder mehrerer Kennarten einer heimischen Assoziation auf deren Vorhandensein im ganzen Areal dieser Einheiten zu schließen, ist es, aus diesen in fossilen Resten gefundenen Arten das frühere Vorkommen der heute lebenden Assoziationen ableiten zu wollen. Solche Schlüsse sind nur mit großen Einschränkungen möglich. Wohl auch aus diesem Grunde findet man in florengeschichtlichen Arbeiten nur selten Angaben über Gesellschaften früherer Zeiten. Immerhin sind aber in großen Zügen ziemlich sicher z.B. lichtliebende, schattenreiche, Nässe oder Trockenheit ertragende, azidophile oder eutrophe Gesellschaften zu erkennen, die, so lange nicht dynamisch wichtige Arten fehlen, nach den erhaltenen Arten mit heutigen Ordnungen oder Verbänden oder in günstigen Fällen selbst mit Assoziationen gleich gesetzt werden können.

In der historischen Pflanzengeographie wird oft von "Vegetation" gesprochen, womit genau genommen die Flora oder auch die Vegetation in physiognomischer Hinsicht gemeint ist. Auch die Verwendung pflanzensoziologisch klingender Namen wie "Quercetum mixtum", "Pinetum silvestris", "Eriophoreto-Callunetum", Eriophoreto-Ericinetum", "Sphagneto Eriophoretum", "Phragmiteto-Scheuchzerietum", "Phragmiteto-Eriophoretum", die vorwiegend in älteren pollenanalytischen Arbeiten vorkommen (jetzt allerdings größtenteils außer Gebrauch sind), darf nicht den Eindruck syntaxonomisch wohl definierter Einheiten erwecken. Wir verwenden hier den Begriff "Vegetation" im pflanzensoziologischen Sinne als die Summe verschiedener oder aller floristisch definierter Pflanzen-Gesellschaften. Andererseits umfaßt das Wort "Vegetation" bei der Verwendung in der historischen

Pflanzengeographie manchmal bereits eine Interpretation der pollenanalytischen Ergebnisse, die deutlich über den Begriff Flora und rein physiognomisch gefaßte Vegetationstypen hinausgeht, aber gewöhnlich vor der Verwendung echter pflanzensoziologischer Begriffe stehen bleibt. In anderen vegetationsgeschichtlichen Arbeiten werden Funde und Untersuchungsergebnisse eindeutig pflanzensoziologisch ausgewertet. In zahlreichen weiteren könnte das leicht geschehen, ist aber von den Verfassern nicht getan worden. Hier ist nur die erste Gruppe Arbeiten berücksichtigt worden, ohne für Vollständigkeit bürgen zu können. Dabei wurde vorwiegend die mitteleuropäische Literatur ausgewertet, weil deren Ergebnisse sich am leichtesten in das hier am weitesten entwickelte System der Pflanzengesellschaften nach BRAUN-BLANQUET [1]) einordnen lassen.

26.2 Karbon

Die am weitesten zurückreichenden palaeosoziologischen Untersuchungen wurden von KELLER (1932), DRÄGERT (1964) und SCHMIDT (1968) mit Hilfe von Großresten im Ruhr-Karbon an Moorgesellschaften durchgeführt, welche die Steinkohlenlager aufgebaut haben. KELLER gab erste Hinweise auf die Anwendbarkeit pflanzensoziologischer Methoden auf die Fossilien der Kohle. DRÄGERT unterscheidet je eine *Articulaten-*, *Pteridophyllen-* und *Lepidophyten-Cordaiten*-Gesellschaft. Die letzte kennzeichnet ein Waldmoor, die erste eine Röhrichtzone im Flachmoor, während die *Pteridophyllen*-Gesellschaft Übergangsmoore anzeigt. Heute lebende Taxa finden sich nicht unter den angegebenen Arten, die ihrerseits alle der heutigen Vegetation fehlen. SCHMIDT weist auf begriffliche und methodische Gesichtspunkte bei der soziologischen Bearbeitung der fossilen Pflanzenreste der *Pteridophyllen*-Gesellschaft. (Weitere Literatur in der Bibliographie von DRÄGERT 1964b.)

26.3 Tertiär

Zwischen Karbon und Tertiär fehlen palaeosoziologische Untersuchungen. Aus dem Pliozän sind von mehreren Stellen in Europa, aber auch in N-Amerika (CHANEY (1956) und AXELROD (1959)) reiche Fundplätze von Pflanzenresten bekannt, die palaeosoziologisch ausgewertet wurden.

A. STRAUS (1935) glaubte nach dem damaligen Stand der

---

[1]) Ein Teil der palaeosoziologischen Literatur wurde in den Bibliographien von DRÄGERT (1964) und von TRAUTMANN (1957) zusammengestellt. Eine weitere Bibliographie (R. TÜXEN) erscheint in Excerpta Botanica B Sociologica.

pflanzensoziologischen Systematik im Pliozän von Willershausen (S-Niedersachsen) eine artenreiche Fagion- Gesellschaft nachweisen zu können, die er "Fagetum subhercynicum pliocenicum" nannte. Auch Quercion wurde für möglich gehalten. In einer späteren Schrift (1954) werden artenreiche Fagion-, Dictamno-Sorbion-, Alnion- und Fraxino-Carpinion-Gesellschaften angenommen. Die Hauptgesellschaft könnte nach Straus ein artenreicher Eichen-Hainbuchenwald gewesen sein. Man darf aber trotz der gleichlautenden Namen gewiß nicht eine allzugroße Ähnlichkeit der tertiären mit den heutigen, sicher viel artenärmeren Gesellschaften erwarten.

W. Szafer (1947, I: 153) konnte in der pliozänen Flora von Krościenko (Polen) mehrere heutige Klassen, Ordnungen und Verbände wieder erkennen (Potamion, Phragmitetalia, Phragmition, Magnocaricion, Querco-Fagetea, Fraxino-Carpinion (vgl. Braun-Blanquet 1949: 200 und 1964: 705). Er ging in seiner soziologischen Gliederung nicht so weit wie Straus (1935), aber viel weiter als E. L. Braun (1935), die von "undifferentiated deciduous forest climax" des Pliozäns gesprochen hatte (Zit. nach Szafer).

W. Szafer (1949) hielt es nach weiteren eingehenden Untersuchungen für möglich, "that a very detailed analysis, both quantitative and qualitative, of the Pliocene remnants of forest flora pursued in particular layers (or lenser) of water deposits, founded upon the detailed knowledge of forest communities living contemporarily in the refugion of Eastern Asia, Atlantic America and Transcaucasia, will enable in future the creation of a real nucleus of paleo-sociology of plants."

Hantke (1954: 91—97) unterschied im Schrotzburger Obermiozän (Schweiz) zwei Auen-Waldgesellschaften: die Podogonium Syellianum-Populus mutabilis-Ass. und den Platanen-reichen Ulmus-Liquidambar-Wald, die in der Gegenwart nicht mehr bestehen. Auch die angedeuteten Wasser- und Röhricht-Gesellschaften weichen stark von den heutigen ab.

Einen stark von rezenten amerikanischen Vorbildern beeinflußten Rekonstruktions-Versuch der Vegetation der niederrheinischen Bucht im Miozän, die zur Bildung der Braunkohle führte, machte Marlis Teichmüller (1958). Die damals herrschenden Busch-Moore, Sumpf- und Trockenwälder haben mancherlei floristische Ähnlichkeit mit rezenten amerikanischen, zeigen aber keine gemeinsamen Arten mit den jetzt in Europa wachsenden Gesellschaften. Dagegen lassen die Wasserpflanzen-Bestände durch *Nymphaea* und *Stratiotes aloides* eine Beziehung mit der heutigen Klasse der Potametea erkennen. Die Gattung *Brasenia*, die damals reichlich in solchen Gesellschaften wuchs, ist in Europa ausgestorben, blieb aber in anderen Erdteilen als *Brasenia purpurea* erhalten.

Die grundlegenden Arbeiten von Straus und Szafer setzte I. Horvat (1959) fort, indem er die fossile südosteuropäische Pliozän-Flora mit den heutigen höheren syntaxonomischen Einheiten SO-Europas und der nordamerikanischen Relikt-Zentren (nach Knapp 1957) verglich. Schon früher hatten mehrere Autoren Gliederungen nach Höhenstufen und Formationen durchgeführt. Während Potamion- und Phragmition-Gesellschaften (vgl. auch Mädler 1939) bis heute sich weiter entwickelten, sind andere tertiäre höhere Einheiten wie die Sumpfwälder der Taxodietea in SO-Europa ausgestorben. Dagegen konnten sich nach Horvat Auwälder (Salicetalia albae, Populion albae), wenn auch stark verarmt, erhalten. In Tieflagen oberhalb der Flußauen wuchsen artenreiche mesophile Waldgesellschaften, die außer den noch nicht vorhandenen Arten der Fagetalia sylvaticae und Querco-Fagetea sylvaticae zahlreiche heute in N-Amerika wachsende Gehölze der Ordnungen Liriodendretalia, Ulmo-Aceretalia sacharini, Fago-Magnolietalia grandiflorae enthielten.

Die heutige Ordnung der Quercetalia pubescentis fand Horvat im so-europäischen Pliozän nur wenig ausgebildet, was allerdings nicht als Beweis für ihr damaliges Fehlen gedeutet werden muß.

Auch die mediterrane Klasse der Quercetalia ilicis war bereits im Pliozän SO-Europas entwickelt, wo sie weiter nördlich vorkam als heute.

Endlich geht nach Horvat auch die Klasse der Vaccinio-Piceetea auf das Pliozän zurück. Sie war damals wesentlich reicher an Gehölzen als heute.

Dagegen waren die Alnetalia in SO-Europa artenärmer, weil ihre volle Ausbildung erst mit der pleistozänen Klima-Verschlechterung erfolgte. Ein ähnliches Schicksal nimmt Horvat für die Caricetalia fuscae, Sphagnetalia medii und Molinietalia-Gesellschaften an. Die Potentilletalia caulescentis und Thlaspeetalia rotundifolii erlitten starke Einbußen, während die Salicetalia herbaceae erst mit dem Pleistozän in SO-Europa auftraten. Seslerietalia, Caricetalia curvulae und Arabidetalia coeruleae erfuhren Bereicherungen, mußten aber auch Verarmung hinnehmen.

Wenn auch diesen palaeosoziologischen Deutungen, die für SO-Europa nach der heutigen Auffassung der Palynologie (Beug) gewisser Berichtigungen bedürfen, nur örtliche Einzelaufnahmen aus dem säkularen Geschehen der Vegetationsentwicklung Mittel- und Südosteuropas zugrunde liegen, so beleuchten sie doch die Synchronologie der hier lebenden Pflanzengesellschaften, bevor die Eiszeiten sie auslöschten, umwandelten oder verdrängten.

In jüngster Zeit hat KNOBLOCH (1969) die tertiären Floren Mährens u.a. auch soziologisch ausgewertet. Er konnte die Vorläufer der heutigen Potametea und eine Phragmites-Gesellschaft erkennen.

Die Zonierung und Sukzession eines Myricaceen-Cyrillaceen-(Betulaceen)-Buschmoores und eines Taxodiaceen-Nyssa-Moorwaldes sowie eines Röhrichts (*Typha*, *Sparganium* u.a.) im Miozän bei Katalinbánya (Ungarn) zeigte SIMONCSICS (1960).

BERGER & ZABUSCH (1953: 258) konnten eine allgemeine Vorstellung von obermiozänen Ufer- und Auwald- und von Strauch-Gesellschaften (die an die Rhamno-Prunetea erinnern) nach Funden aus der Türkenschanze bei Wien vermitteln.

BERGER (1955a: 426) beschrieb eine Uferwaldgesellschaft aus *Salix varians* und *S. macrophylla* mit *Myrica lignitum* aus einem wärmeren Klima des Jungtertiär (Obertorton) aus Wolfsberg in Ostkärnten.

Im Pannon des mährischen Teils des Wiener Beckens wies KNOBLOCH (1969: 163) eine heute unbekannte Sumpfwald-Gesellschaft nach, an der vor allem die Arten der Gattungen *Glyptostrobus*, *Nyssa*, *Byttneriophyllum* und *Alnus* beteiligt waren. Sie zeigt gewisse ökologische Beziehungen zu den heutigen Alnetea glutinosae (vgl. p. 281).

BERGER (1955b: 108—110) gab ein Bild der Vegetation in der Umgebung des Laaerberges in Wien während des Altpliozän (Pannon), ohne allerdings wegen der Mischung der allochthonen Reste syntaxonomisch auswertbare Angaben machen zu können.

BERGER (1952: 112) entwarf ein Bild der Pflanzengesellschaften aus einer stillen flachen Bucht des großen pannonischen Süßwassersees aus dem mittleren Unterpliozän bei Brunn-Vösendorf. Er konnte eine Ceratophyllum-, eine Phragmites-, Salix-Nerium-Gesellschaften, einen feuchten farnreichen Auenwald mit *Carya*, *Zelkova*, *Platanus*, *Taxodium* u.a. und einen stark gemischten Laubwald von *Castanea*, *Fagus*, *Carpinus* u.a. erkennen mit Lianen- und Sträuchern, die an Rhamno-Prunetea-Waldmäntel erinnern (*Rosa*, *Vitis*, *Clematis*).

Auf feuchten Böden (ohne Sumpfcharakter) wuchs im Pannon des mährischen Teiles des Wiener Beckens ein Wald mit *Ginkgo*, *Carpinus*, *Betula*, *Fagus*, *Liquidambar*, *Ulmus* und *Platanus*, der in ähnlicher Zusammensetzung auch von BERGER (1955a, 1956b: 17) aus dem österreichischen Wiener Becken angegeben wird und für den ein Gegenstück heute nicht mehr besteht. (Knobloch 1969: 163). Auch Andréansky's Untersuchungen in Ungarn (1963) deuten in die gleiche Richtung.

Auf nord-amerikanische Arbeiten (z.B. von Chaney (1956) und Axelrod (1959) kann hier nur verwiesen werden.

26.4 Pleistozän und Holozän

Lona & Follieri (1958: 97) konnten während Günz II—III und nach Günz III in den Warmzeiten eine periodische diskontinuierliche Substitution des alten Caryetum durch das jüngere Querco-Carpinetum feststellen.

Nach Watts (1958: 147) "seems it possible that a dry-heath community similar to the modern one (d.i. Blechno-Quercetum) may have occurred" (im Mindel-Riss-Interglazial in Kilbeg, Irland).

Eiszeitliche Florenreste sind lange bekannt und pflanzengeographisch wie florengeschichtlich vielfach untersucht worden. Dagegen ist ihre soziologische Auswertung noch im Rückstand. Sie wurde wohl dadurch erschwert (vgl. auch Braun-Blanquet 1964: 707), daß die syntaxonomische Gliederung der heutigen arktischen Pflanzengesellschaften erst nach und nach durchgeführt und immer noch nicht abgeschlossen werden konnte. Ältere Vegetationseinheiten niederen Ranges lassen sich genau genommen syntaxonomisch nur dann erkennen, wenn die vergleichbare rezente Vegetation systematisch untersucht worden ist. Aus diesem Grunde bleiben unsere Beispiele auf solche gut studierten Gebiete oder auf höhere syntaxonomische Einheiten beschränkt.

Wir wollen versuchen für mitteleuropäische Gesellschaftsklassen einige Beispiele aus ihrer Synchronologie zu geben, soweit verwertbare Angaben vorliegen oder zugänglich sind. Dabei haben wir allgemeine nicht soziologisch definierte Formationsangaben hier nicht oder nur ausnahmsweise berücksichtigt, obwohl ihr Wert für das Verständnis der Vegetationsgeschichte durchaus nicht gering einzuschätzen ist.

Gewiß ging die Wiederbewaldung vereister Gebiete nicht in der Weise vor sich, daß sich die zurückgewichenen Zonen der Vegetation und ihre Gesellschaften als Ganzes wieder nach Norden in die vereist gewesenen Gebirge vorschoben, sondern neue Gesellschaften bauten sich aus den wieder oder erstmalig in das vom Eise verlassene Gebiet einwandernden Arten auf [vgl. Lang (1967: 26), Braun-Blanquet (1964: 709), Frenzel (1955, 1968)].

Strandgesellschaften der Meeresküsten (z.B. Zosteretea,

Thero-Salicornietea, Cakiletea, Bolboschoenetea maritimi, Honckenyo-Elymetea, Ammophiletea u.a.) dürften in ihrer Zusammensetzung über lange Zeiträume nahezu gleich geblieben sein und mit Klimaänderungen entlang der Küstenlinien sich harmonika-artig verschoben haben, wenn jedoch auch dabei die sie aufbauenden Arten selbständig reagierten, so daß in der Feingliederung der Ordnungen, Verbände und wohl auch Assoziationen auch hier im Laufe der Zeit mancherlei örtliche durch Gestein und Relief bedingte Differenzierungen auftraten. Andere Einheiten entstanden aber auch hier spät und sind noch in der Entfaltung begriffen (Spartinetea).

Die meisten übrigen Gesellschaften aber zerfallen bei entscheidender Umwelt-Änderung bis auf Relikte oder verschwinden nach und nach ganz, oder sie bauen sich um oder neu wieder aus einzelnen Arten, die zur Verfügung stehen und zusagende exogene sowie endogene Lebensbedingungen finden.

Während die Spezialisten-Gesellschaften also tertiären Ursprungs sein können, wie übrigens auch die Relikt-Gesellschaften an Fels-Standorten oder in Schneetälchen der Hochgebirge z.B. S-Europas solche Wurzeln haben dürften, später jedoch in verschiedener Weise sich entwickelten (z.B. Asplenietea rupestria, Thlaspeetea rotundifolii, Elyno-Seslerietea, Salicetea herbaceae) mußten sich die meisten der auf ausgeglicheneren Standorten wachsenden Gesellschaften aus den einzeln, oft aus verschiedenen Richtungen zuwandernden Arten aufbauen (vgl. E. Schmid 1936). Alle Übergänge zwischen diesen beiden Extremen der Gesellschafts-Bildung und -Verlagerung sind denkbar.

Die mit zunehmender Erwärmung nach der Eiszeit artenreicher werdenden Gesellschaften fügten sich unter laufender Entstehung neuer syntaxonomischer Einheiten zu einem durch Standorts- und endogene Gesellschafts-Wirkungen immer feiner gegliederten Mosaik zusammen.

Viele der heute lebenden Pflanzengesellschaften sind aber erst durch den Menschen als Substitions- oder Ersatz-Gesellschaften erzeugt worden und darum viel jünger als die meisten natürlichen, die in Mittel- und Nordeuropa, ebenso wie in Nord-Asien und Nord-Amerika, allerdings zum größten Teil auch erst nach der Eiszeit entstanden.

Nach Pawłowski (1966: 534) haben sich die meisten der heutigen polnischen Pflanzengesellschaften in verschiedenen Klima-Perioden des Holozän entwickelt. Die ältesten polnischen Wald-Gesellschaften dürften Pino-Vaccinietum myrtilli und Pino-Vaccinietum uliginosi sein, die vielleicht schon seit dem Ende des Pleistozän in Polen wuchsen. Alle anderen sind jünger.

26.5 SUMMARY

Synchronology during the Particular Geological Periods in Europe

It is thinkable to conclude from the species composition of fossil plant deposits on the eventual occurrence of certain vegetation units in former geological periods in cases of a partial identity or near relation of the fossil floras concerned with species living in present times. These conclusions seem to be possible, since the relations of species to certain associations, alliances, orders and classes are well investigated in the recent vegetation. But such conclusions from fossil floras with partly extinct taxa on recent vegetation are always more or less hypothetical, at least in details.

As far as in the literature on historical plant geography "vegetation" is mentioned, apparent confusions with "flora" or discussions restricted to vegetational physiognomy are mostly prevailing. In the present explanations the term "vegetation" is used phytosociologically, designating the sum of plant communities defined by syntaxonomical methods using mainly criteria of fidelity, namely characteristic and differential species. The present review is restricted to discussions of publications dealing with "vegetation" in the meaning just mentioned and of predominantly Central European literature.

The palaeosociology of the rich fossil plant deposits of the carbon age needs special methods, since these plants living at this period are not identical with any recent species.

But the similarity of the tertiary floras, namely of its later parts, with recent plants (at least at the generic level) affords parallelizations with recent vegetation units. In many cases, the European tertiary vegetation has apparently more affinities to recent plant communities of Southeastern North America than with Central European vegetation of present times (e.g. occurrence of forests with *Carya*, *Liquidambar*, *Nyssa* etc.).

During the pleistocene interglacials, only few fossil deposits have been found appropriate for palaeosociological conclusions. The phytosociological treatment of the glacial deposits known since long periods is impeded by deficiencies of the syntaxonomy of the recent arctic vegetation.

In the course of the postglacial climatic changes, many coastal plant communities of salt marshes and of sand dunes remained presumably relatively stable in their species composition during long periods. But other coastal vegetation units are of late origin or are even forming in present times (e.g. European Spartinetea).

The most plant communities of other sites were much more influenced by the decisive environmental alterations occurring since the last glacial period.

A great part of the present Central European plant communities were originating apparently in postglacial times. A considerable percentage of the Central European vegetation is a result of human activities. These anthropogenic vegetation units are generally younger than the natural plant communities.

27 SYNCHRONOLOGIE EINZELNER VEGETATIONS-EINHEITEN IN EUROPA

R. Tüxen

27 SYNCHRONOLOGIE EINZELNER VEGETATIONS-EINHEITEN IN EUROPA

Je reicher fossile Funde sich darbieten, je vollkommener sie konserviert wurden und je sauberer sie getrennt sind, desto feinere pflanzensoziologische Einheiten kann man aus ihnen feststellen. Höhere syntaxonomische Rangstufen, wie Klassen, Ordnungen oder Verbände, lassen sich leichter erkennen als Assoziationen. Dabei bleibt als Voraussetzung aber die Annahme einer gewissen Konstanz der Arten-Verbindung über längere Zeiträume, die dann wohl in der Tat gegeben ist, wenn keine oder nur wenige andere Arten als heute vorkommen.

In syntaxonomischer Anordnung geben wir die uns zugänglichen Angaben über die Geschichte der Gesellschaften aus den einzelnen Klassen, wobei insbesondere die Verhältnisse in Europa im jüngeren Tertiär, Pleistozän und Holozän berücksichtigt seien.

Lemnetea

Troels-Smith erkannte das Vorkommen von *Lemna* in Verbindung mit einer Kulturschicht der Pfahlbauten bei Weiher (Schweiz). (Vgl. a. Welten 1955 : 77.) Die gleiche Koinzidenz fand Jørgensen (1954 : 170) bei Aamosen (Seeland-Dänemark). Weitere Angaben über subfossile *Lemna*-Funde bei Godwin 1956, Rohlmann 1958, Dorofeev 1963, Knörzer sowie Burrichter (vgl. dazu R. Tüxen 1973a, b).

Bidentetea tripartiti

Die Angaben von Szafer (1931 : 25) lassen Bidention-Gesellschaften in Polen schon im Tertiär erkennen. Willerding (1960 : 467, 1967 : 74) wies das Bidention an Hand von Großrest-Funden von 6 Arten aus dem Göttinger Leinetal für die ältere bis jüngere Nachwärmezeit nach. Gesellschaften desselben Verbandes wurden von Udelgard Körber-Grohne (1967: 108) von der Unterweser (Feddersen Wierde) in den ersten Jahrhunderten nach Christi Geburt und von Behre (1970 : 36) von der unteren Ems aus der älteren Eisenzeit festgestellt.

Stellarietea mediae

Willerding (1960 : 467, 1967 : 74) fand Großreste von 25 Arten der Stellarietea mediae in Ablagerungen der älteren bis zu der jüngeren Nachwärme-Zeit im Göttinger Leine-Tal. Er warnt aber (1970 : 300, 325) vor zu weit getriebenen soziologischen Schlüssen aus den vorgefundenen Unkrautsamen wegen Verschiebung der Konkurrenzverhältnisse. Er weist zugleich (brieflich) darauf hin, daß man die niedrig wüchsigen Arten, z.B. aus der Gattung *Veronica* oder *Anagallis*, mit einiger Sicherheit als Acker-Unkräuter erst aus einer Zeit nachweisen könnte, in der das Getreide nahe dem Boden geschnitten und nicht nur die Ähren geerntet wurden. Helbaek (1954 : 257) gab eine Zusammenstellung der urgeschichtlichen Funde unter anderem von Arten der Stellarietea mediae aus Dänemark, die das Vorkommen dieser Klasse seit der frühen römischen Eisenzeit in den beiden ersten nachchristlichen Jahrhunderten erkennen läßt. Lüdi (1954 : 19) erwähnt zahlreiche Stellarietea-Arten aus bronzezeitlichen Funden im schweizerischen Alpenvorland.

Aus bronzezeitlichen Pfahlbauten von Valeggio am Mincio konnte Margita Villaret-von Rochow (1957 : 112) eine fragmentarische Liste einer Secalinetalia-Gesellschaft zusammenstellen.

Wenn auch die irischen Pollenanalysen von Smith (1964) keine näheren Angaben enthalten, die pflanzensoziologisch auszuwerten wären, geht doch daraus hervor, daß Wildkrautgesellschaften der Äcker auch dort nicht älter als neolithisch sein können.

Gesellschaften der Klasse der Stellarietea mediae wurden von Udelgard Körber-Grohne (1967) auf der Feddersen Wierde (NW-Deutschland) aus den ersten Jahrhunderten nach Christi Geburt und von Behre (1970) von der unteren Ems (NW-Deutschland) schon seit der älteren Eisenzeit erkannt.

Aus römischen Abfall-Resten konnte Knörzer (1962 : 263, 1968a : 100, 1968b : 168) sowohl Polygono-Chenopodion-, als auch Aperetalia spica-venti-Gesellschaften (Aphanion, Papaveretum argemone) nachweisen. Knörzer (1971) gelang die Kennzeichnung einer neolithischen Getreide-Wildkrautgesellschaft, die er Bromo secalini-Lapsanetum praehistoricum nennt. Sie dürfte etwa zwei Jahrtausende bestanden haben und aus Wald- und Saumpflanzen entstanden sein, denen sich Arten beimengten, die mit dem Saatgut eingeschleppt wurden.

Die Gesellschaft wurde in der Eisenzeit durch *Scleranthus annuus-Spergula arvensis*-reiche Bestände ersetzt. Chenopodietalia und Aperetalia-Gesellschaften waren noch nicht deutlich getrennt.

Mit Hilfe der prähistorischen und historischen Siedlungsforschung beleuchtete J. Tüxen (1958) aus dem Verhältnis von Archaeophyten und Neophyten die prähistorische Entwicklung von Chenopodietalia albi-Gesellschaften auf Hackfrucht-Äckern in NW-Deutschland.

Isoëto-Nanojuncetea

Eine Nanocyperion-Gesellschaft fand Udelgard Körber-Grohne (1967) an der Unterweser in der Zeit um Christi Geburt. Auch eine *Juncus ranarius*-Gesellschaft wurde von dort gemeldet.

Litorelletea

Nach Niklewski & van Zeist (1970:751) wuchs eine Isoëtes-Gesellschaft mit *Myriophyllum* während der Eiszeit und noch im Postglazial in NW-Syrien.

Am Ende des Mindel-Riss-Interglazials wuchs in Irland (Kilbeg) nach Watts (1958: 147) eine Littorellion-Gesellschaft mit *Eriocaulon septangulare* (ohne *Lobelia dortmanna*), die dem heutigen Eriocauleto-Lobelietum nahezu gleicht.

Eine Isoëtes-Litorella-Myriophyllum alterniflorum-Gesellschaft, die eine Potamion-Gesellschaft ablöste, konnte Andersen (1967 : 110) schon im dänischen Eem-Interglazial erkennen. Perring (1967 : 258) erwähnt eine ähnliche Artengruppe aus interglazialen Ablagerungen Irlands. Zagwijn (1963 : 55) gibt aus dem niederländischen Tiglian-Interglazial das häufige Vorkommen von *Isoëtes lacustris* an. Lang (1955, 1967 : 29) berichtet von einer vergleichbaren spätglazialen Isoëtes-Gesellschaft, die in See-Ablagerungen des Süd-Schwarzwaldes gefunden wurde. Aus Ostfriesland meldete Behre (1966 : 82) die Isoëtes echinospora-Gesellschaft, die sich in der ersten Alleröd-Hälfte offenbar aus dem Nupharetum nach der Oligotrophierung des Seewassers einstellte und lange erhielt. H. Müller (1970 : 44) konnte am Otterstedter See (NW-Deutschland) eine durch Eutrophierung bedingte Sukzession von einer Isoëtes echinospora- über eine Isoëtes lacustris- zu einer reinen Litorella lacustris-Gesellschaft nachweisen, wie sie den heute zu beobachtenden Trophie-Ansprüchen dieser Gesellschaften entspricht. Udelgard Körber-Grohne (1967 : 108) fand Arten der Litorelletalia nach Christi Geburt an der Unterweser.

Potametea

Nachweise von Gesellschaften dieser Klasse sind häufiger gelungen. Nach Mädler (1939) wuchsen schon im Pliozän in Glässum bei Frankfurt /Main ähnlich wie nach Szafer (1925 : 291, 1930 : 462) in Samostrzelniki bei Grodno (Polen) *Brasenia schröteri*, *B. nehringi*, *Najas maior*, *N. flexilis*, *Ceratophyllum demersum*, *Stratiotes aloides*, *Trapa natans*, *T. muzanensis*, *Nuphar luteum*, *Nymphaea candida*, *Potamogeton* sp. variae und in Krościenco (Polen) nach Szafer (1947) *Euryale*, *Nuphar*, *Potamogeton-* und *Ceratophyllum*-Arten, die mit wenigen Ausnahmen (*Brasenia, Euryale*) heute Potametea-Gesellschaften eigen sind.

Im mährischen Teil des Wiener Beckens fand KNOBLOCH (1969:163) eine aus dem Pannon stammende Wasserpflanzen-Gesellschaft mit *Trapa und Brasenia*, die ebenfalls als Vorläufer der heutigen Potametea aufgefaßt werden darf.

Auch I. HORVAT (1959 : 394) gibt für die Seen und Teiche der südosteuropäischen Pliozän-Landschaft Potamion-Gesellschaften an.

Die Untersuchungen fossiler Großreste und die Pollenanalyse interglazialer Ablagerungen in Polen durch SZAFER (1931 : 38) und in Brandenburg durch STARK, FIRBAS & OVERBECK (1932 : 125) ergaben das Vorhandensein reich entwickelter Potametalia-Gesellschaften in dieser Zeit und ihr Verschwinden im Laufe der Verlandung durch Phragmitetalia-Bestände.

ANDERSEN (1967 : 110) konnte in Jütland die Entwicklung von interglazialen Potametalia- (Myriophyllo-Nupharetum und Stratiotetum) zu einer Isoëtes-Gesellschaft zeigen. Auch im Interglazial von Tidofeld (NW-Deutschland fand PFAFFENBERG (1934:125) eine Potamion-Gesellschaft mit *Brasenia purpurea*, die im Laufe der Verlandung ihres Gewässers in eine Magnocaricion-Gesellschaft mit *Dulichium* überging.

Eine aus der Spät-Weichsel-Zeit stammende Liste einer *Myriophyllum alterniflorum-Gesellschaft* teilte BIRKS (1970 : 836) von Abernethy Forest in Inverness-shire (Schottland)mit.

Nach KOCH (1934a : 127) könnte sich der Eindruck verstärken, daß die Gesellschaften der kleinen *Potamogeton*-Arten in der pflanzensoziologischen Systematik, wie das von einigen Autoren (vgl. Parvopotametalia Den Hartog & Segal 1964 em. Westhoff 1968) vorgeschlagen wurde, selbständige Einheiten bilden, weil ohne die großen *Potamogeton*-Arten in den ältesten verlandenden Gewässern des Emslandes (NW-Deutschland) im Praeboreal *Potamogeton filiformis*, *P. pectinatus*, *P. pusillus*, *P. obtusifolius*, *P. gramineus* mit *Myriophyllum alterniflorum* und *M. verticillatum* gefunden wurden. RYBNIČEK & RYBNIČKOVÁ (1968 : 140) weisen ebenfalls auf einen arktischen Typus einer Gesellschaft der Potametalia Koch 1926 mit *Potamogeton filiformis* und *Hippuris vulgaris* usw. hin.

In Süd-Böhmen konnte VLASTA JANKOVSKÁ (1970 : 48) in der jüngeren Dryaszeit und im Praeboreal eine Wasserpflanzengesellschaft erkennen, die dem Nupharetum pumili Oberd. 1957 nahesteht. TROELS-SMITH (1969 : 588) wies das Nupharetum aus dem frühen Neolithikum in W-Seeland (Dänemark) nach.

Im Neolithikum war nach PFAFFENBERG (1947 : 79) das Potamion im Dümmer (NW-Deutschland) reich entwickelt. WELTEN

(1955 : 81) fand Potamion im Burgäschsee (Schweiz) im Subboreal.

PFAFFENBERG (1954 : 483) wies eine Verlandungssukzession vom Nupharetum über Phragmition zum Scheuchzerietum und Sphagnetum in Süddeutschland nach. Ein Potametum lucentis fand PFAFFENBERG (PFAFFENBERG & DIENEMANN 1964 : 82/3) in der Kalkmudde des Dümmer (NW-Deutschland). Diese Gesellschaft ging in ein Nupharetum mit *Najas marina* über. (Vgl. auch PFAFFENBERG 1947 : 79.)

Nach WILLERDING (1960 : 467, 1967 : 74) kommen in nachwärmezeitlichen Ablagerungen des Leinetals bei Göttingen Großreste von mehreren Potametalia-Arten vor. RYBNIČEK & RYBNIČKOVÁ (1968) konnten durch kombinierte Analysen von Großresten und Pollen die Sukzession Potametalia → Magnocarion → Scheuchzerio-Caricetea fuscae → Oxycocco-Sphagnetea in einem von ihnen untersuchten Moor in SO-Böhmen aufklären.

In bronzezeitlichen Pfahlbau-Resten der Nord-Schweiz fand LÜDI (1954 : 17) Arten des Potamion-Verbandes, während UDELGARD KÖRBER-GROHNE (1967 : 108) Potametalia-Arten von der Unterweser aus der Zeit um Christi Geburt angibt.

Bolboschoenetea maritimi

Das Bolboschoenetum (Scirpetum) maritimi wuchs nach OVERBECK (1950 : 87) schon in der mittleren Wärmezeit im Brackwasser-Bereich der nordwestdeutschen großen Stromtäler, nach MENKE (1968 : 214) in den Marschrand-Gebieten W-Holsteins vom Atlantikum bis zum frühen Subatlantikum und in der Zeit um Christi Geburt nach UDELGARD KÖRBER-GROHNE (1966 : 182, 1967 : 109 ff.) an der Niederweser (Feddersen Wierde).

Phragmitetea

Im Frankfurter Gebiet gab es nach MÄDLER (1939) im Miozän *Sparganium-*, *Scirpus-*, *Dulichium-* und *Carex-*Arten, durch die Phragmitetea-Gesellschaften für diese Zeit nachgewiesen sein dürften, deren Zusammensetzung im einzelnen allerdings nicht bekannt ist.

SZAFER (1947, I : 153) berichtet aus dem Pliozän von Kroscienko (Polen) von Phragmitetalia- (Phragmition- und Magnocaricion-)Gesellschaften (vgl. BRAUN-BLANQUET 1951 : 518; 1964 : 706). Phragmitetalia-Gesellschaften haben im Interglazial in Brandenburg Potametalia-Bestände im Laufe der Verlandung von Gewässern verdrängt, wie aus Pollen- und Großrest-Analysen von STARK, FIRBAS & OVERBECK (1932 : 125) zu folgern ist.

Eine ufernahe Röhricht-Gesellschaft mit *Phragmites* fand KNOBLOCH (1969 : 163) im Pannon des mährischen Teils des Wiener Beckens.

Wichtige Kenntnisse über die Zusammensetzung und Erhaltung früherer Röhrichte (auch mit *Cladium mariscus*) vermittelte GROSSE-BRAUCKMANN (1962-1967) aus dem Gebiet zwischen Weser und Elbe (ohne Zeitangaben) durch seine Torfanalysen aus Großresten. Er beschränkt seine Deutungen in der Regel auf die höheren Einheiten (Klassen und Ordnungen).

Im Leinetal bei Göttingen fand WILLERDING (1960 : 467, 1967 : 74) in nachwärmezeitlichen Ablagerungen 7 Phragmitetalia- und Phragmition-Arten.

Das Scirpo-Phragmitetum war am Dümmer (NW-Deutschland) nach PFAFFENBERG (in PFAFFENBERG & DIENEMANN 1964: 83/4) in der Frühen Wärmezeit (Boreal) ausgebildet. PFAFFENBERG (1947 : 80) konnte diese Assoziation nach Großresten aus dem Dümmer auch für die Jüngere Steinzeit belegen. In W-Holstein verfolgte MENKE (1968 : 214) das Scirpo-Phragmitetum vom Atlantikum bis zum frühen Subatlantikum.

LÜDI (1954 : 17) wies das Scirpo-Phragmitetum in bronzezeitlichen Ablagerungen im Schweizer Alpenvorland nach.

UDELGARD KÖRBER-GROHNE (1966 : 182, 1970 : 113) erkannte diese Assoziation nach Christi Geburt an der Unterweser (Feddersen Wierde), und BEHRE (1970) stellte sie an der unteren Ems fest.

PFAFFENBERG (1939 : 362) konnte das Cladietum marisci im Interglazial des Emslandes belegen, das BEUG (1964 : 434) etwa aus dem 13. Jhd. vor Chr. aus dem Gardasee-Gebiet nachwies.

Diese Gesellschaft erreichte im Postglazial am Dümmer nach PFAFFENBERG (in PFAFFENBERG & DIENEMANN 1964 : 83/4) zwischen 7000 und 5500 v. Chr. ihr Optimum. Auch im Neolithikum kam sie noch dort vor (PFAFFENBERG 1947 : 80). MENKE (1968 : 214) fand sie in Westholsteinischen Marschrand-Gebieten vom Atlantikum bis zum frühen Subatlantikum. Über die Geschichte des Cladietum in Norwegen berichtet HAFSTEN (1965).

Der Magnocaricion-Verband wurde von SZAFER (1947) im Pliozän Polens (Kroscienko) erkannt (vgl. a. BRAUN-BLANQUET 1951 : 518; 1964 : 706).

Aus dem Interglazial von Haren (Emsland) nennt PFAFFENBERG (1939) ebenfalls nach Großresten ein "Magnocaricetum", in dem allerdings *Cladium mariscus* den Ton angibt (s. oben).

RYBNIČEK & RYBNIČKOVÁ (1968 : 140) geben in SO-Böhmen nach dem Alleröd Verlandungsgesellschaften an, die dem Magnocaricion ähnlich sind.

Seit 7000 Jahren wächst das Caricetum rostratae nach

Birks (1970 : 837) in Inverness-shire (Abernethy forest) in Schottland.

Overbeck (1950:69) gibt als "Magnocaricetum" eine Liste von Großresten aus dem Praeboreal (Vorwärmezeit) von Huxfeld bei Bremen, die zum Caricetum rostratae gestellt werden müssen.

Der wärmezeitliche Fund von Tichomirow (1938, zit. nach Frenzel 1955 : 44) vom Ufer des Nord-Jamal am Weißen Meer mit *Carex rostrata, Cicuta virosa, Equisetum limosum,* und *Menyanthes trifoliata* kann kaum anders als Caricetum rostratae gedeutet werden.

Jonas (1933 : 195) identifizierte Vorlauftorfe der Emsmoore zwischen älterem und jüngerem *Sphagnum*-Torf mit einem *Sphagnum cuspidatum*-reichen Caricetum rostratae.

Das Caricetum inflato-vesicariae nennt Udelgard Körber-Grohne (1967 : 113) aus den ersten Jahrhunderten nach Christi Geburt von der Unterweser. Behre (1970 : 33) fand an der Unteren Ems zur gleichen Zeit ebenfalls Magnocaricion. Dieser Verband war im Neolithikum auch am Dümmer (NW-Deutschland) nach Pfaffenberg (1947 : 80) reich entwickelt.

Die Cicuta virosa-Carex pseudocyperus-Ass. erkannte Pfaffenberg (Pfaffenberg & Dienemann 1964 : 83/4) in der Leber- und Torfmudde der Frühen Wärmezeit des Dümmer.

Aus dem Schweizerischen Alpen-Vorland wies Lüdi (1954 : 17) aus der Bronzezeit das Caricetum elatae und das Caricetum gracilis nach.

Asplenietea rupestria

Die Potentilletalia caulescentis sind nach Horvat (1959 : 400) in SO-Europa schon im Pliozän nachweisbar, erfuhren aber in den Eiszeiten Einbußen.

Thlaspeetea rotundifolii

Auch die Thlaspeetea rotundifolii, die nach Horvat (1959 : 400) im Pliozän in SO-Europa reich entwickelt waren, erlitten im Pleistozän dort Verluste.

Violetea calaminariae

Ernst (1969 : 399) konnte eine Armerion halleri-Gesellschaft in Wales durch Pollenanalyse bis etwa 1100 n. Chr. zurückverfolgen.

Asteretea tripolium

(Juncetea maritimi)

Menke (1968 : 214) erkannte Gesellschaften dieser Klasse in

Marschrand-Gebieten W-Holsteins vom Atlantikum bis zum Subatlantikum. Die Großrest-Analysen der Wurt-Grabung Feddersen Wierde an der Unterweser von UDELGARD KÖRBER-GROHNE (1966, 1967, 1970) ergaben das Vorhandensein des Juncetum gerardi in mehreren Subassoziationen in der Zeit vom 1. bis zum 5. Jh. nach Chr., was von BEHRE (1970 : 39) von der unteren Ems bestätigt werden konnte.

Auch das Puccinellietum maritimae wurde an der Unterweser um diese Zeit gefunden (KÖRBER-GROHNE 1970 : 210, 219).

Epilobietea angustifolii

WILLERDING 1960 : 468, 1967 : 74 gelang es in nachwärmezeitlichen Ablagerungen des Leinetals bei Göttingen mehrere Arten der Epilobietea zu finden.

Artemisietea vulgaris

WILLERDING (1960 : 467, 1967 : 74) gibt aus nachwärmezeitlichen Ablagerungen im Leinetal bei Göttingen etwa 10 Arten der Artemisietea (darunter Convolvuletalia sepii) nach Großrest-Funden an.

Von Convolvuletalia sepii-Gesellschaften fand KNÖRZER (1962 : 263) Großreste in römischen Abfallhaufen am Niederrhein, während UDELGARD KÖRBER-GROHNE (1967 : 113) die Sonchus paluster-Archangelica-Assoziation nach Christi Geburt von der Niederweser nachweisen konnte.

Plantaginetalia majoris

Aus dem Auftreten von *Plantago major* (nicht *P. lanceolata*) in Pollendiagrammen darf auf die anthropogene Erzeugung von Plantaginetalia maioris-Gesellschaften geschlossen werden (vergl. z. B. IVERSEN 1941 : 40/41, 1949 : 9, JANSSEN 1960 : 85, STRAKA 1970 : 250). In Dänemark treten diese Gesellschaften im Subboreal mit dem Neolithikum auf.

Nach Großrestfunden bei der Grabung Feddersen-Wierde ist das Vorkommen einer Plantaginetalia-Trittgesellschaft an der Unterweser in den ersten nachchristlichen Jahrhunderten anzunehmen (UDELGARD KÖRBER-GROHNE 1967 : 108).

Agropyro-Rumicion

Nach Großrest-Funden in Ablagerungen des Göttinger Leinetales wuchsen dort in der Nachwärmezeit mehrere Arten des Agropyro-Rumicion-Verbandes (WILLERDING 1960: 468, 1967: 74).

BEHRE (1970) konnte das Juncetum compressi von der

unteren Ems seit der älteren Eisenzeit nachweisen (vergl. UDELGARD KÖRBER-GROHNE 1967 : 108).

Molinio-Arrhenatheretea

Wirtschaftswiesen-Gesellschaften sind bisher nur selten aus früheren Zeitabschnitten nachgewiesen worden, was für das geringe Alter dieser meist anthropogen bedingten Gesellschaften spricht. Nur das Filipendulion und das Molinion konnten, wenn auch nur zum Teil fragmentarisch, erkannt werden. Diese Hochstaudengesellschaften sind stellenweise natürlich.

HORVAT (1959 : 40) nimmt das Vorhandensein von Molinietalia-Gesellschaften in SO-Europa schon im Pliozän an.

RYBNIČEK & RYBNIČKOVÁ (1968 : 140, Tab. 2) fanden nach dem Alleröd Bestände mit *Filipendula ulmaria* und *Lysimachia vulgaris* in SO-Böhmen.

Auch in Schottland zeichnet sich das Filipendulion ulmariae nach BIRKS (1970 : 837) schon seit mehr als 7000 Jahren (Zone AF-2) ab.

Nach MENKE (1968) wuchsen seit dem Atlantikum Filipendulion-Gesellschaften und auch natürliche Molinion-Wiesen im Marschen-Randgebiet der schleswigholsteinischen Westküste.

Das Bestehen von Mähwiesen der Molinietalia seit 3000 v. Chr. wurde von JANSSEN (1960 : 85) pollenanalytisch für Süd-Limburg (Niederlande) erschlossen.

Aus dem schweizerischen Alpenvorland nennt LÜDI (1954 : 17) aus der Bronzezeit das Geranio-Filipenduletum, von dem Pollen mehrerer Arten in Pfahlbau-Siedlungen gefunden wurden. (Vergl. a. WELTEN 1955 : 81.)

WILLERDING (1960 : 468, 1967 : 74) gab einige Molinio-Arrhenatheretea-Arten aus den nachwärmezeitlichen Ablagerungen im Göttinger Leinetal an.

UDELGARD KÖRBER-GROHNE (1967 : 108) fand an der Unterweser in den ersten nachchristlichen Jahrhunderten zahlreiche Arten der Molinietalia.

Sedo-Scleranthetea

MENKE (1969 : 110) eichte Oberflächen-Pollenproben von Strandwällen in W-Holstein auf Pflanzengesellschaften und wies in fossilen Strandwällen auf diese Weise das Vorhandensein von Festuco-Sedetalia um 200 n. Chr. nach.

In eisenzeitlichen Friedhöfen fand R. TÜXEN (1960) die humosen Bodenprofile von Festuca ovina-Rasen, die zu den Festuco-Sedetalia gerechnet werden und die als anthropogene Ersatzgesellschaften des Querco roboris-Betuletum auf Friedhofsgelände durch die häufige Öffnung neuer Gräber als Sukzessionsstadium vom Corynephoretum und Agrostidetum aridae sich einstellten und lange als Trockenrasen erhalten blieben.

Elyno-Seslerietea

Die Seslerietalia waren nach Horvat (1959 : 400) schon im Pliozän in SO-Europa ausgeprägt, Sie erfuhren in den Eiszeiten sowohl Verluste als auch Bereicherungen.

Das "Seslerietum" bildete nach Zólyomi (1953 : 398) im frühen Postglazial in den ungarischen Mittelgebirgen mit subalpinen klimatisch anspruchslosen Arten und an geschützten Stellen mit interglazialen und Tertiär-Relikten durchsetzt einen Komplex mit Pinus silvestris-Wäldern (vgl. a. Zólyomi 1951 : 58).

Caricetea curvulae

Die Eiszeiten brachten den nach Horvat (1959 : 400) schon im Pliozän in SO-Europa erkennbaren Caricetalia curvulae Verarmungen, aber auch neuen Arten-Zuwachs.

Das alpine Nardetum wird von Welten (1958 : 273) auf Grund seiner Pollenanalysen als "ein Charakteristikum der Zeit der regressiven Höhenstufen" gewertet.

Festuco-Brometea

Über die Geschichte der Gesellschaften aus dieser Klasse liegen kaum palynologische Nachrichten vor.

Als submediterraner Karst-Rasen drang nach Zólyomi (1953 : 398) das Festuco-Brometum aus dem Süden in der zweiten Phase der Wärmezeit nach Ungarn ein.

Scheuchzerietea

Das Scheuchzerietum (Caricetum limosae) wurde in einer Sukzessionsreihe vom Nupharetum bis zum Sphagnetum von Pfaffenberg (1954 : 483) aus dem Wurzacher Ried, einem Moor in Süddeutschland, ohne genaue Zeit-Angabe vermutet.

Auch Jonas (1934 : 12, 51, 69) erwähnt aus Emsland-Mooren wiederholt fossiles Scheuchzerietum (Caricetum limosae) (vgl. auch p. 278).

Pfaffenberg (1942 : 83) konnte die Lagerstätte einer Moorleiche aus der letzten Hälfte des vorchristlichen Jahrtausends in einer Moorschlenke im nördlichen Oldenburg als ein Rhynchosporetum sphagnetosum cuspidati deuten. Janssen (1960 : 81, 89) erkannte das Rhynchosporion in den Zonen VIII-X (Firbas 1949), d.h. in der späten Wärmezeit bis zur jüngeren Nachwärmezeit in einem der von ihm pollenanalytisch untersuchten s-limburgischen Moore (Niederlande). Zum Rhynchosporion vgl. Koch (1930, 1934).

Caricetea fuscae

Caricetalia fuscae-Gesellschaften waren schon im Pliozän in SO-Europa entwickelt (Horvat 1959 : 400), erfuhren aber im Pleistozän eine Bereicherung.

In bronzezeitlichen Pfahlbau-Ablagerungen wurden nach Lüdi (1954 : 17) Arten der Caricetea fuscae gefunden.

Die Großreste der Wurten-Grabung Feddersen-Wierde aus der Unterweser lieferten Arten der Caricetalia fuscae aus den ersten nach-christlichen Jahrhunderten. Udelgard Körber-Grohne (1967 : III) vermutet das Eleocharetum uniglumis als zugehörige Gesellschaft.

Menke (1968) konnte Caricetalia fuscae-Gesellschaften aus seinen Großrest-Analysen der Marschrandgebiete W-Holsteins erkennen.

Eine Carex lasiocarpa-Sphagnum sect. Subsecunda- und andere dem Verband Eriophorion gracilis nahestehende Gesellschaften fanden Rybníček & Rybníčková (1968 : 139) aus SO-Böhmen vom ausklingenden Boreal bis zum Atlantikum (Zone V-VI der Moorentwicklung).

Moravec & Rybníčková (1964 : 390) klärten im Böhmerwald-Vorgebirge (ČSSR) die Entstehung des Valeriano dioicae-Caricetum davallianae durch die menschlich bedingte Entwaldung in der Umgebung von Quellen im jüngeren Subatlantikum.

In Inverness-shire fand Birks (1970 : 838) in vermutlich spätweichselzeitlichen Ablagerungen das Carici-Saxifragetum aizoidis aus dem Caricion bicoloris-atrofuscae-Verband.

Salicetea herbaceae

Für die Geschichte der arktisch-alpinen Gesellschaften der Salicetea herbaceae wird die soziologische Auswertung mancher mittel- und nordeuropäischen eiszeitlicher fossilführender Ablagerungen aufschlußreich sein, wie schon aus der noch vorwiegend florengeschichtlichen Übersicht von Braun-Blanquet (1923 : 155 ff.) hervorgeht.

Nach Horvat (1959 : 400) haben die Salicetalia herbaceae ihre volle Ausbildung und ihre gegenwärtigen Grenzen in SO-Europa in den Eiszeiten erreicht.

Auf die schon im Pliozän vorhandenen Arabidetalia coeruleae haben sich in SO-Europa dagegen die Eiszeiten sowohl negativ als positiv ausgewirkt (Horvat 1959 : 400).

Oxycocco-Sphagnetea

Nach HORVAT (1959 : 400) erfuhren die schon im Pliozän bestehenden Sphagnetalia mediae während des Pleistozäns in SO-Europa eine Bereicherung.

KOLUMBE & BEYLE (1938 : 24) gaben Artenlisten von Sphagnion-Gesellschaften aus Holstein für die letzten Jahrhunderte vor und die ersten nach Christi Geburt.

Auch JONAS (z.B. 1933, 1934a, b) machte Angaben über solche Gesellschaften der Emsland-Moore und ihrer Sukzessionen.

H. KOCH (1930 : 523, 1934a : 133, 1934b : 103) wies bei dem Vergleich fossiler Sphagnion und Rhynchosporion-Gesellschaften darauf hin, "daß die fossilen Hochmoor-Pflanzengesellschaften von den rezenten der nw-deutschen Moore gründlich verschieden sind." (Vgl. a. GROSSE-BRAUCKMANN 1962a.)

Auf den gründlichen stratigraphischen Untersuchungen von OVERBECK (1941, 1950, 1964), KUBITZKI (1960) und anderen fußend hat JAHNS (1969 : 60; dort weitere Literatur) die Entwicklung der Sphagnion-Gesellschaften in der Esterweger Dose (Nordwestdeutschland) studiert. Daraus kann gefolgert werden, daß sowohl das Caricetum limosae (Cuspidato-Scheuchzerietum) als auch das Erico-Sphagnetum magellanici nebeneinander bestanden und sich nicht rhythmisch oder zyklisch abgelöst haben, wie es die alte Hypothese von L. v. POST & R. SERNANDER (1910) annahm und von vielen Autoren seitdem vertreten wurde, und wie noch KOCH (1934a : 134) in den Mooren des Emslandes und des Hümmling (NW-Deutschland) in Anlehnung an OSVALDS Befunde im schwedischen Komosse den "Wechsel von Bult- und Schlenkenkomponenten als eine typische Regenerationsfolge" gedeutet hatte (vgl. a. OVERBECK 1950 : 43).

Die Untersuchungen von CASPARIE (1969) bestätigen eindeutig die Konstanz des Erico-Sphagnion-Caricetum limosae-Mosaiks in den Hochmooren der Niederlande und NW-Deutschlands.

Sukzessionen von Potametea über Phragmitetea und Scheuchzerietea zu Oxycocco-Sphagnetea wurden von PFAFFENBERG (1934), MENKE (1968) aus N-Deutschland und von RYBNIČEK und RYBNIČKOVÁ (1968 : 140) aus SO-Böhmen nachgewiesen.

GROSSE-BRAUCKMANN (1962b : 114) ordnete die im Unterwesergebiet gefundenen Großreste in Torfen nach soziologischen Gesichtspunkten (ohne Zeitangaben). Dabei tritt deutlich das Erico-Sphagnetum magellanici hervor.

BIRKS (1970) fand Erico-Sphagnion schon vor 7000 Jahren in Schottland.

Nachdem schon OVERBECK (1950 : 87) das Ericetum tetralicis für das Atlantikum und Subboreal nachgewiesen hatte, gab MENKE (1963) mit Hilfe der Pollenanalyse gestützt auf Großrest-Funde einen weiteren Beitrag zum Alter und zur Frage der Entstehung des Ericetum tetralicis seit der späten Wärmezeit (Subboreal) (vgl. JANSSEN 1970 : 192). Diese Assoziation ist nach MENKE z.T. natürlich, z.T. durch den Menschen erzeugt.

Das Bodenprofil aus "braunem Flugsand", das JONAS (1934a : 157) pollenanalytisch untersuchte, läßt sich unschwer als dasjenige eines fossilen (natürlichen) Ericetum tetralicis deuten, was mit dem Pollen-Befund in Einklang steht.

R. TÜXEN (1957 : 22) zeigte die morphologischen Merkmale im Bodenprofil des Ericetum tetralicis an Hand von Lackprofilen.

Auf grund der pollenanalytischen Befunde von MENKE (1963) und anderen Autoren erwog J. TÜXEN (1967 : 6) die Entstehung seit dem Beginn der späten Wärmezeit und dieVorläufer-Gesellschaften des Ericetum tetralicis, die er als Bruchwälder aus Moorbirke und Kiefer oder Nanocyperion deutete.

Nardo-Callunetea

Das Alter des Calluno-Genistetum, das R. TÜXEN (1930 : 13, 1932 : 34, 1937 : 117, 1938) als eine Ersatzgesellschaft von Quercion robori-petraeae-Wälder erkannte (vgl. DIMBLEBY 1954), wurde in NW-Deutschland sowohl pollenanalytisch (z.B. OVERBECK & SCHMITZ 1931) als auch durch vergleichende Makromorphologie der Bodenprofile mit Hilfe prähistorischer Datierungen nachgewiesen. Schon ZOTZ (1930) deutete Calluna-Heide-Profile unter bronzezeitlichen Grabhügeln. (Vgl. dazu FIRBAS 1952 : 161.) Die pollenanalytische Untersuchung frühbronzezeitlicher Grabhügel in den südlichen Niederlanden durch VAN ZEIST (1967a, b) wies deren Anlage auf der Heide erneut nach. (Vgl. 1967b, Abb. p. 49.)

Die makromorphologischen Merkmale der Bodenprofile des Calluno-Genistetum wurden von R. TÜXEN (1957, vgl. seine Sammlung von Lackabzügen z.Zt. im Staatlichen Museum für Naturkunde und Vorgeschichte Oldenburg) beschrieben und pflanzensoziologisch gedeutet. R. TÜXEN (1957) fand Profile des Calluno-Genistetum unter neolithischen Grab-Anlagen, die als die ältesten Nachweise dieser vom Menschen erzeugten Gesellschaft und des von ihr gebildeten Ortsteins gelten dürfen (vgl. DIMBLEBY 1954). Die Bodenprofile geben zugleich Aufschluß über die Ausgangsgesellschaft der Heide, die entweder Querco roboris-Betuletum oder Fago-Quercetum war. Auch die Subassoziationen des Calluno-Genistetum lassen sich aus der Morphologie der Profile erschließen, wobei allerdings vorausgesetzt wird, daß die prähistorischen Zusammensetzung der Heidegesellschaft etwa die gleiche wie heute war.

Über die Entstehung der britischen Calluna-Heiden vom Neolithikum bis zur Bronzezeit machte DIMBLEBY (1954 : 75) Angaben unter Berücksichtigung der Bodenprofile.

JONAS verteidigte in Anlehnung an BEIJERINCK in mehreren Arbeiten (1933, 1934a, 1935) auf Grund seiner Pollenzählungen in Sand-Profilen das natürliche Dasein und die Entwicklung von "Heide"-Gesellschaften in Nordwestdeutschland. Weil aber seine soziologischen Angaben zu wenig bestimmt sind, lassen sich seine Funde und die daraus gezogenen Schlüsse nur z.T. auswerten.

In Jütland bestätigte JONASSEN (1950 : 126) die Entstehung der Calluna-Heide seit dem Neolithikum und in verstärkten, Maße seit der Bronzezeit durch die Tätigkeit des Vieh züchtenden und Ackerbau treibenden Menschen. Das Alter des durch die Heide erzeugten Ortsteins wurde entgegen der früher von P. E. MÜLLER (1924) geäußerten Hypothese als verschieden anerkannt. Diese Ergebnisse der Pollenanalyse decken sich vollständig mit den von R. TÜXEN pedologisch-urgeschichtlich gewonnenen Erkenntnissen in NW-Deutschland. IVERSEN (1969) zeigte erneut pollenanalytisch die Erzeugung von Calluna-Heide durch den Menschen in der Wikinger-Zeit und ihre Erhaltung durch Brand bis vor etwa 200 Jahren.

Seit 1200 n. Chr. anthropogen geschaffene Calluna-Heide wies MIKKELSEN (1954 : 220) pollenanalytisch auf Bornholm nach. MOTHES et al. (1937) fanden Calluna-Heide, die seit der Mitte des 17. Jhd. vom Menschen erzeugt worden war.

Auf eine im späten Weichsel-Stadium in den Cairngorms (Schottland) weit verbreitete Empetrum-Heide weist BIRKS (1970 : 838) hin.

Loiseleurio-Vaccinietea

Die Sukzession des Alnetum viridis zum Rhodoreto-Vaccinietum im Laufe der (klimatisch bedingten) Waldgrenzen-Verschiebung wird durch die Pollenanalysen von WELTEN (1958 : 272) beleuchtet.

Betulo-Adenostyletea

Über die Rolle des Alnetum viridis im Laufe der klimabedingten Verschiebung der Waldgrenze (Relikt einer früheren Höhenstufe höher gelegener Alpenwälder) in den Schweizerischen Alpen unterrichten Pollenanalysen von WELTEN (1958 : 272).

Salicetea albae

Das Saliceto-Populetum des Alföld blieb nach Zólyomi (1953 : 398) in der postglazialen Wärmezeit auf die Überschwemmungs-Auen der Flüsse beschränkt, ohne infolge des herrschenden Steppenklimas sich dicht schließen zu können.

Nach Willerding (1960 : 466, 1967 : 74) wuchsen in der Nachwärmezeit im Göttinger Leinetal nach massenhaften Blatt- und Zweigfunden von Weidenarten das Salicetum albo-fragilis und das Salicetum triandro-viminalis. Beide Gesellschaften konnte Willerding (1969 : 995) auch für die Eisenzeit dort nachweisen.

Rhamno-Prunetea (Crataego-Prunetea)

Zoller (1960b : 1967) rechnet im Boreal auf Schutthalden, an Steilhängen und in Runsen mit größeren Hasel-Gebüschen.

Die Querco-Fagetea-Arten, die Willerding (1960 : 468, 1967 : 74) aus nachwärmezeitlichen Ablagerungen im Leinetal bei Göttingen mitteilte, deuten z.T. eher auf Prunetalia spinosae wie *Corylus avellana*, *Prunus spinosa*, *Cornus sanguinea* und auch *Sambucus nigra*. Damit wäre das Vorkommen von Prunetalia-Waldmänteln zu jener Zeit wahrscheinlich gemacht.

Für die Bronzezeit können Prunetalia-Gesellschaften aus der Liste der Samen und Früchte abgeleitet werden, die Margita Villaret-von Rochow (1958 : 97, 112) von Valeggio am Mincio (Prov. Verona) mitteilte. (Vgl. dazu Zoller 1962 : 181, 182.)

Alnetea glutinosae

Nach Horvat (1959 : 400) waren Alnetalia glutinosae-Gesellschaften schon im Pliozän in SO-Europa reichlich vertreten. Sie erfuhren eine weitere Entwicklung im Pleistozän.

Die Annahme einer Kontinuität dieser und anderer Gesellschaften im gleichen Gebiet, von der Horvat offenbar ausgeht, hat sich nach Beug (brieflich) nicht bestätigen lassen. Die Alnetalia glutinosae und wohl auch andere tertiäre Gesellschaften waren in den Kalt-Zeiten vollständig aus SO-Europa verschwunden.

Im Interglazial des Emslandes (Haren) fand Pfaffenberg (1939 : 362) Bruchwaldtorf von *Alnus glutinosa*, der von einer Alnion glutinosae-Gesellschaft gebildet sein dürfte.

In der älteren Nachwärmezeit wuchs Erlenbruchwald im Münsterland (Trautmann 1969 : 116).

Am Dümmer (NW-Deutschland) erkannte Pfaffenberg (1947: 75, 78) farnreiches Carici elongatae-Alnetum glutinosae anhand von Großresten, sowie von Pollen und Sporen im Neolithikum durch die Datierung einer darin angelegten Siedlung.

BEHRE (1970 : 42) belegte eine Sukzession vom Betuletum pubescentis zum Alnetum glutinosae infolge des Zuflusses von nährstoffenreichem Wasser an der unteren Ems im frühen Subboreal. Mit weiterem Anstieg des Wassers setzte eine regressive Sukzession zum Phragmition und später zum Potamion ein. Im Dümmergebiet wurde nach PFAFFENBERG (PFAFFENBERG & DIENEMANN 1964 : 86) seit dem Atlantikum das Alnetum immer wieder von Betuletum pubescentis oder von Seggen-Gesellschaften abgelöst.

Auch JONAS (1934 : 64) gab Hinweise auf fossiles Alnetum glutinosae.

Von der Unterweser erwähnt UDELGARD KÖRBER-GROHNE (1967 : 108) Alnion glutinosae aus den ersten nachchristlichen Jahrhunderten nach Großfunden.

Im Darss stellte sich nach FUKAREK (1961 : 266, 275) das Carici-Alnetum am Ende von Verlandungsreihen, die mit dem Nupharetum begannen, in jüngerer Zeit ein. *Salix*-Gebüsche (wohl zu den Alnetalia glutinosae gehörig) fand BIRKS (1970 : 837) an einem Bach und in feuchten Vertiefungen in Invernessshire (Schottland) vor 7000 Jahren (Zone AF-2 und AF-3).

In S-Limburg trat nach JANSSEN (1960 : 32) in Zone VIII, wohl unter dem Einfluß des Menschen, das Saliceto-Franguletum an die Stelle des Thelypterideto-Alnetum.

Betuletum pubescentis

MENKE's pollenanalytische Untersuchungen eines fossilen Moores bei Ostrohe (Schleswig-Holstein) (1967 : 134) lassen ein Betuletum pubescentis im Jüngeren Eem erkennen, das aus einem eutraphenten Erlen-Bruch (Alnetum glutinosae) hervorging.

Am Dümmer trat nach PFAFFENBERG (PFAFFENBERG & DIENEMANN 1964 : 86) das Betuletum pubescentis seit dem späten Atlantikum auf.

Ein Betuletum pubescentis von langer Wuchsdauer konnte FUKAREK (1961 : 271) im Darss durch Pollenanalyse belegen.

Vaccinio-Piceetea

Die schon im Pliozän in SO-Europa reich vertretene Ordnung der Vaccinio-Piceetalia erfuhr im Pleistozän nach HORVAT (1959 : 400) weitere Bereicherung.

Nach den Pollenanalysen von ELIŠKA RYBNIČKOVÁ (1966 : 298) wuchsen Vaccinio-Piceion Gesellschaften in den Orlické hory-Bergen (ČSSR) im älteren Subatlantikum nur an isolierten Standorten auf Torf oder nassen Böden. Ihre Ausdehnung auf das heutige

Maß begann im 16.-17. Jhdt. durch menschliche Eingriffe in das natürliche Abieto-Fagetum.

Auf bosnische Relikt-Gesellschaften (Pino-Betuletum pubescentis Stef. 1962, Sphagno-Piceetum montanum (prov. 1964), Quercetum roboris montanum Stef. 1968) weist STEFANOVIĆ (1970 : 79) hin.

Angaben zur postglazialen Geschichte des Piceetum montanum, des Piceetum transalpinum, des Erico- und des Vaccinio-Pinetum im Bereich des Südabfalls der mittleren Ostalpen und der Nachbargebiete macht H. MAYER (1969a : 147-163).

Pinion

Nach ZÓLYOMI (1953 : 398) wurde das "Pinetum silvestris" in Ungarn mit dem Anbruch der postglazialen Wärmezeit im westlichen Transdanubien aus den Mittelgebirgen auf die meisten sauren, schottrigen Böden zurückgedrängt, wo es bis dahin mit Sesleria-Rasen einen Komplex gebildet hatte und nur noch im westlichen Teil als Relikt-Föhrenwald eine Zeitlang erhalten blieb. Die schottischen *Pinus silvestris* var. *scotica*-Wälder (vgl. BIRKS 1970: 842/3) erreichten im Cairngorm-Gebiet Höhen von fast 800 m NN, bevor das "blanket bog" (wohl Erico-Sphagnion) sie hier verdrängte.

H. MAYER (1970 : 60) faßt die natürlichen Relikt-Kiefernwälder (Vaccinio-Pinetum cladonietosum) der Buckligen Welt südlich von Wiener Neustadt (Österreich) in 400-600 m Meereshöhe als Reste der frühpostglazialen Kiefern-Heidewälder auf.

Im Pustertal ist nach H. MAYER (1969a : 20) das Rhododondro-Pinetum als Reliktgesellschaft der postglazialen *Pinus*-Wälder zu werten.

Quercetea robori-petraeae

Eingehende Beschreibungen der makromorphologischen Merkmale nordwestdeutscher Bodenprofile von Quercion roboripetraeae-Gesellschaften gab R. TÜXEN zu verschiedenen Malen (z.B. 1939, 1957, 1964). Diese Profile zeigen die ehemalige Verbreitung des natürlichen Querco robori-Betuletum und des Fago-Quercetum im nw-deutschen Altmoränengebiet an.

KOLUMBE & BEYLE (1938 : 18) erkannten nach dem Bodenprofil das Querco-Betuletum im Bereich des Wittmoores (Holstein) für das Subatlantikum.

PFAFFENBERG (1939b, 1952 : 33, 1964 : 95) konnte das natürliche Vorkommen des Querco roboris-Betuletum seit dem

Beginn des Atlantikum durch Pollenanalyse neben anderen an Kleinst-Mooren beweisen (vgl. dazu JANSSEN 1960 : 94). FUKAREK (1955 : 57) bestätigte das Bestehen des "Eichen-Birkenwaldes" seit der Jüngeren Steinzeit an Hand von Holzkohle-Resten einer Siedlung und des Bodenprofils bei Wahlitz am Elbetal (Mecklenburg).

Auch auf dem Darss (Mecklenburg) sind nach FUKAREK (1961 : 262/3) "heidekrautreiche Eichen-Birkenwälder, die einen gewissen Anteil von Buche und Kiefer aufweisen", seit der Älteren Nachwärmezeit natürlich.

TRAUTMANN (1969: 124) weist darauf hin, daß in NW-Deutschland "in der Regel mehr oder weniger breite Zonen azidophiler Wälder (Eichen-Birkenwald, Buchen-Eichenwald)" an deren Entstehung nicht der Mensch beteiligt ist, die Hochmoore umgeben.

BURRICHTER (1970) bestätigte das natürliche Vorkommen des Querco-Betuletum in verschiedenen Subassoziationen als potentiell natürliche Waldgesellschaft im Sand-Münsterland seit der Einwanderung der Eiche,

Nach LOSERT (1953 : 142) ist ein "ausgedehntes Vorkommen armer azidophiler Eichenwälder (Eichen-Birkenwälder)" auf der Geest östlich des mittleren Leinetals "mit den pollenanalytischen Befunden verträglich, durch sie allein aber nicht sicher zu beweisen" (vgl. dazu FIRBAS 1954 : 196).

Im Subboreal wandelte sich das Querco-Betuletum im Rheiderland an der unteren Ems (NW-Deutschland) nach BEHRE (1970 : 42) infolge des Meeresanstiegs durch stagnierendes hohes Grundwasser zum Betuletum pubescentis und zum Kiefern-Bruch.

ELSBETH LANGE & W. HEINRICH (1970 : 82) konnten pollenanalytisch im Gebiet von Frankenberg (Sachsen) Fichten-Birken und Eichen-Kiefern-Birkenwald erkennen, die von SCHLÜTER in die Gruppe des Stieleichen-Birkenwaldes (?) gestellt werden und im Kontakt mit Melampyro-Fagetum-Wälder als potentiell natürliche Waldgesellschaften gelten.

In der Eifel ergaben nach TRAUTMANN (1962 : 262/3) Pollenanalysen ausgedehnte Vorkommen des natürlichen Fago-Quercetum, dessen Ausdehnung in der Jüngeren Nachwärmezeit vom Menschen gefördert wurde, was ebenfalls pollenanalytisch im Sauerland schon von BUDDE (1938, 1949) erkannt worden war und später pflanzensoziologisch von SOFIE MEISEL-JAHN (1955) und SEIBERT (1955) bestätigt wurde.

TRAUTMANN (1966b : 23) fand natürliche Buchen-Eichen-Wälder (Fago-Quercetum) durch Pollenanalyse in großer Ausdehnung im Münsterland, was mit den pollenanalytischen und pflanzensoziologischen Ergebnissen von BUDDE & RUNGE (1940 :

18) in Einklang steht (vgl. dazu FIRBAS 1952 : 159) und von BURRICHTER (1970) bestätigt wurde, (vgl. dazu BUDDE & RUNGE (1940 : 18), TRAUTMANN (1969 : 111, 121)).

Im Dravet, einen naturnahen Wald in SW-Jütland (Dänemark), belegte IVERSEN (zuletzt 1969) durch Pollenanalysen kleinster Moore die säkulare Sukzession von Eichen-Hasel-Wald (Carpinion?) zum Fago-Quercetum (Quercion robori-petraeae).

J. J. MOORE (1967 : 101) klärte die Geschichte des Blechno-Quercetum in Irland mit Hilfe der Pollenanalyse auf.

Diese Tatsachen sind anscheinend STRAKA (1970) entgangen, der alle Quercion robori-petraeae-Gesellschaften als menschlich bedingte Degradationsstadien von Buchenwäldern (ohne Angabe von deren syntaxonomischer Stellung) bewertet.

ZOLLER (1967) gelang der pollenanalytische Nachweis der anthropogenen Entstehung der Kastanien-Wälder am Südfuß der Alpen, die größtenteils zum Quercion robori-petraeae gestellt werden (vgl. a. BEUG 1964 : 431).

Zunehmende Bodenversauerung, die durch die postglaziale Klimaverschlechterung bedingt war, hält ZÓLYOMI (1953 : 399) für die Ursache der Ausbreitung des Querco-Luzuletum in Ungarn.

Querco-Fagetea

JIRINA SLAVIKOVÁ (1960) wies das Vorkommen von Tilio-Acerion-Gesellschaften und des Taxo-Fagetum für das Subatlantikum (100 v. Chr.) von der mittleren Moldau nach. ZÓLYOMI (1953 : 399) zeigte das Aufkommen des "Fagetum" in der Nachwärmezeit im ungarischen Mittelgebirge und in Transdanubien als Buchenzone. Die Buche drang in dieser Zeit auch in die Baumschicht der gemischten Felsenwälder der Nordhänge ein und baute dort das Fago-Ornetum und das Fagetum seslerietosum auf.

Die Geschichte des Abietetum, des Abieti-Fagetum und des Fagetum im Bereich des Südabfalls der mittleren Ostalpen wird nach den Pollenanalyse dieses Gebietes in seine Nachbarbezirke von H. MAYER (1969b : 144-166) dargestellt.

Im Orlické hory-Gebirge (ČSSR) bildete nach Pollenanalysen von ELIŠKA RYBNIČKOVÁ (1966 : 298) das Abieti-Fagetum im Subatlantikum die natürliche Vegetation bis es durch den Menschen im 16.-17. Jhdt. in Vaccinio-Piceion-Gesellschaften umgewandelt wurde.

Das Rhododendro-Abietetum Kuoch 1954 erhielt nach

Zoller (1967 : 263) seinen heutigen Charakter erst in der zweiten Hälfte des Subboreals. (Vgl. a. Mayer 1969 : 160.)

Zoller (1962 : 356) erkannte mit Hilfe der Pollenanalyse im Schweizerischen Mittelland, daß sich das Querco-Abietetum sphagnetosum an der untersuchten Stelle erst in sehr junger Zeit nach einer Periode des Ackerbaues und der Beweidung seit dem ausgehenden Mittelalter als sekundäre Folgegesellschaft entwickelt hat. Die natürliche Vegetation gehörte hier vorher zum Fagion.

Das Acero-Fagetum besiedelte nach Pollenanalysen von Eliška Rybničková (1966 : 298) im Subatlantikum steinige und felsige Böden der Abhänge und Rücken des Orlické hory-Gebirges (ČSSR).

Am südlichen Alpenrand entstanden die heutigen Eu-Fagion- und Luzulo-Fagion-Gesellschaften nach Zoller (1967 : 262) in der zweiten Hälfte des Subboreals.

Zur Entwicklung der Buchen- und Buchen-Weißtannen-Wälder (Fagion) in den südlichen Alpenketten machen Zoller & Kleiber (1971 : 146) neue Angaben (vgl. a. Zoller 1960 : 200 und Zoller & Kleiber 1967).

Faegri (1954) wies pollenanalytisch die Entstehung eines wahrscheinlich künstlich begründeten Buchenwaldes in Alversund (W-Norwegen) in der Zeit zwischen 500 und 1000 n. Chr. nach.

Zahlreiche Arbeiten beschäftigen sich mit der Frage nach der anthropogenen Entstehung von Eichenwäldern durch Degradation von Buchenwäldern, wobei die Pollenanalyse natürlich nicht die syntaxonomische Stellung dieser Buchenwälder entscheiden kann (Jonas 1934a : 45/6; 1934b; Couteaux 1962, Trautmann 1957, 1962, 1966 u.a.; Budde 1938, 1949; Janssen 1960 u.a.). Danach müssen viele heutige Querco-Carpineten, Fago-Querceten und "Eichen-Birkenwälder" als menschlich erzeugte Ersatzgesellschaften ehemaliger Fagion-Gesellschaften bewertet werden.

Für den pollenanalytischen Nachweis des Querco-Carpinetum ist nicht so sehr das Mengen-Verhältnis von *Fagus* zu *Carpinus* (oder gar die Menge der Hainbuche) maßgebend, sondern das Vorhandensein einer gewissen Menge von *Quercus* mit Fagetalia- (oder auch Prunetalia-)Arten, vor allem von *Corylus*.

Die Frage der Natürlichkeit oder anthropogenen Bedingtheit des Querco-Carpinetum in den Niederlanden wird sehr eingehend und sorgfältig von Janssen (1960 : 82) diskutiert. Er konnte neben der menschlich verursachten Ausbreitung des Querco-Carpinetum auch das natürliche Vorkommen dieser Assoziation in S-Limburg wahrscheinlich machen und auch schon früh Alno-Ulmion (Ulmion) nachweisen (vgl. Westhoff 1966 : 289).

KOLUMBE & BEYLE vermuteten (1938 : 18) in Holstein (Wittmoor) das Querco-Carpinetum im Subatlantikum.

STEINBERG (1944 : 578) ließ die Frage offen, ob die pollenanalytisch nachgewiesenen sehr häufigen von Natur aus hainbuchenreichen Wälder des Eichsfeldes zum Fagion oder zum Carpinion gestellt werden müssen (vgl. FIRBAS 1952).

SELLE (1936 : 408) bekundete pollenanalytisch das natürliche Vorkommen des Querco-Carpinetum stachyetosum silvaticae im Allertal (NW-Deutschland).

PFAFFENBERG (1939b : 145) belegte das Vorkommen des natürlichen Querco-Carpinetum am Lengener Moor (NW-Deutschland) durch Pollenanalysen. MÜLLER (1970 : 39/40) wies mit der gleichen Methode im älteren Subatlantikum in der Umgebung des Otterstedter Sees (NW-Deutschland) herrschendes natürliches Querco-Carpinetum nach, das durch Ausbreitung des Kulturlandes im Mittelalter zurückgedrängt wurde.

Im Eggegebirge (östlich Paderborn) konnte TRAUTMANN (1957: 290) neben Fago-Quercetum das natürliche Querco-Carpinetum für die ältere Nachwärmezeit feststellen.

Für das Münsterland wies TRAUTMANN (1966, 1969 : 123) durch pollenanalytische Untersuchung eines kleinen Moores das Vorkommen ausgedehnter Eichen-Hainbuchenwälder (Querco-Carpinetum) zu Beginn der menschlichen Besiedlung nach, was von BURRICHTER (1970) bestätigt wurde (vgl. dazu BUDDE & RUNGE 1940 : 18).

WILLERDING (1965 : 60) fand in einer bandkeramischen Siedlung bei Rosdorf (Göttingen) Holzkohle von Eiche, Linde, Ahorn, Esche und Erle, die einer Carpinion-Gesellschaft jener Zeit entstammen. Das Vorhandensein des Querco-Carpinetum (mit reichlichen Buchen-Gehalt) in der Bronze- und Eisenzeit auf von Löß bedeckten Hügeln nahe Göttingen konnte WILLERDING (1966b : 61) nicht widerlegen, wenn auch *Carpinus* selten gefunden wird (vgl. a. WILLERDING 1969 : 395 und ELLENBERG 1963 : 208 ff.).

ZEIDLER (1956) wies in Franken in mehreren Mooren pollenanalytisch natürliche Eichen-Hainbuchenwälder nach.

Im Vorland des Orlické hory (ČSSR) wuchsen nach den Pollenanalysen von ELIŠKA RYBNIČKOVA im Subatlantikum Carpinion-Gesellschaften.

Anklänge an das Querco-Carpinetum fand LÜDI (1954 : 15) in den Großrest-Funden der bronzezeitlichen Siedlungen im Schweizerischen Alpenvorland.

Mit dem Ende der postglazialen Wärmezeit entwickelte sich in der ungarischen Tiefebene nach ZÓLYOMI (1951 : 58, 1953 : 399) das Querco-Carpinetum.

In Irland konnte die Geschichte des Corylo-Fraxinetum durch Pollenanalysen von J. J. MOORE (1967 : 100) beleuchtet werden.

Zur Geschichte des Querco-Ulmetum sei auf die kritische Darstellung von TROELS-SMITH (1961) verwiesen. BEHRE (1970 : 42) konnte um die Wende Subboreal/Subatlantikum (700 v. Chr.) im Rheiderland (Ostfriesland) auf einem Ems-Uferwall das Fraxino-Ulmetum nachweisen.

Im Leinetal wuchsen nach WILLERDING (1960 : 465, 1967 : 75) in der Nachwärmezeit Fraxino-Carpinion bzw. Alno-Ulmion-Wälder, darunter ein erlenreicher Auewald, der nach unserer Meinung am ehesten dem Stellario-Alnetum, vielleicht auch dem Pruno-Fraxinetum entsprechen dürfte. Dagegen konnte das Fraxino-Ulmetum dort nicht nachgewiesen werden.

Das Auftreten kleinflächiger Bestände des Pruno-Fraxinetum in der älteren Nachwärmezeit im Münsterland machte TRAUTMANN (1969 : 124) wahrscheinlich. Dieses wurde aber schon in der jüngeren Nachwärmezeit ebenso wie das Querco-Carpinetum und das Fago-Quercetum zum großen Teile vernichtet, um Grünland und Äckern Platz zu machen.

Im Alföld (Ungarn) war nach ZÓLYOMI (1953 : 398) in der postglazialen Wärmezeit das Ulmeto-Fraxineto-Roboretum als Auewald ausgebildet, konnte sich jedoch erst in der zweiten Hälfte der Wärmezeit ausbreiten.

In der postglazialen Wärmezeit bildeten sich nach ZÓLYOMI (1953 : 398) im ungarischen Hügel- und Bergland auch das Acero-Fraxinetum und das Tilio-Fraxinetum aus, die sich mit der Klima-Verschlechterung auf Blockhalden und Felskämme zurückzogen (vgl. auch ZÓLYOMI 1951 : 58).

Über die Geschichte des Orno-Ostryetum als anthropogene Degradationsphase des Quercus pubescens-Waldes am Südabfall des mittleren Ostalpengebietes unterrichtet H. MAYER (1969b : 166).

In Ungarn nahm in der postglazialen Nachwärmezeit nach ZÓLYOMI (1953 : 399) das Querco-Potentilletum albae (meist auf Kosten des Querco-Lithospermetum) seinen heutigen Platz ein.

Nach ZÓLYOMI (1951 : 58, 1953 : 398) wanderte in der zweiten Hälfte der postglazialen Wärmezeit in Ungarn das submediterrane Querceto-Cotinetum aus dem Süden ein, das mit der Klimaverschlechterung auf die exponiertesten Südhänge und Grate zurückgedrängt wurde und dort zugleich die Rasen- und Steppengesellschaften stark einengte.

OPRAVIL (1961) konnte in Böhmen das Quercion pubescentis an Hand von Holzkohle-Funden bis zur Hallstatt-Periode zurück verfolgen.

Quercetea ilicis

NIKLEWSKI & VAN ZEIST (1970) verfolgten an Hand von Pollenanalysen das Schicksal des Ceratonio-Pistacietum lentisci und des Pistacio-Quercetum calliprini, des Cedretum libani, einer Crataegus-Amygdalus-Prunus-Vorwald-Gesellschaft, sowie der Steppe durch die letzte Periode der Eiszeit und das Postglazial in NW-Syrien.

H. MAYER berichtet über die Geschichte des extrazonalen Quercetum ilicis im Bereich der S-Alpen (Ledro-See).

SUMMARY

Synchronology of Particular Vegetation Units in Europe

In the following sentences, vegetation units recognizable in tertiary, pleistocene, and holocene deposits are arranged syntaxonomically. Only the vegetational classes represented in several findings are mentioned in this summary. Literature references and classes apparent until now only once, twice or three times in fossil deposits are to be found in the German text.

The occurrence of the Lemnetea can be concluded from several fossil and subfossil findings (e.g. in Switzerland and in Denmark).

The Bidentetea tripartiti seem to be recognizable in tertiary deposits already in Poland. In the younger warm periods of post-glacial and later deposits, this class is investigated in some Northwest German valleys (Leine, Lower Weser, Lower Ems).

Fossil weed communities of the Stellarietea mediae are described in several publications during the last years. The references start with few neolithic deposits, but become much more numerous in the course of the excavations of the bronze and iron ages. More detailed parallelisations with recent weed communities are somewhat questionable owing to shiftings of the competition balance. The conditions for weeds in former periods with harvest only of the ears were apparently different from the situation under the present methods of cutting of the grain crops near the soil surface.

Plant communities of the Litorelletea growing in or near oligotrophic water were recognized in many deposits, partially already of interglacial origin (Denmark, Netherlands, Ireland). They were represented even in N.W. Syria during glacial and early post-glacial periods. Late glacial and post-glacial German lake deposits contain plant remainders demonstrating successions within the Litorelletea in connexion with oligo- and eu-trophication of the water.

The Potametea, demanding more eu-trophic water, are apparent in many fossil lake deposits. During the pliocene and pannon, genera, now restricted to climatically warmer areas, were associated with species growing in the recent Potametea vegetation of Central Europe. Studies on interglacial deposits in Poland, Brandenburg and N.W. Germany show successions from Potametea aquatic vegetation to reed and sedge marshes. The late glacial and post-glacial fossils permit more detailed differentiations of alliances, associations and successional stages within the Potametea.

The Bolboschoenetum (Bolboschoenetea maritimi) was represented already in the atlanticum and sub-atlanticum in areas with brackish water of the lower parts of streams and edges of marshes in N.W. Germany.

Several species of reed and sedge marshes (Phragmitetea) occur already in late tertiary deposits (miocene, pliocene, pannon) in Central Europe. The post-glacial macrofossils from numerous parts of Europe can be attached to various associations and alliances (e.g. Phragmition, Magnocaricion) of the Phragmitetea. These findings suggest in some cases syndynamical processes from Potametea to Phragmition and from Phragmition to Magnocaricion.

Salt marshes of the Asteretea tripolium were living apparently in coastal areas of N.W. Germany (Holstein, Lower Weser, Lower Ems) during the atlanticum, sub-atlanticum and in later periods. Associations and even subassociations are well recognizable from the deposits.

The Plantaginetalia majoris are referable to anthropogenic influences. The first records in Denmark are neolithic.

Plant fossils referable to Molinio-Arrhenatheretea are rarely found in earlier periods. This fact confirms the anthropogenic character of the most communities of the meadows and pastures belonging to this class. Only, the Filipendulion with partially natural tall forb communities could be recognized already in the pliocene (S.E. Europe) and Alleroed (Bohemia). Molinietalia meadows, presumably cut for hay, grew in the Netherlands since 3000 B.C.

Communities of the Caricetea fuscae are supposed to be existing already in the tertiary of S.E. Europe. Many findings of deposits of bronze and younger periods refer to this syntaxonomical class.

Also the acidophilous bog communities of the Oxycocco-Sphagnetea were presumably already in existence in the pliocene in S.E. Europe. These vegetation units and their syndynamical relations were found at several places in connexion with the studies on the history of bogs during the late glacial and post-glacial periods. In N.W. Germany and in the Netherlands, the Caricetum limosae and the Erico-Sphagnetum magellanici were living side by side for long periods. This demonstrates that these both plant communities are not always superseding mutually in rhythmical or cyclical vegetational changes.

The dwarf shrub heather vegetation of the Calluneto-Genistetum can replace anthropogenically acidophilous forest vegetation (Quercion robori-petraeae) since the neolithicum. This replacement is proved by pollen-analytical methods and by comparative macro-morphology of soil profiles in N.W. Germany and Jutland (Denmark).

The Alnetalia glutinosae were assumed already in the pliocene in S.E. Europe and found in the inter-glacial of the Emsland. In holocenic deposits, these *Alnus* communities were recognizable in several cases. Also conclusions on progressive and regressive successions starting or ending with Alnetalia are possible for these periods.

Traces of conifer forests of the Vaccinio-Piceetea are supposed already for the pliocene in S.E. Europe. Synchronological investigations proved the relic character of forests with *Picea abies* and namely with *Pinus sylvestris* (Pinion sylvestris) and their retrogression since the beginning of the post-glacial warm periods in many parts of Europe.

The former distribution of the acidophilous deciduous forests of the Quercetea robori-petraeae in Northern Germany, Jutland and Ireland could be concluded by studies of soil profiles and by pollen analysis. Pollen-analytical results proved the anthropogenic origin of the *Castanea vesca* forests on former sites of vegetation with dominant oaks (Quercion robori-petraeae) in hill areas south of the Alps.

The development and immigration of the forests with *Fagus sylvatica* (Eu-Fagion, Luzulo-Fagion etc.) during the subboreal or during other periods was investigated in various countries, e.g. in Hungary, in parts of the Alps and adjacent regions. Several authors studied the anthropogenic replacement of forests with dominant *Fagus sylvatica* by vegetation rich in *Quercus petraea* and *Qu. robur*. Since mainly or only the changes of the tree species composition of the forests can be recognized by pollen-analytical data, no syntaxonomical evidences emerge from such studies. The partial natural character of the Querceto-Carpinetum becomes probable by pollen-analytical studies in various parts of Germany and in the Netherlands.

The bottomland forests with *Quercus*, *Fraxinus*, *Ulmus* and partially also *Carpinus* and *Alnus* could be stated mainly for the post-glacial warm periods and for the later times.

The occurrence of various communities of the Quercetalia pubescenti-petraeae (thermophilous woodlands) is supposed for the post-glacial warm periods and later times in Hungary and at low altitudes in warm parts of the southeastern Alps and in Bohemia.

Fossil remainders of the sclerophyllous evergreen woodlands of the Quercetea ilicis were found near the great lakes at the southern edge of the Alps (e.g. at the Lake Ledro) and in deposits of the last glacial and post-glacial periods in N.W. Syria.

BIBLIOGRAPHY

The main objective of this bibliography is the presentation of the references of the 27 articles within the volume 'Vegetation Dynamics'. Therefore, also publications quoted in these articles had to be included not dealing primarily with vegetational changes. But beyond the purpose mentioned, this bibliography contains much more publications on vegetation dynamics than any other literature reference list. Thus, it is the amplest bibliography on vegetation dynamics published until now. But completeness in syndynamical papers could not be aspired; otherwise, the size of this bibliography would be much larger and exceed the space available.

Some examples of earlier literature lists concentrating on successional research work are in the papers of LÜDI (1930), WHITTAKER (1953), TÜXEN (1961b, containing 113 references), GREGORY & WALKER (1966, with 87 references) and in some articles of the 'Botanical Reviews' (e.g. CHURCHILL & HANSON 1958, COUPLAND 1958, SELLECK 1960). The annual reports on the progress of vegetation science (in 'Fortschritte der Botanik') regularly include publications on syndynamics (ELLENBERG 1965, 1966 and earlier, ELLENBERG & KNAPP 1967, KNAPP 1968–1973).

The authors writing mainly in the Russian language submitted their reference lists for the present volume in English translation or transcription; therefore, most of the references of publications written originally in Russian are included in this form. The titles of further papers published mainly in other languages are cited in English, French or German, if such a summary or authorized title translation is existing.

It may be mentioned that the authors' epithets to the scientific names of taxonomic or syntaxonomic units are not regarded to be literature references; consequently, the papers concerned are normally not included in the bibliography.

AARIO, L. – 1940 – Waldgrenzen und subrezente Pollenspektren in Petsamo-Lappland. *Ann. Acad. Sci. Fenn.* A 54(8): 1–120.

ABRAMOVA, L. I. – 1968 – The formation of vegetation on rototilled peat fields. *Vestn. Mosk. Univ.* VI 6: 77–82.

ADRIANI, M. J. – 1937 – Synökologische Beiträge zur Frage der Bedeutung von *Fagus silvatica* in einigen niederländischen Waldassoziationen. *Mitt. Flor.-soz. Arb.-gem. Niedersachsen* 3: 185–192.

ADRIANI, M. J. & MAAREL, E. VAN DER – 1968 – Breakers on Voorne. 104 pp. St. Wetensch. Duinonderzoek, Oostvoorne.

AICHINGER, E. – 1951 – Soziationen, Assoziationen, Waldentwicklungstypen. *Angew. Pfl.-soz.* (*Wien*) 1: 21–68.

AICHINGER, E. – 1952 – Fichtenwälder und Fichtenforste als Waldentwicklungstypen. *Angew. Pfl.-soz.* (*Wien*) 7: 1–178.

AICHINGER, E. – 1954 – Statische und dynamische Betrachtung in der pflanzensoziologischen Forschung. *Veröff. Geobot. Inst. Rübel* 29: 9–28.

AICHINGER, E. – 1957 – Die Zwergstrauchheiden als Vegetationsentwicklungstypen. *Angew. Pfl.-soz.* (*Wien*) 12: 1–124; 13: 1–84; 14: 1–171.

AICHINGER, E. – 1962 – Wald- und Wiesenentwicklungstypen als Grundlage der Produktivitätsleistung. In: LIETH, H. (Ed.): Die Stoffproduktion der Pflanzendecke. pp. 83–89. Stuttgart.

AICHINGER, E. – 1967 – Pflanzen als forstliche Standortsanzeiger. XXVIII+366 pp. Wien.

AICHINGER, E. – 1969 – Von der Erfassung des Arrhenatheretum elatioris Br.-Bl. 1925 im ostalpin-dinarischen Raum. *Acta Bot. Croat.* 28: 21–30.

ALBERDA, T. & WIT, C. T. DE – 1961 – Dry matter production and light interception of crop surfaces. *Jaarb. I.B.S.* 1961: 37–44.

ALBERTSON, F. W., RIEGEL, A. & LAUNCHBAUGH, J. – 1953 – Effects of different intensities of clipping on short grasses in west-central Kansas. *Ecology* 34: 1–20.

ALBERTSON, F. W. & WEAVER, J. E. – 1944 – Effects of drought, dust, and intensity of grazing on cover and yield of short-grass pastures. *Ecol. Monogr.* 14: 1–29.

ALEKHIN (= ALECHIN), V. V. – 1934 – Les steppes de la région centrale du tchernosiom . Voronesh.

ALEKHIN, V. V. – 1936 – Vegetation of the U.S.S.R. in zones. In: WALTER, H. & ALEKHIN, V. V.: Osnovy botan. geograf. Moscov, Leningrad.

ALEKSANDROVA, V. D. – 1956 – The vegetation of the southern Island of Novaya Zemlya between 70°56′ N.L. and 72°12′ N.L. *Rastit. Krain. Sev. SSSR i ee osv.* 2.

ALEKSANDROVA, V. D. – 1961– The plant community in the light of some cybernetic ideas. *Bull. MOIP* 66, (3).

ALEKSANDROVA, V. D. – 1962a – The problem of development in phytocoenology. *Bjull. Mosk. O. I. Prir. O. Biol.* 67(2): 86–107.

ALEKSANDROVA, V. D. – 1962b – On the underground structure of some plant communities in the arctic tundra of the Greater Lyakhov Island. *Problemy Bot.* 6.

ALEKSANDROVA, V. D. – 1963 – A sketch of the flora and vegetation of the Greater Lyakhov Island. *Trudy Arct. Antarct. Inst.* 224.

ALEKSANDROVA, V. D. – 1964 – Study of change in the plant cover. *Field Geobotany (Leningrad)* 3: 300–447.

ALEKSANDROVA, V. D. – 1966 – On the postglacial history of vegetation on the New Siberian Islands. *Bot. Zh.* 51(11).

ALEXANDER, T. R. & DICKSON, J. H. – 1971 – Vegetational changes in the National Key Deer Refuge. *Quart. J. Florida Acad.* 33: 81–89.

ANDERSEN, S. T. – 1967a – Interglaziale Pflanzensukzessionen aus Dänemark und ihr Verhältnis zu Umweltfaktoren. In: TÜXEN, R. (Ed.): Pflanzensoziologie und Palynologie. *Ber. Int. Sympos. Stolzenau/W.* 1962: 106–118.

ANDERSEN, S. T. – 1967b – Tree-pollen rain in a mixed deciduous forest in South Jutland (Denmark). *Rev. Palaeobot. and Palynol.* 3: 267–275.

ANDERSEN, S. T. – 1969 – Bestimmung der Pollenproduktion im Walde mit Hilfe von Oberflächenproben. *Ber. Dtsch. Bot. Ges.* 81: 488.

ANDERSEN, S. T. – 1970 – The relative pollen productivity and pollen representation of north European trees, and correction factors for tree pollen spectra determinated by surface pollen analyses from forests. *Danmarks Geol. Unders.* II. 96: 99 pp.

ANDREÁNSKY, G. – 1963 – Das Trockenelement der jungtertiären Flora Mitteleuropas. *Vegetatio* 11: 155–172.

ANDREEV, V. N. – 1930 – Geobotanical exploration of the tundra reindeer pastures. *Sov. Sever. No.* 5.

ANDREEV, V. N. – 1931 – The vegetation of the Northern Kanin tundra. *Olenyi pastb. Severn. Kraya (Arkhangelsk)* 1.

ANDREEV, V. N. – 1933 – The forage sources of the reindeer breeding in the western part of the Bolshesemelskaya tundra. *Olenyi pastb. Severn. Kraya* 2.

ANDREEV, V. N. – 1938 – Investigations of the tundra reindeer pastures by means of the aeroplane. *Trudy Leningrad Inst. Polarn. Zemled. Ser. Olenevodstvo* 1.

ANDREW, L. E. & RHOADES, H. F. – 1947(1948) – Soil development from calcareous glacial material in eastern Nebraska during 75 years. *Soil Sci. Soc. Amer. Proc.* 12: 407–408.

ANT, H. – 1967a – Die Pioniergesellschaften der Schlammflächen trockengefallener Talsperrensohlen. *Decheniana* 118: 139–144.

ANT, H. – 1967b – Zum räumlichen und zeitlichen Gefüge der Vegetation trock-

engefallener Talsperrenböden. *Arch. Hydrobiol.* 62: 439–452.
ANTONOVICS, J. – 1968 – Evolution in closely adjacent populations. V. Evolution of self fertility. *Heredity* 23: 219–238.
APINIS, A. E., CHESTERS, C. G. C. & TALIGOOLA, H. K. – 1972 – Colonisation of *Phragmites communis* leaves by fungi. *Nova Hedwigia* 23: 113–124.
ARBER, A. – 1954 – The mind and the eye. Cambridge Univ. Press, London.
ARORA, R. K. – 1967 – The vegetation of South Kanara district, IV. Succession in plant communities. *J. Indian Bot. Soc.* 46: 15–24.
ASAI, T. – 1952 – Die Vegetation auf den Kegeln des Aso-Vulkans. *Kumamoto J. Sci.* 1: 1–31.
ASHBY, W. C. & WEAVER, G. T. – 1970 – Forest regeneration on two old fields in south-western Illinois. *Amer. Midland Natural.* 84: 90–104.
ASHTON, P. S. – 1969 – Speciation among tropical forest trees: some deductions in the light of recent evidence. *Biol. J. Linn. Soc.* 1: 155–196.
ASSMANN, E. – 1961 – Waldertragskunde. XV+490 pp. München, Bonn, Wien.
AUBRÉVILLE, A. – 1938 – Les forêts de l'Afrique occidentale française. *Ann. Acad. Sci. Col.* 9: 1–245.
AUSTIN, M. P. – 1972 – Models and analysis of descriptive vegetation data. *Sympos. Brit. Ecol. Soc.* 12: 61–86.
AUSTIN, M. P. & GREIG-SMITH, P. – 1968 – The application of quantitative methods to vegetation survey. II. Some methodological problems of data from rain forests. *J. Ecol.* 56: 827–844.
AXELROD, D. J. – 1959 – Evolution of the Madro-Tertiary geoflora. *Bot. Rev.* 24: 433–509.
BABCOCK, E. – 1947 – The genus *Crepis*. *Univ. Cal. Publ. Bot.* 21/22: 1–1030.
BACHTHALER, G. – 1968 – Die Entwicklung der Ackerunkrautflora in Abhängigkeit von veränderten Feldbaumethoden. *Z. Acker- u.Pfl.-bau* 127: 149–170, 327–358.
BACHTHALER, G. – 1969 – Entwicklung der Unkrautflora in Deutschland in Abhängigkeit von den veränderten Kulturmethoden. *Angew. Bot.* 43: 59–69.
BACHTHALER, G. & DIERCKS, R. – 1968 – Chemische Unkrautbekämpfung auf Acker- und Grünland. 2. Aufl. BLV, München.
BAER, K. – 1838 – Végétation et climat de Novaia Zemlia. *Bull. Sci. Publ. Acad. Imp. Sci. S.-Pétersb.* 3.
BAKER, H. G. – 1972 – Migration of weeds. In: VALENTINE, D. H. (Ed.): Taxonomy, Phytogeography and Evolution, pp. 327–347. Academic Press, London, New York 1972.
BAKKER, D. – 1951 – Oecologie van Klein Hoefblad, *Tussilago farfara* L., en de bestrijding van deze plant in de Noordoostpolder. *Diss. Groningen.*
BAKKER, D. – 1960 – *Senecio congestus* (R. Br.) DC. in the Lake Ijssel Polders. *Acta Bot. Neerl.* 9: 235–259.
BARADZIEJ, E. – 1969 – The disappearance of dead material . . . xerothermic grassland of the Ojców National Park. *Fragm. Florist. Geobot.* (*Kraków*) 15: 469–485.
BARCLAY-ESTRUP, P. & GIMINGHAM, C. H. – 1969 – Vegetational change in relation to the *Calluna cycle*. *J. Ecol.* 57: 737–758.
BARKMAN, J. J. – 1958 – On the ecology of cryptogamic epiphytes with special reference to the Netherlands. *Belmontia* 2.
BARKMAN, J. J. – 1968 – Das synsystem. Problem der Mikrogesellschaften innerhalb der Biozönosen. In: TÜXEN, R. (Ed.): Pflanzensoziologische Systematik. *Ber. Intern. Sympos. Stolzenau/W.* 1964: 41–53.
BARKMAN, J. J. – 1969 – The influence of air pollution on bryophytes and lichens. *Proc. Europ. Congr. Infl. Air Pollut. on Plants and Anim.* (*Wageningen*) 1: 197–241.
BARKMAN, J. J. – 1970 – The epiphytic vegetation of 'Speulder'- and 'Sprielder-

bosch', the communities, ecology and succession. Comparison with other woods. *Belmontia* 16(96): 1–13.

Barrow, M. D., Costin, A. B. & Lake, P. – 1968 – Cyclical changes in an Australian fjaeldmark community. *J. Ecol.* 56: 89–96.

Bartholomew, W. V., Meyer, J. & Laudelout, H. – 1953 – Mineral nutrient immobilization under forest and grass fallow in the Yangambi (Belgian Congo) region. *Publ. I.N.E.A.C. Sér. Sci.* 57: 1–27.

Barton, L. V. – 1961 – Seed preservation and longevity. XVIII, 216 pp. New York and London.

Bartsch, A. F. – 1969 – Accelerated eutrophication of lakes, ecological response to human activities. *Abstr. Intern. Bot. Congress* 11: 9.

Bastin, B. – 1964 – Recherches sur les relations entre la végétation actuelle et le spectre pollinique récent dans la Forêt de Soignes (Belgique). *Agricultura* 2e *Sér.* 12(2): 341–373.

Bateman, K. G. – 1959 – The genetic assimilation of four venation phenocopies. *J. Genet.* 56: 443–474.

Batzli, G. O. & Pitelka, F. A. – 1970 – Influence of meadow mouse populations on California grassland. *Ecology* 51: 1027–1039.

Baule, H. & Fricke, C. – 1967 – Die Düngung von Waldbäumen. 259 pp. München.

Baxter, D. V. & Middleton, J. T. – 1961 – Geofungi in forest successions following retreat of the Alaskan glaciers. *Recent Adv. Bot.* 14: 1514–1517.

Bazzaz, F. A. – 1968 – Succession on abandoned fields in the Shawnee Hills, southern Illinois. *Ecology* 49: 924–936.

Beadle, N. C. W. – 1948 – The vegetation and pastures of western New South Wales, with special reference to soil erosion. 281 pp. Govt. Printer, Tennant, Sidney.

Beadle, N. C. W. – 1951 – The misuse of climate as an indicator of vegetation and soils. *Ecology* 32: 343–345.

Beadle, N. C. W. & Costin, A. B. – 1952 – Ecological classification and nomenclature. *Proc. Linn. Soc. N. S. W.* 77: 61–82.

Beard, J. S. – 1944 – Climax vegetation in tropical America. *Ecology* 25: 127–158.

Beard, J. S. – 1945 – The natural vegetation of Trinidad. *Oxford Forest. Mem.* 20.

Beard, J. S. – 1945a – The progress of plant succession on the Soufrière of St. Vincent. *J. Ecol.* 33: 1–9.

Beard, J. S. – 1953 – The savanna vegetation of northern tropical America. *Ecol. Monogr.* 23: 149–215.

Beard, J. S. – 1969 – The vegetation of the Boorabbin and Lake Johnston areas, Western Australia. *Proc. Linn. Soc. N. S. W.* 93: 239–269.

Becher, R. – 1963 – Entwicklungs-Möglichkeiten der Wald-Vegetation nach Einwirkung bestimmter experimenteller Beeinflussungen ... *Geobot. Mitteilungen* 19: 1–241.

Becking, R. W. – 1957 – The Zürich-Montpellier school of phytosociology. *Bot. Rev.* 23: 411–488.

Becking, R. W. – 1968 – Vegetational response to change in environment and change in species tolerance with time. *Vegetatio* 16: 135–158.

Behre, K.-E. – 1966 – Untersuchungen zur spätglazialen und frühpostglazialen Vegetationsgeschichte Ostfrieslands. *Eisz. Gegenw.* 17: 69–84.

Behre, K.-E. – 1970 – Die Entwicklungsgeschichte der natürlichen Vegetation im Gebiet der unteren Ems und ihre Abhängigkeit von den Bewegungen des Meeresspiegels. *Probleme Küstenforschung im südl. Nordseegebiet* 9: 14–47.

Beidemann, I. N. – 1957 – Observations of the changes of the coastal vegetation and the salinization of the sea bottom when recessing the Caspian. *Acta Inst. Bot. Acad. Sci. URSS, Ser. III, Geobot., fas.* 2.

Beidemann, I. N. & Preobrazhensky, A. S. – 1957 – The interdependence of the development of soils and vegetation in the Kura-Araxian Plain. *Acta Inst. Bot. Acad. Sci. URSS, Ser. III, Geobot., fas.* 2.

Beijerinck, W. – 1929 – De subfossiele plantenresten in de Terpen van Friesland en Groningen, I. Vruchten, jaden en bloemen. Wageningen.

Beijerinck, W. – 1933 – Die mikropaläontologische Untersuchung äolischer Sedimente und ihre Bedeutung für die Quartär-Stratigraphie. *Proc. Kon. Akad. Wetensch. Amsterdam.*

Beijerinck, W. – 1934 – Humusortstein und Bleichsand als Bildungen entgegengesetzter Klimate. *Proc. Kon. Akad. Wetensch. Amsterdam* 37: 93–98.

Bell, D. T. – 1969 – Allelopathic effects of *Brassica nigra* on annual grasslands in Southern California. *Abstr. Intern. Bot. Congress* 11: 12.

Bennett, H. H. & Allison, R. V. – 1928 – The soils of Cuba. Washington.

Benninghoff, W. S. – 1952 – Interaction of vegetation and soil frost phenomena. *Arctic* 5: 34–44.

Beran, N. – 1964 – Experimentelle Untersuchungen, Einflüsse von Blattstreu und andere gegenseitige Beeinflussungen bei einigen Pflanzenarten auf Abraumhalden und ähnlichen Substraten. *Diss. Giessen.*

Berger, W. – 1952 – Die altpliozäne Flora der Congerienschichten von Brünn-Vösendorf bei Wien. *Palaeontographica* 92B.

Berger, W. – 1955a – Die altpliozäne Flora des Laaerberges in Wien. *Palaeontographica* 97B.

Berger, W. – 1955b – Neue Ergebnisse zur Klima- und Vegetationsgeschichte des europäischen Jungtertiärs. *Ber. Geobot. Inst. Rübel* 1954: 12–29.

Berger, W. & Zabusch, F. – 1953 – Die obermiozäne (sarmatische) Flora der Türkenschanze in Wien. *Neues Jb. Geol. Pal. Abh.* 98: 226–276.

Berger-Landefeldt, U. – 1967 – Der Energieumsatz am Standort. *Bot. Jahrb.* 86: 402–448.

Berghen, C. van den – 1959 – Etude sur la végétation des dunes et des landes de la Bretagne. *Vegetatio* 8: 193–208.

Berglund, B. E. – 1962, 1963 – Vegetation of the island of Senoren. *Bot. Not.* 115, 116.

Bergman, H. F. & Stallard, H. – 1916 – The development of climax formations in northern Minnesota. *Minn. Bot. Stud.* 4: 333–378.

Berthold, H.-J. – 1954 – Erfahrungen bei der Aufforstung von Halden auf dem Gelände des westdeutschen Steinkohlenbergbaues. *Forschung u. Beratung* 1: 146–150.

Bertsch, K. – 1935 – Der deutsche Wald im Wechsel der Zeiten. 91 pp. Tübingen.

Betz, R. F. & Cole, M. H. – 1969 – The Peacock Prairie; a study of a virgin Illinois mesic black-soil prairie forty years after initial study. *Trans. Ill. Acad.* 62: 44–53.

Beug, H. J. – 1964 – Untersuchungen zur spät- und postglazialen Vegetationsgeschichte im Gardaseegebiet. *Flora* 154: 401–444.

Bews, J. W. – 1916 – An account of the chief types of vegetation in South Africa, with notes on the plant succession. *J. Ecol.* 4: 129–159.

Billings, W. D. – 1941 – Quantitative correlations between vegetational changes and soil development. *Ecology* 22: 448–456.

Billings, W. D. – 1969 – Vegetational pattern near alpine timberline as affected by fire-snowdrift interactions. *Vegetatio* 19: 192–207.

Birger, S. – 1906 – Die Vegetation einiger 1882–1886 entstandener schwedischer Inseln. *Bot. Jb.* 38(3).

Birks, H. J. B. – 1970 – Inwashed pollen spectra at Loch Fada, Isle of Skye. *New Phytol.* 69: 807–820.

Bittmann, E. – 1953 – Das Schilf (*Phragmites communis* Trin.) und seine Verwendung im Wasserbau. *Angew. Pfl. soz.* (Stolzenau) 7: 1–44.
Black, W. H. & Clark, V. I. – 1942 – Yearlong grazing of steers in the northern Great Plains. *U.S. Dept. Agr. Circ.* 642.
Blackburn, W. H. & Tueller, P. T. – 1970 – Pinyon and juniper invasion in black sagebrush communities in east-central Nevada. *Ecology* 51: 841–848.
Bledsoe, L. J. & Dyne, G. M. van – 1971 – A compartment model simulation of secondary succession. *Systems Analysis and Simulation in Ecology* 1: 479–511.
Bledsoe, L. J. & Jameson, D. A. – 1969 – Model structure for a grassland ecosystem. *Range Science Series* (*Colorado State Univ.*) 2.
Bliss, L. C. – 1966 – Plant productivity in alpine microenvironments on Mt. Washington, New Hampshire. *Ecol. Monogr.* 36: 125–155.
Bliss, L. C. & Cantlon, J. E. – 1957 – Succession on river alluvium in Northern Alaska . *Amer. Midl. Natural.* 58: 452–469.
Bliss, L. C. & Linn, R. M. – 1955 – Bryophyte communities associated with old field succession in the North Carolina piedmont. *Bryologist* 58: 120–131.
Blum, J. L. – 1956 – Application of the climax concept to algal communities of streams. *Ecology* 37: 603–604.
Blum, U. & Rice, E. L. – 1969 – Inhibition of symbiotic nitrogen fixation by gallic and tannic acid, and possible roles in old-field succession. *Bull. Torr. Bot. Club* 96: 531–544.
Bobrov, E. G. – 1972 – Introgressive hybridization, speciation and succession of vegetative cover. *Bot.* Zh. 57: 865–879.
Boch, M. S. – 1968 – On the method of studying the temporal changes of mire vegetation. In: Materialy po dinamike rast. pokrova. Vladimir.
Bodard, K. S. – 1969 – Cytogenetic mechanisms associated with speciation in tropical tree species. *Abstr. Intern. Bot. Congr. Seattle* 11: 10.
Bode, H. R. – 1940 – Über die Blattausscheidungen des Wermuts und ihre Wirkungen auf andere Pflanzen. *Planta* 30: 567–589.
Bodrogközy, G. & Harmati, I. – 1969 – Die mit Bewässerung verknüpfte Wirkung der verschiedenen Nitrogen- und Phosphordarreichungen für die Achilleo-Festucetum pseudovinae-Artenzusammensetzung der Trokkenweiden im Donautal. In: Tüxen, R. (Ed.): Experimentelle Pflanzensoziologie. Ber. Intern. Sympos. Rinteln 1965: 75–92.
Boeker, P. – 1965 – Die Entwicklung einer Ansaat auf einem Grundwasserstandsversuch in der Boker Heide. *Netherl. J. Agr. Sci.* 13: 164–170.
Boeker, P. – 1966 – Leistungsvergleich zwischen Wiese und Weide, dargestellt am Beispiel eines Grundwasserstandsversuches. *Bayer. Landw. Jahrb.* 43: 223–230.
Bogdanovskaya-Gienef, I. D. – 1938 – Reindeer pastures and natural conditions of the reindeer bieeding on the Kolguev Island. *Trudy Inst. Polarn. Zemled. Ser. Olenevodstvo* 2.
Boguslawski, E. v. – 1961 – Zur Zielsetzung der Forschungsarbeit in Rauisch-Holzhausen. *Ber. Oberhess. Ges. N. F. Naturwiss. Abt.* 31: 46–57.
Bonner, J. – 1950 – The role of toxic substances in the interactions of higher plants. *Bot. Rev.* 16: 51–65.
Bonner, J. & Galston, A. W. – 1944 – Toxic substances from the culture media of guayule which may inhibit growth. *Bot. Gaz.* 106: 185–198.
Borchert, J. R. – 1950 – The climate of the central North American grassland. *Ann. Ass. Amer. Geogr.* 40: 1–39.
Borhidi, A. – 1970 – Ökologie, Wettbewerb und Zönologie des Schilfrohres. *Acta Bot. Acad. Sci. Hung.* 16: 1–12.
Borhidi, A. & Balogh, M. – 1970 – Die Entstehung von dystrophen Schaukelmooren in einem alkalischen (Szik-) See. *Acta Bot. Acad. Sci. Hung.* 16: 13–31.

Borisova, I. V. & Popova, T. A. – 1971 – Age phases of bunch formation in steppe grasses. *Bot. Zh.* 56: 619–626.
Borisova, I. V. & Popova, T. A. – 1972 – The dynamics of quantity and age composition of bunch grass coenopopulations in desert steppes of Central Kazakhstan. *Bot.* Zh. 57: 779–793.
Bormann, F. H. – 1953 – Factors determining the role of loblolly pine and sweetgum in early old-field succession in the piedmont of North Carolina. *Ecol. Monogr.* 23: 339–358.
Bormann, F. H. – 1969 – Abstr. Intern. Bot. Congr. 11: 18.
Bormann, F. H., Siccama, T. G., Likens, G. E. & Whittaker, R. H. – 1971 – The Hubbard Brook eco-system study: composition and dynamics of the tree stratum. *Ecol. Monogr.* 40: 373–388.
Börner, H. – 1956 – Die Abgabe organischer Verbindungen aus den Karyopsen, Wurzeln und Ernterückständen. . . *Beitr. Biol. Pfl.* 33: 33–83.
Börner, H. – 1960 – Liberation of organic substances from higher plants and their role in the soil sickness problem. *Bot. Rev.* 26: 383–424.
Bornkamm, R. – 1961a – Zur Konkurrenzkraft von *Bromus erectus*. *Bot. Jahrb.* 80: 466–479.
Bornkamm, R. – 1961b – Vegetation und Vegetations-Entwicklung auf Kiesdächern. *Vegetatio* 10: 1–24.
Bornkamm, R. – 1963 – Erscheinung der Konkurrenz zwischen höheren Pflanzen und ihre begriffliche Fassung. *Ber. Geobot. Inst. ETH St. Rübel* 34: 83–107.
Bornkamm, R. – 1969 – Interactions and their influence on protein content of plants. *Abstr. Intern. Bot. Congress* 11: 18.
Bothmer, H. J. G. – 1953 – Der Einfluss der Bewirtschaftung auf die Ausbildung der Pflanzengesellschaften niederrheinischer Dauerweiden. *Z. Acker- u. Pfl.-bau* 96: 457–476.
Botkin, D. B., Janak, J. F. & Wallis, J. R. – 1972a – Some ecological consequences of a computer model of forest growth. *J. Ecol.* 60: 849–872.
Botkin, D. B., Janak, J. F. & Wallis, J. R. – 1972b – The rationale, limitations and assumptions of a northeast forest growth simulator. *I.B.M. Res. Dev.* 16: 101–116.
Böttcher, H. – Some remarks on the vegetation of South-Icelandic cultivated hayfields and their damages by "winterkilling,,. *Res. Inst. Nedri Ás, Iceland, Rep.* 9: 1–28.
Bourdo, E. A. – 1956 – A review of the General Land Office Survey and of its use in quantitative studies of former forests. *Ecology* 37: 754–768.
Bourne, R. – 1934 – Some ecological conceptions. *Empire Forest. J.* 13: 15–30.
Bournérias, M. – 1959 – Le peuplement végétal des espaces nus. *Bull. Soc. Bot. France Mém.* 1959: 1–300.
Bowen, R. N. C. – 1958 – The exploration of time. London.
Bradshaw, A. D. – 1969 – An ecologist's viewpoint. *Brit. Ecol. Soc. Sympos.* 9: 415–427.
Bradshaw, A. D. – 1971 – Plant evolution in extreme environments. In: Creed, R. (Ed.): Ecological genetics and evolution. pp. 20–50. Blackwell, Oxford and Edinburgh.
Bradshaw, A. D., McNeilly, T. S. & Gregory, R. P. G. – 1965 – Industrialization, evolution, and development of heavy metal tolerance in plants. *Brit. Ecol. Soc. Symp.* 5: 327–343.
Braun, E. L. – 1935 – The undifferentiated deciduous forest climax and the association-segregate. *Ecology* 16: 514–519.
Braun, E. L. – 1950 – Deciduous forests of eastern North America. 596 pp. Blakiston, Philadelphia.

Braun, E. L. – 1956 – The development of association and climax concepts: Their use in interpretation of the deciduous forest. *Amer. J. Bot.* 43: 906–911.
Braun-Blanquet, J. – 1923 – L'origine et développement des flores du Massif Central de France. Zurich et Paris.
Braun-Blanquet, J. – 1928 – Pflanzensoziologie. 1. Aufl. 330 pp. Berlin.
Braun-Blanquet, J. – 1951 – Pflanzensoziologie. 2. Aufl. 631 pp. Wien.
Braun-Blanquet, J. – 1964 – Pflanzensoziologie. 3. Aufl. 865 pp. Wien.
Braun-Blanquet, J. – 1932 – Plant sociology. Transl. by Fuller, G. D. & Conard, H. S. 439 pp. Mc Graw-Hill, New York.
Braun-Blanquet, J. – 1933 – L'association végétale climatique et le climax du sol dans le Midi méditerranée. *Bull. Soc. Bot. Fr.* 80(9).
Braun-Blanquet, J. – 1949 – W. Szafer: The pliocene flora of Kroscienko in Poland (Referat). *Vegetatio* 1: 199–200.
Braun-Blanquet, J. – 1955 – Die Vegetation des Piz Languard, ein Maßstab für Klima-Änderungen. *Svensk Bot. Tidskr.* 49: 1–8.
Braun-Blanquet, J. – 1957 – Ein Jahrhundert Florenwandel am Piz Linard. *Bull. Jard. Bot. Et. Bruxelles Vol. Jub. Robyns*: 221–232.
Braun-Blanquet, J. – 1961 – Die inneralpine Trockenvegetation. *Geobot. Selecta* 1: 1–273.
Braun-Blanquet, J. et al. – 1931 – Pflanzensoziologisch-pflanzengeographische Studien in Südwestdeutschland. *Beitr. Naturdenkmalpfl.* 14(3): 217–292.
Braun-Blanquet, J. & De Leeuw, W. C. – 1936 – Vegetationsskizze von Ameland. *Nederl. Kruidk. Arch.* 46: 359–393.
Braun-Blanquet, J. & Jenny, H. – 1926 – Vegetationsentwicklung und Bodenbildung in der alpinen Stufe der Zentralalpen. *Denkschr. Schweiz. Naturf. Ges.* 63: 181–349.
Braun-Blanquet, J., Pallmann, H. & Bach, R. – 1954 – Pflanzensoziologische und bodenkundliche Untersuchungen im schweizerischen Nationalpark und seine Nachbargebieten. II. *Ergebn. Wiss. Untersuch. Schweiz. Nationalpark* 4(28): 1–200.
Braun-Blanquet, J., Wikus, E., Suter, R. & Braun-Blanquet, G. – 1958 – Lagunenverlandung und Vegetationsentwicklung an der französischen Mittelmeerküste bei Palavas, ein Sukzessionsexperiment. *Veröff. Geobot. Inst. Rübel* 33: 9–22.
Bray, J. R. – 1960 – The chlorophyll content of some native and managed plant communities in Central Minnesota. *Canad. J. Bot.* 38: 313–333.
Brereton, A. J. – 1971 – The structure of the species populations in the initial stages of salt-marsh succession. *J. Ecol.* 59: 321–338.
Brock, T. D. & Brock, M. L. – 1969 – Recovery of a hot spring community from a catastrophe. *J. Phycol.* 5: 75–77.
Brown, R. T. – 1950 – Forests of the central Wisconsin sand plains. *Ecol. Soc. Amer. Bull.* 31: 56.
Brown, R. T. & Curtis, J. T. – 1952 – The upland conifer-hardwood forests of northern Wisconsin. *Ecol. Monogr.* 22: 217–234.
Buchwald, K. & Losert, H. – 1953 – Pflanzensoziologische und pollenanalytische Untersuchungen im "Blanken Flat" bei Vesbeck. *Mitt. Flor.-soz. Arbeitsgem. N.F.* 4: 124–146.
Budd, A. C., Chepil, W. S. & Doughty, J. C. – 1954 – The influence of crops and fallows on the weed seed population of the soil. *Canad. J. Agr. Sci.* 34: 18–27
Budde, H. – 1938 – Pollenanalytische Untersuchung eines Sauerländischen Moores bei Lützel. *Decheniana* 97B: 169–187.
Budde, H. – 1949 – Die Waldgebiete Westfalens während der älteren Nachwärmezeit, etwa 500 vor bis 1000 nach Chr. *Natur u. Heimat* 9(1): 26–33.

BUDDE, H. & RUNGE, F. – 1940 – Pflanzensoziologische und pollenanalytische Untersuchung des Venner Moores, Münsterland. *Abh. Landesmus. Naturkd. Prov. Westf.* 11(1): 3–28.
BUDOWSKI, G. – 1965 – Distribution of tropical American rain forest species in the light of successional processes. *Turrialba* 15: 40–42.
BUDOWSKI, G. – 1969 – The use of plant indicators to recognize successional steps. *Abstr. Intern. Bot. Congress* 11: 24.
BUDYKO, M. I. & YEFIMOVA, N. A. – 1968 – The utilization of solar energy by the natural vegetational cover over the territory of the USSR. *Bot. Zh.* 53: 1384–1389.
BUELL, M. F., BUELL, H. F., SMALL, J. A. & SICCAMA, T. G. – 1971 – Invasion of trees in secondary succession in the New Jersey piedmont. *Bull. Torr. Bot. Club* 98: 67–74.
BURG, P. F. J. VAN & ARNOLD, G. H. (Ed.), – 1966 – Nitrogen and grassland. 234 pp. Wageningen.
BURGES, A. & DROVER, D. P. – 1953 – The rate of podzol development in the sands of the Woy Woy district, N. S. W. *Austral. J. Bot.* 1: 83–94.
BURGESS, R. L. – 1964 – Ninety years of vegetational change in a township in southeastern North Dakota. *Proc. North Dakota Acad. Sci.* 18: 84–94.
BUROLLET, P. A. – 1927 – Le Sahel de Sousse. 276 pp. Tunis.
BURRICHTER, E. – 1960 – Die Therophyten-Vegetation an nordrhein-westfälischen Talsperren im Trockenjahr 1959. *Ber. Dtsch. Bot. Ges.* 73: 24–37.
BURRICHTER, E. – 1970 – Beziehungen zwischen Vegetations- und Siedlungsgeschichte im nordwestlichen Münsterland. *Vegetatio* 20: 199–209.
BURSCHEL, P. & SCHMALTZ, J. – 1965 – Die Bedeutung des Lichtes für die Entwicklung junger Buchen. *Allg. Forst- u. Jagdzg.* 136: 193–210.
BUSCH, K. K. – 1971 – Some problems of biometric analysis of anthropogenous successions on drained forest areas. *Bot. Zh.* 56: 22–30.
BYKOV, B. A. – 1955 – Vegetation and forage resources of the Western Kazakhstan. Acad. Sci. Kazakh SSR Press, Alma-Ata.
BYKOV, B. A. – 1957 – Geobotany. 2nd ed. Alma-Ata.
BYKOV, B. A. – 1966 – On the way to the genetic classification of vegetation. *Isv. Akad. Nauk Kaz. SSR. Ser. Biol.* 4.
BYKOV, B. A. (ed.) – 1968 – Bio-ecological basis of exploitation and improvement of pastures in the area of the north of the Aral Sea. "Nauka", Alma-Ata.
CAIN, S. A. – 1939 – The climax and its complexities. *Amer. Midl. Natural.* 21: 146–181.
CAIN, S. A. – 1947 – Characteristics of natural areas and factors in their development. *Ecol. Monogr.* 17: 185–200.
CAJANDER, A. K. – 1909 – Über Waldtypen. *Acta Forest. Fenn.* 1(1): 1–175.
CAJANDER, A. K. – 1926 – The theory of forest types. *Acta Forest. Fenn.* 29(3): 1–108.
CANTLON, J. E. – 1969 – Continuum concept of vegetation: responses. *Bot. Rev.* 34: 255–258.
CAPUTA, J. – 1948 – Untersuchungen über die Entwicklung einiger Gräser und Kleearten in Reinsaat und Mischung. 127 pp. *Diss. Zürich.*
CASPARIE, W. A. – 1969 – Bult- und Schlenkenbildung in Hochmoortorf (zur Frage des Moorwachstums-Mechanismus). *Vegetatio* 19: 146–180.
CHAMPNESS, ST. S. & MORRIS, K. – 1948 – The population of buried viable seeds in relation to contrasting pasture and soil types. *J. Ecol.* 36: 149–173.
CHANDAPILLAI, M. M. – 1970 – Variation in fixed dune vegetation at Newborough Warren, Anglesey. *J. Ecol.* 58: 193–201.
CHANEY, R. W. – 1956 – Paleobotany. *Carnegie Inst. Washington Year Book* 55.
CHAPMAN, V. J. – 1960 – Salt marshes and salt deserts of the world. XVI, 392

pp. London, New York.
Charter, C. F. – 1941 – Reconnaissance survey of the soils of British Honduras. Trinidad.
Chaudhri, I. I. & Quadir, S. A. – 1958 – Sand dunes vegetation of coastal regions of Karachi. *Pakistan J. Forest.* 8: 337–341.
Chesters, C. G. C. – 1950 – On the succession of microfungi associated with the decay of logs and branches. *Trans. Lincs. Nat. Un.* 12: 129–135.
Christiansen, W. – 1937 – Beobachtungen an Dauerquadraten auf der Lotseninsel Schleimünde. *Schr. Naturw. Ver. Schlesw.-Holst.* 21: 19–57.
Churchill, E. D. & Hanson, H. C. – 1958 – The concept of climax in arctic and alpine vegetation. *Bot. Rev.* 24: 127–191.
Clarke, S. E., Tisdale, E. W. & Skoglund, N. A. – 1943 – The effects of climate and grazing practices on short-grass prairie vegetation in southern Alberta and southwestern Saskatchewan. *Canad. Dept. Agr. Tech. Bull.* 46.
Clements, F. E. – 1904 – The development and structure of vegetation. *Bot. Surv. Nebr. Univ.* 3: 1–175.
Clements, F. E. – 1905 – Research methods in ecology. 334 pp. Univ. Publ. Co., Lincoln.
Clements, F. E. – 1910 – The life history of lodgepole burn forests. *U.S. Forest Service Bull.* 79.
Clements, F. E. – 1916 – Plant succession: An analysis of the development of vegetation. *Carnegie Inst. Wash. Publ.* 242: 1–512.
Clements, F. E. – 1928 – Plant succession and indicators: A definitive edition of plant succession and plant indicators. 453 pp. Wilson, New York.
Clements, F. E. – 1936 – Nature and structure of the climax. *J. Ecol.* 24: 252–284.
Clements, F. E. – 1934 – The relict method in dynamic ecology. *J. Ecol.* 22: 39–68.
Clements, F. E. & Hanson, H. C. – 1929 – Plant competition. *Carnegie Inst. Wash. Publ.* 398, 1–340.
Cleve, K. van, Viereck, L. A. & Schleutner, R. L. – 1970 – Accumulation of nitrogen in alder ecosystems developed on the Tanana River flood plain near Fairbanks, Alaska. *Arctic and Alpine Res.* 3: 101–114.
Cleve, K. van, Viereck, L. A. & Schleutner, R. L. – 1972 – Distribution of selected chemical elements in even-aged alder (*Alnus*) ecosystems near Fairbanks, Alaska. *Arctic and Alpine Res.* 4: 239–255.
Cline, A. C. & Spurr, S. H. – 1942 – The virgin upland forest of central New England. *Bull. Harvard Forest* 21: 1–58.
Coaz, J. – 1887 – Erste Ansiedlung phanerogamer Pflanzen auf vom Gletscher verlassenen Böden. *Mitt. Natf. Ges. Bern* 1886: 3–12.
Comendar, V. I. – 1967 – The successions of the associations formed by Pinus mughus Scop. on the stony substrates in the Ukrainan Carpathians. *Bot. Zh.* 52: 1170–1176.
Conway, G. R. & Murdie, G. – 1972 – Population models as a basis for pest control. *Sympos. Brit. Ecol. Soc.* 12: 195–213.
Cooper, C. F. – 1960 – Changes of vegetation structure and growth of southwestern pine forests since white settlement. *Ecol. Monogr.* 30: 129–164.
Cooper, C. F. – 1961 – The ecology of fire. *Sci. Amer.* 204: 150–156, 158.
Cooper, C. F. – 1969 – Ecosystem models in watershed management. In: The ecosystem concept in natural resource management. Academic Press, New York.
Cooper, W. S. – 1913 – The climax forest of Isle Royale, Lake Superior, and its development. *Bot. Gaz.* 55: 1–44, 115–140, 189–235.
Cooper, W. S. – 1916 – Plant succession in the Mount Robson Region, British Columbia. *Plant World* 19: 211–238.

COOPER, W. S. – 1923 – The recent ecological history of Glacier Bay, Alaska. *Ecology* 4: 93–128, 223–246, 355–365.

COOPER, W. S. – 1926 – The fundamentals of vegetation change. *Ecology* 7: 391–413.

COOPER, W. S. – 1931 – A third expedition to Glacier Bay, Alaska. *Ecology* 12: 61–95.

COOPER, W. S. – 1939 – A fourth expedition to Glacier Bay, Alaska (1938). *Ecology* 20: 130–155.

COSTELLO, D. F. – 1944 – Natural revegetation of abandoned plowed land in the mixed prairie association of northeastern Colorado. *Ecology* 25: 312–326.

COSTIN, A. B., WIMBUSH, D. J., BARROW, M. D. & LAKE, P. – 1969 – Development of soil and vegetation climaxes in the Mount Kosciusko area, Australia. *Vegetatio* 18: 273–288.

COTTAM, G. – 1949 – The phytosociology of an oak wood in southwestern Wisconsin. *Ecology* 30: 271–287.

COTTAM, G. & CURTIS, J. T. – 1949 – A method for making rapid surveys of woodlands by means of pairs of randomly selected trees. *Ecology* 30: 101–104.

COUPLAND, R. T. – 1958 – The effects of fluctuations in weather upon the grasslands of the Great Plains. *Bot. Rev.* 24: 273–317.

COUPLAND, R. T. – 1959 – Effects of changes in weather conditions upon grasslands in the northern Great Plains. *A.A.A.S. Symposium volume on Grasslands.*

COUPLAND, R. T., SKOGLUND, N. A. & HEARD, A. J. – 1960 – Effects of grazing in the Canadian Mixed Prairie. *Proc. Intern. Grassl. Congress* 8: 212–215.

COUTEAUX, M. – 1962 – Etude palynologique de la tourbière du Buchelbusch à Bonnert et de la tourbière du Heideknapp à Tontelange. *Bull. Soc. Bot. Belg.* 94: 261–278.

COUTEAUX, M. – 1967 – Etude palynologique de tourbières exigues et de sols sableux dans une optique phytosociologique. In: TÜXEN, R. (Ed.): Pflanzensoziologie und Palynologie. Ber. Int. Sympos. Stolzenau/W. 1962: 193–202.

COWLES, H. C. – 1899 – The ecological relations of the vegetation of the sand dunes of the Lake Michigan. *Bot. Gaz.* 27: 95–117, 167–202, 281–308, 361–391.

COWLES, H. C. – 1901 – The physiographic ecology of Chicago and vicinity: A study of the origin, development and classification of plant societies. *Bot. Gaz.* 31: 73–108, 145–182.

COWLES, H. C. – 1910 – The fundamental causes of succession among plant associations. *Rept. Brit. Assoc. Adv. Sci.* 1909: 668–670.

COWLES, H. C. – 1911 – The causes of vegetative cycles. *Bot. Gaz.* 51: 161–183.

COX, C. F. – 1933 – Alpine plant succession on James Peak, Colorado. *Ecol. Monogr.* 3: 299–372.

CRAMPTON, C. B. – 1911 – The vegetation of Caithness considered in relation to the geology. *Ed. Committee for the Survey and Study of British Vegetation*, 132 pp.

CRAMPTON, C. B. – 1912 – The geological relations of stable and migratory plant formations. *Scot. Bot. Rev.* 1: 1–17, 57–80, 127–146.

CRANKSHAW, W. B., QADIR, S. A. & LINDSEY, A. A. – 1965 – Edaphic controls of tree species in presettlement Indiana. *Ecology* 46: 688–698.

CRAWFORD, R. M. M., WISHART, D. & CAMPBELL, R. M. – 1970 – A numerical analysis of high altitude scrub vegetation in relation to soil erosion in the Eastern Cordillera of Peru. *J. Ecol.* 58: 173–191.

CROCKER, R. L. – 1952 – Soil genesis and the pedogenic factors. *Quart. Rev. Biol.* 27: 139–168.

CROCKER, R. L. & DICKSON, B. A. – 1957 – Soil development on the recessional moraines of the Herbert and Mendenhall Glaciers, southwestern Alaska. *J. Ecol.* 45: 169–185.

CROCKER, R. L. & MAJOR, J. – 1955 – Soil development in relation to vegetation

and surface age of Glacier Bay, Alaska. *J. Ecol.* 43: 427–448.
CROCKER, R. L. & WOOD, J. G. – 1947 – Some historical influences on the development of the South Australian vegetation communities and their bearing on concepts and classification in ecology. *Trans. R. Soc. Austral.* 71: 91–136.
CROCKER, W. – 1938 – Life-span of seeds. *Bot. Rev.* 4: 235–274.
CUATRECASAS, J. – 1964 – Cacao and its allies. *Contr. U.S. Nat. Herb.* 35: 379–614.
CURTIS, J. T. – 1955 – A prairie continuum in Wisconsin. *Ecology* 36: 558–566.
CURTIS, J. T. – 1959 – The vegetation of Wisconsin: An ordination of plant communities. 657 pp. Univ. Wisc. Press, Madison.
CURTIS, J. T. & MCINTOSH, R. P. – 1951 – An upland forest continuum in the prairie-forest border region of Wisconsin. *Ecology* 32: 476–496.
DAHL, E. – 1957 – Rondane: Mountain vegetation in South Norway and its relation to the environment. *Skr. Norske Vid. Akad. Oslo, Mat.-Nat. Kl.* 1956(3): 1–374.
DALY, G. T. – 1966 – Nitrogen fixation by nodulated *Alnus rugosa*. *Canad. J. Bot.* 44: 1607–1621.
DAMMAN, A. W. H. – 1971 – Effect of vegetation changes on the fertility of a Newfoundland forest site. *Ecol. Monogr.* 41: 253–270.
DANSEREAU, P. – 1946 – L'érabliere laurentienne. II. Les successions et leurs indicateurs. *Canad. J. Res. C* 24: 235–291.
DANSEREAU, P. – 1954a – Climax vegetation and the regional shift of controls. *Ecology* 35: 575–579.
DANSEREAU, P. – 1954b – Studies on Central Baffin vegetation. I. Bray Island. *Vegetatio* 5, 6.
DANSEREAU, P. – 1956a – Le régime climatique régional de la végétation et les contrôles édaphiques. *Rev. Canad. Biol.* 15: 1–71.
DANSEREAU, P. – 1956b – Le coincement, un processus écologique. *Acta Biotheoretica* 11(3–4): 157–178.
DANSEREAU, P. – 1957 – Biogeography: an ecological perspective. XIII. 394 pp. Ronald Press Comp., New York.
DANSEREAU, P. – 1959 – Phytogeographia laurentiana. II. The principal plant associations of the Saint Lawrence Valley. *Contrib. Inst. Bot. Univ. Montréal* 75: 1-147.
DANSEREAU, P. – 1966 – Ecological impact and human ecology. In: DARLING, F. F. & MILTON, J. P. (Ed.): Future environments of North America. pp. 425–462. Nat. Hist. Press, New York.
DANSEREAU, P. – 1969 – Espoir de l'écologie humaine. *Bull. Comm. Canad. UNESCO* 12(1–2) Suppl.: 14 pp.
DANSEREAU, P. – 1970 – Macaronesian studies. IV. Natural ecosystems of the Azores. *Rev. Géogr. Montréal* 24(1): 21–42.
DANSEREAU, P. & BUELL, P. F. – 1966 – Studies on the vegetation of Puerto Rico. *Inst. Carib. Sci. Spec. Publ.* 1: 287 pp.
DAPPER, H. – 1966 – Zur Stoffproduktivität der Grossen Brennnessel (*Urtica dioica* L.) an einem Ruderalstandort. *Verh. Bot. Ver. Prov. Brandenburg* 103: 54–64.
DAUBENMIRE, R. – 1968 – Plant communities. 300 pp. Harper and Row, New York, London.
DAVIS, C. A. – 1907 – Peat, essays on its origin, uses and distribution in Michigan. *Mich. Geol. Surv. Ann. Report* 1906.
DAVIS, R. M. & CANTLON, J. E. – 1969 – Effect of size area open to colonization on species composition in early old-field succession. *Bull. Torr. Bot. Club* 96: 660–673.
DAWKINS, H. C. – 1971 – Techniques for long-term diagnosis and prediction in forest communities. *Sympos. Brit. Ecol. Soc.* 11: 33–44.

DAY, R. J. – 1972 – Stand structure, succession, and use of southern Alberta's Rocky Mountain forest. *Ecology* 53: 472–478.
DEBRECZY, Z. – 1968 – Die Rolle dei Moosarten in der Vegetationssukzession eines Gebietes des Balatonoberlandes. *Fragm. Bot. Mus. Hist.-Nat. Hung.* 6: 59–66.
DECKER, H. F. – 1966 – Plants. In: MIRSKY, A. (Ed.): Soil development and ecological succession in a deglaciated area of Muir Inlet, southeast Alaska. *Ohio State Univ. Report* 20: 73–96.
DEEVEY, E. S. – 1970 – Commentary: Science and education for environmental management. *Ecology* 51: 363–364.
DELEUIL, G. – 1950 – Mise évidence de substances toxiques pour les thérophytes dans les associations du Rosmarino-Ericion. *C. R. Acad. Sci.* (*Paris*) 230: 1362–1364.
DELEUIL, G. – 1951 – Origine des substances toxiques du sol des associations sans thérophytes du Rosmarino-Ericion. *C. R. Acad. Sci.* (*Paris*) 232: 2038–2039.
DELPECH, R. – 1971 – Observations expérimentales sur l'évolution de la végétation de trois types de pelouses subalpines sous l'influence de facteurs anthro-pogènes. *Cahiers Géogr. Besançon* 21: 63–79.
DEWERS, F. – 1926 – Der Einfluß der Vegetation auf Schichtung und Schichtgrenzen der oberflächlichen diluvialen Ablagerungen. *Abh. Nat. Wiss. Ver. Bremen* 26(2).
DEWERS, F. – 1933 – Die geologische Bedeutung der Pflanzenwurzeln. *Natur und Museum* (8).
DEWERS, F. – 1934/35 – Probleme der Flugsandbildung in Nordwestdeutschland. *Abh. Nat.-wiss. Ver. Bremen* 29(3/4).
DICKSON, B. A. & CROCKER, R. L. – 1953, 1954 – A chronosequence of soils and vegetation near Mt. Shasta, California. *J. Soil Sci.* 4: 123–141, 142–154, 173–191.
DICKSON, C. A. – 1970 – The study of plant macrofossils in British quarternary deposits. In: WALKER, D.&WEST, R. G.: Studies in the vegetational history of the British Isles. pp. 233–254. Cambridge, London.
DIEKJOBST, H. & ANT, H. – 1967 – Die Pioniergesellschaften der Schlammflächen trockengefallener Talsperrensohlen. *Decheniana* 118: 139–144.
DIEMONT, W. H. – 1938 – Zur Soziologie und Synökologie der Buchen- und Buchenmischwälder der nordwestdeutschen Mittelgebirge. *Mitt. Flor.-soz. Arbeitsgem. Niedersachsen* 4: 1–182.
DIEREN, J. W. VAN – 1934 – Organogene Dünenbildung. 304 pp. Den Haag.
DIERSCHKE, H. – 1968 – Syndynamik, ein wichtiger Forschungszweig der Vegetationskunde. *Vegetatio* 15: 388–397.
DIMBLEBY, G. W. – 1954 – The origin of heathland podzols and their conversion by afforestation. *C. R. 8. Congr. Intern. Bot. Paris, Sect.* 13.
DINESMANN, L. G. – 1968 – The study of the history of biogeocoenoses based on burrows of animals. *Bot. Zh.* 53: 214–222.
DOCHMANN, G. I. – 1954 – The vegetation of Mugojary. Geography Press, Moscow.
DOCTERS VAN LEEUWEN, W. M. – 1936 – Krakatau, 1883–1933. *Ann. Bot. Buitenzorg* 46/47: 1–506.
DOING KRAFT, H. – 1954 – L'analyse des carrées permanents. *Acta Bot. Néerl.* 3: 421–424.
DOKUCHAEV, V. V. – 1885 – Russian chernozem. Moscow.
DOKUCHAEV, V. V. – 1892 – Our steppe in past and nowadays. St. Petersburg.
DOKUCHAEV, V. V. – 1899 – On zones of the nature. St. Petersburg.
DOMIN, K. – 1923 – Is the evolution of the earth's vegetation tending towards a

small number of climatic formations? *Acta Bot. Bohem.* 2: 54–60.
DONALD, C. M. – 1963 – Competition among crops and pasture plants. *Adv. Agron.* 15: 1–118.
DOROFEEV, P. I. – 1963 – Neue Daten über die pleistozäne Flora von Weißrußland und der Region um Smolensk. In: SUKACHEV, V. N. (Ed.): Materialien zur Geschichte der Flora und Vegetation der USSR.
DOUGLASS, A. E. – 1936 – Climatic cycles and tree growth, III. *Carneg. Inst. Washingt. Publ.*
DRÄGERT, K. – 1964a – Pflanzensoziologische Untersuchungen in den mittleren Essener Schichten des nördlichen Ruhrgebietes. *Forschungsber. Land Nordrhein-Westfalen* 1963.
DRÄGERT, K. – 1964b – Bibliographia phytosociologica: Palaeosociologica. *Excerpta Botanica* B6: 320.
DRESCHER, W. – 1965 – Aus der Bestands- und Ertragsgeschichte von Beständen des südlichen Hochschwarzwaldes. *Schr.-r. Landesforstverw. Baden-Württembg.* 19: 1–58.
DRIVER, C. H. & GINNES, J. H. – 1969 – Ecology of slash pine stumps: fungal colonization and infection by *Fomes annosus*. *Forest Sci.* 15: 2–10.
DRURY, W. H. – 1956 – Bog flats and physiographic processes in the upper Kuskokwim River region, Alaska. *Contr. Gray Herbarium* 178: 1–130.
DUNCAN, T. & STUCKEY, R. L. – 1970 – Changes in the vascular flora of seven small islands in western Lake Erie. *Mich. Bot.* 9: 175–200.
DURIETZ, G. E. – 1921 – Zur methodologischen Grundlage der modernen Pflanzensoziologie. 267 pp. Wien.
DURNO, S. E. – 1965 – Pollen analytical evidence of "landnam" from two Scottish sites. *Trans. Bot. Soc. Edinb.* 40: 13–19.
DUVIGNEAUD, P. – 1967 – Ecosystèmes et biosphère. 2me éd. 135 pp. Ministère Education Nat. et Culture, Bruxelles (1 éd. 1961).
DUVIGNEAUD, P. – (in press) – Symposium on the productivity of the forest ecosystems of the world in Brussels, Belgium, 1969. UNESCO Publ., Paris.
DYDINA, R. A. – 1954 – Meadowing of the forest- and tundralands. *Doklady VI Sess. Uchen. Sov. Inst. Selsk. Khos. Krayn. Severa* 3.
DYDINA, R. A. – 1957 – The improvement of the botanical composition of the Ob-Irtysch meadows of the Far North. *Bull. Naucho-tekhn. Inform. Inst. Selsk. Khos. Krayn. Severa* 2.
DYKSTERHUIS, E. J. – 1949 – Condition and management of range land based on quantitative ecology. J. *J. Range Mangt.* 2: 104–115.
DYNE, G. M. VAN (Ed.) – 1969 – The ecosystem concept in natural resource management. Academic Press, New York, London.
DYNE, G. M. VAN – 1969b – Grassland management, research, and training viewed in a systems context. *Range Science Series* (*Colorado State Univ.*) 3.
DYNE, G. M. VAN – 1972 – Organization and management of an integrated ecological research program. *Sympos. Brit. Ecol. Soc.* 12: 111–172.
DYRNESS, C. T. – 1969 – Early plant succession following logging and slash burning in *Pseudotsuga* forests in Oregon. *Abstr. Intern. Bot. Congress* 11: 50.
DZIUBALTOWSKI, S. – 1925 – Les associations steppiques sur le plateau de la Petite Pologne et leurs successions. *Acta Soc. Bot. Pol.* 3(2).
EDMONDSON, W. T. – 1969 – The response of Lake Washington to large changes in its nutrient income. *Abstr. Intern. Bot. Congress* 11: 51.
EGGLER, W. A. – 1969 – Comparison of vegetation on sixteen young lava flows on Kilauea and Mauna Loa volcanoes, Hawaii. *Abstr. Intern. Bot. Congress* 11: 52.
EGLER, F. E. – 1942 – Indigene versus alien in the development of arid Hawaiian vegetation. *Ecology* 23: 14–23.

EGLER, F. E. – 1947 – Arid southeast Oahu vegetation, Hawaii. *Ecol. Monogr.* 17: 383–435.

EGLER, F. E. – 1951 – A commentary on American plant ecology, based on the textbooks of 1947–1949. *Ecology* 32: 673–695.

EGLER, F. E. – 1967 – Vegetation management: a breakthrough. *Ecology* 48: 1056.

EGUNJOBI, J. K. – 1969 – Dry matter and nitrogen accumulation in secondary successions involving gorse (*Ulex europaeus* L.) and associated shrubs and trees. *N. Z. J. Sci.* 12: 175–193.

EHRENDORFER, F. – 1965 – Über stammesgeschichtliche Differenzierungsmuster bei den *Dipsacaceae. Ber. Dtsch. Bot. Ges.* 77: (83)–(94).

EHRENDORFER, F. – 1968 – Geographical and ecological aspects of intraspecific differentiation. *Bot. Soc. Brit. Isl. Conf. Rep.* 10: 261–296.

EHRENDORFER, F. – 1970 – Evolution: Pathways and patterns. Evolutionary patterns and strategies in seed plants. *Taxon* 19: 185–195.

EHWALD, E. – 1960 – Das Problem der Bodenentwicklung. *Ber. u. Vortr. Dtsch. Akad. Landw.-wiss. Berlin* 4: 68–97.

EICKE-JENNE, J. – 1960 – Sukzessionsstudien in der Vegetation des Ammersees in Oberbayern. *Bot. Jahrb.* 79: 447–520.

EILART, J. & MASING, V. – 1961 – Taimkatte detailse suur eodulise kaardistamise juhised. *Eesti Loodus* 4,6.

EINARSSON, E. – 1967 – The colonization of Surtsey, the new volcanic island, by vascular plants. *Aquilo Ser. Bot.* 6: 172–182.

EINARSSON, E. – 1970 – Plant ecology and succession in some nunataks in the Vatnajökull glacier in south-east Iceland. *Proc. Helsinki Sympos. UNESCO*: 247–256.

ELLARSON, R. S. – 1949 – The vegetation of Dane County, Wisconsin in 1835. *Trans. Wisc. Acad. Sci. Arts and Letters* 39: 21–45.

ELLENBERG, H. – 1952 – Auswirkungen der Grundwassersenkung auf die Wiesengesellschaften am Seitenkanal westlich Braunschweig. *Angew. Pflanzensoz. (Stolzenau)* 6: 1–46.

ELLENBERG, H. – 1953a – Führt die alpine Vegetations- und Bodenentwicklung auch auf einem reinen Karbonatgestein zum Krummseggenrasen (Caricetum curvulae)? *Ber. Dtsch. Bot. Ges.* 66: 241–246.

ELLENBERG, H. – 1953b – Physiologisches und ökologisches Verhalten derselben Pflanzenarten. *Ber. Dtsch. Bot. Ges.* 65: 350–361.

ELLENBERG, H. – 1954 – Über einige Fortschritte der kausalen Vegetationskunde. *Vegetatio* 5/6: 199–211.

ELLENBERG, H. – 1959 – Typen tropischer Urwälder in Peru. *Schweiz. Z. Forstwes.* 110: 169–187.

ELLENBERG, H. – 1959 – Kausale Vegetationskunde und Grünlandwirtschaft. *Tg.-ber. Dtsch. Akad. Landw.-wiss. Berlin* 18: 43–48.

ELLENBERG, H. – 1963 – Vegetation Mitteleuropas mit den Alpen in kausaler, dynamischer und historischer Sicht. 943 pp. Ulmer, Stuttgart.

ELLENBERG, H. – 1965, 1966 – Vegetationskunde (Soziologische Geobotanik). *Fortschritte Bot.* 27, 28: 289–296.

ELLENBERG, H. – 1968a – Wald- und Feldbau im Knyphauser Wald, eine Heide-Aufforstung in Ostfriesland. *Ber. Naturhist. Ges. Hannover* 112: 17–90.

ELLENBERG, H. – 1968b – Leistung und Haushalt von Lebensgemeinschaften. *Umschau* 1968: 481–485.

ELLENBERG, H. (Ed.) – 1971 – Integrated experimental ecology. *Ecol. Studies* 2: 1–214.

ELLENBERG, H. – 1972 – Welche Ursachen bewirken das Verkahlen von Kulturwiesen auf Island. In: TÜXEN, R. (Ed.): Grundfragen u. Methoden in der

Pflanzensoziologie. Ber. Sympos. Int. Rinteln 1970: 451–463.
ELLENBERG, H. (Ed.) – 1973 – Ökosystemforschung. 280 pp. Springer, Berlin, Heidelberg, New York.
ELLENBERG, H. & KNAPP, R. – 1967 – Vegetationskunde (Soziologische Geobotanik). *Fortschritte Bot.* 29: 352–365.
ELLENBERG, H. & LEBRUN, J. – 1970 – Natural vegetation and its management for rational land use. *Natural Resources Research* (*UNESCO*) 10: 105–122.
ELLENBERG, H. & MUELLER-DOMBOIS, D. – 1966(1967) – Tentative key to a physiognomic-ecological classification of plant formations of the earth. *Ber. Geobot. Inst. E. T. H. St. Rübel* 37: 21–55.
ELLENBERG, H. & REHDER, H. – 1962 – Natürliche Waldgesellschaften auf aufzuforstenden Kastanienflächen im Tessin. *Schweiz. Z. Forstwes.* 113: 128–142.
ELLISON, L. – 1954 – Subalpine vegetation of the Wasatch Plateau. *Ecol. Monogr.* 24: 89–148.
ERDTMAN, G. E. – 1921 – Pollenanalytische Untersuchungen von Torfmooren und marinen Sedimenten in SW-Schweden. *Arkiv Bot.* 17(10): 1–173.
ERNST, A. – 1907 – Die neue Flora der Vulkaninsel Krakatau. *Vierteljahrsschr. Naturf. Ges. Zürich* 52.
ERNST, W. – 1969 – Pollenanalytischer Nachweis eines Schwermetallrasens in Wales. *Vegetatio* 18: 393–400.
ESKUCHE, U. – 1955 – Vergleichende Standortsuntersuchungen an Wiesen im Donauried bei Herbertingen. *J – ber. Ver. Vaterl. Naturkd. Württemberg* 109: 33–135.
EUROLA, S. – 1971 – Die Vegetation einer Sturzhalde (Sveagruva, Spitzbergen). *Aquilo Ser. Bot.* 10: 8–28.
EVANS, F. C. & CAIN, S. A. – 1952 – Preliminary studies on the vegetation of an old-field community in southeastern Michigan. *Contr. Lab. Vertebr. Biol. Univ. Michigan* 51: 1–17.
EVENARI, M. – 1961 – Chemical influences of other plants (allelopathy). *Handb. Pflanzenphysiol.* 16: 691–736.
EWING, J. – 1924 – Plant successions of the brush-prairie in northwestern Minnesota. *J. Ecol.* 12: 238–266.
FAEGRI, K. – 1933 – Über die Längenvariationen einiger Gletscher des Jostedalsbre und die dadurch bedingten Pflanzensukzessionen. *Bergens Mus. Aarbok Naturv. Rekke* 1933(7): 1–255.
FAEGRI, K. – 1954 – On age and origin of the beech forest (*Fagus sylvatica* L.) at Lygerfjorden near Bergen (Norway). *Danm. Geol. Unders.* II 80: 230–249.
FARRON, C. – 1968 – Contribution à la taxonomie des *Ourateaceae* (*Ochnaceae*) d'Afrique. *Candollea* 23: 177–228.
FASSETT, N. C. – 1944 – Vegetation of the Brule basin, past and present. *Trans. Wisc. Acad. Sci. Arts and Letters* 36: 33–64.
FEDOROV, A. A. – 1966 – The structure of the tropical rain forest and speciation in the humid tropics. *J. Ecol.* 54: 1–11.
FEDOROV, V. D. – 1970 – Community of phytoplanktonic organisms and the seasonal variations in their structure. *Bot. Zh.* 55: 626–637.
FEDOSEYEV, A. P. – 1964 – The climate and the pasture herbage of Kazakhstan. Hydrometeorological Press, Leningrad.
FEEKES, W. – 1936 – De ontwikkeling van de natuurlijke vegetatie in de Wieringermeerpolder. *Nederl. Kruidk. Arch.* 46: 1–295.
FEEKES, W. & BAKKER, D. – 1954 – De ontwikkeling van de natuurlijke vegetatie in de Noordoostpolder. *Van Zee tot Land* (*Zwolle*) 6: 1–92.
FIALA, K. & KVĚT, J. – 1971 – Dynamic balance between plant species. *Brit. Ecol. Soc. Sympos.* 11: 241–269.

Finley, D. & Potzger, J. E. – 1952 – Characteristics of the original vegetation in some prairie counties of Indiana. *Butler Univ. Bot. Studies* 10: 114–118.
Firbas, F. – 1949, 1952 – Spät- und nacheiszeitliche Waldgeschichte Mitteleuropas nördlich der Alpen. 1. Band: Allgemeine Waldgeschichte. 480 pp. Jena 1949. 2. Band: Waldgeschichte der einzelnen Landschaften. 256 pp. Jena 1952.
Firbas, F. – 1954 – Über die nachwärmezeitliche Ausbreitung einiger Waldbäume. *Forstwiss. Cbl.* 73, 1/2.
Fishwick, R. W. – 1970 – Sahel and Sudan Zone of Northern Nigeria, North Cameroons and the Sudan. (Afforestation.) *Monogr. Biol. Junk, The Hague* 20: 59–85.
Fitzpatrick, E. A. & Nix, H. A. – 1969 – A model for simulating soil water regime in alternating fallow-crop systems. *Agr. Met.* 6: 303–319.
Florschütz, F. – 1941 – Resultaten van microbotanisch onderzoek van het complex Loodzand-Oerzand en van daaronder en daarboven gelegen afzettingen. Besprekingen over het heidepodsolprofiel gehouden op de bijeenkomst der Sectie Nederland van de Intern. Bodenkundige Vereeniging, pp. 36–56.
Flower-Ellis, J. G. K. – 1971 – Age structure and dynamics in stands of bilberry (*Vaccinium myrtillus* L.). *Avd. Skogsekol. Skogshögskolan Rapp. Upps.* 9: 1–108.
Ford, E. D. & Newbould, P. J. – 1970 – Stand structure and dry weight production through the sweet chestnut (*Castanea sativa* Mill.) coppice cycle. *J. Ecol.* 58: 275–296.
Formozov, A. N., Khadasheva, K. S. & Popova, B. A. – 1954 – The effect of rodents on the vegetation of pastures and hay meadows of typical semideserts of the Volga-Ural watershed. In: The problems of improvement of the forage reserve in the steppe, semidesert and desert zones of the USSR. Acad. Sci. USSR Press, Moscow, Leningrad.
Forrest, W. & Ovington, J. D. – 1970 – Organic matter change in an age series of *Pinus radiata* plantations. *J. Appl. Ecol.* 7: 177–186.
Franke, H. W. – 1969 – Methoden der Geochronologie 132. pp. Springer, Berlin, Heidelberg, New York.
Frankland, J. C. – 1969 – Fungal decomposition of bracken petioles. *J. Ecol.* 57: 25–36.
Freise, F. W. – 1938 – Inselberge und Inselberg-Landschaften in Granit- und Gneisgebiete Brasiliens. *Z. Geomorph.* 10: 137–168.
Freist, H. – 1962 – Untersuchungen über den Lichtungszuwachs der Rotbuche und seine Ausnutzung im Forstbetrieb. *Beih. Forstwiss. Cbl.* 17: 78 p.
Frenzel, B. – 1955 – Die Vegetationszonen Nord-Eurasiens während der postglazialen Wärmezeit. *Erdkunde* 9: 40–53.
Frenzel, B. – 1968 – Grundzüge der pleistozänen Vegetationsgeschichte Nordeurasiens. 326 pp. Wiesbaden.
Frey, E. – 1959 – Die Flechtenflora und -vegetation des Nationalparks im Unterengadin. II. Die Entwicklung der Flechtenvegetation auf photogrammetrisch kontrollierten Dauerflächen. *Ergebn. Wiss. Unters. Schweiz. Nationalpark N.F.* 6(41).
Friedel, H. – 1938a – Die Pflanzenbesiedlung im Vorfeld des Hintereisferners. *Z. Gletscherkde.* 26: 215–239.
Friedel, H. – 1938b – Boden- und Vegetationsentwicklung im Vorfeld des Rhonegletschers. *Ber. Geobot. Inst. Rübel* 1937: 65–76.
Fukarek, F. – 1955 – Die Holzkohlenfunde der Wahlitzer Grabungen I. *Wiss. Abh. Akad. Landw.-wiss. Berlin* 15.
Fukarek, F. – 1961 – Die Vegetation des Darss und ihre Geschichte. *Pflanzensoziologie (Jena)* 12: 1–321.
Fuller, G. D. – 1914 – Evaporation and soil moisture in relation to the succes-

sion of plant associations. *Bot. Gaz.* 58: 193–234.
FURNIVAL, G. M. & WILSON, R. W. – 1971 – Systems of equations for predicting forest growth and yield. *Statistical Ecology* 3: 43–57.
FURRER, E. – 1914 – Vegetationsstudien im Bormiesischen. *Vierteljahrschr. Naturforsch. Ges. Zürich* 59.
FURRER, E. – 1922 – Begriff und System der Pflanzensukzession. *Vj.-Schr. Naturf. Ges. Zürich* 67: 132–156.
FURRER, E. – 1962 – Vegetationsforschung in der Schweiz seit 1900. *Geogr. Helvet.* 17(1).
GAMS, H. – 1918 – Prinzipienfragen der Vegetationsforschung. *Vj.-Schr. Naturf. Ges. Zürich* 63: 293–493.
GAMS, H. – 1936 – Zur Geschichte, klimatischen Begrenzung und Gliederung der immergrünen Mittelmeerstufe. *Veröff. Geobot. Inst. Rübel* 12: 163–204.
GASHWILER, J. S. – 1970a – Further study of conifer seed survival in a western Oregon clearcut. *Ecology* 51: 849–854.
GASHWILER, J. S. – 1970b – Plant and mammal changes on a clearcut in west-central Oregon. *Ecology* 51: 1018–1026.
GAUSSEN, H. – 1951 – Le dynamisme des biocénoses végétales. *Année Biol. 3e Sér.* 27: 89–102.
GAUSSEN, H. & DURRIEU, G. – 1971 – Etude d'une prairie de montagne *à* 48 ans d'intervalle. *Webbia* 25: 465–479.
GEERING, J., FREI, E. & LANINI, F. – 1966 – Risultati di esperimenti di concimazione minerale su prati e pascoli nel Ticino. *Schweiz. Landw. Forsch.* 5: 88–152.
GEHRIS, C. W. – 1971 – Plant community development in Bergen Swamp. *Proc. Rochester Acad.* 12: 98–109.
GÉHU, J.-M. & GÉHU, J. – 1969 – Les associations végétales des dunes mobiles et des bordures de plages de la côte atlantique française. *Vegetatio* 18: 122–166.
GÉHU, J.-M. & PLANCHAIS, N. – 1965 – Evolution de la végétation de quelques landes littorales bretonnes d'après l'analyse pollinique des sols. *Pollen et Spores* 7: 339–360.
GEISLER, S. – 1967 – Verbreitungskarten von Pflanzen im Hohen Westerwald und südlichen Siegerland. *Geobot. Mitteilungen* 44: 1–32.
GERSPER, P. L. – 1968 – Effect of stemflow water on soil formation under beech trees. *Ph. D. Thesis Ohio State Univ., Diss. Abstr.* 29B(3): 840–841.
GERSPER, P. L. & HOLOWAYCHUCK, N. – 1970 – Effects of stemflow water on a Miami soil under a beech tree. *Proc. Amer. Soil Sci.* 34: 779–794.
GERSPER, P. L. & HOLOWAYCHUCK, N. – 1971 – Some effects of stem flow from forest canopy trees on chemical properties of soils. *Ecology* 52: 691–702.
GESSNER, F. – 1932 – Die Entstehung und Vernichtung von Pflanzengesellschaften an Vogelnistplätzen. *Beih. Bot. Cbl.* 49 *Erg.-bd.*: 113–128.
GESSNER, H. & SIEGRIST, R. – 1925 – Über die Auen des Tessinflusses. *Veröff. Geobot. Inst. Rübel* 3: 127–169.
GHOSH, H. R. B. – 1968 – *Caryologia* 21: 111–114.
GIGON, A. – 1971 – Vergleich alpiner Rasen auf Silikat- und auf Karbonatboden. *Veröff. Geobot. Inst.* E T H *St. Rübel* 48: 1–159.
GILLISON, A. N. – 1969 – Plant succession in an irregularly fired grassland area: Doma Peaks region, Papua. *J. Ecol.* 57: 415–428.
GIMINGHAM, C. H. – 1951 – Contributions to the maritime ecology of St. Cyrus, Kincardineshire, II. The sand dunes. *Trans. Bot. Soc. Edinb.* 35: 387–414.
GLAUSER, R. – 1968 – The ecosystem approach to the study of Mt. Shasta mudflows. *Ph. D. Thesis, Univ. California* (1967), *Diss. Abstr. Sect.* B28(10): 3966–3967.
GLEASON, H. A. – 1926 – The individualistic concept of the plant association. *Bull. Torr. Bot. Club* 53: 7–26.

GLEASON, H. A. – 1927 – Further views on the succession-concept. *Ecology* 8: 299–326.
GLEASON, H. A. – 1939 – The individualistic concept of the plant association. *Amer. Midland Natural.* 21: 92–110.
GODWIN, H. – 1929 – The sub-climax and deflected succession. *J. Ecol.* 17: 144–147.
GODWIN, H. – 1936 – Studies in the ecology of Wicken Fen. III. The establishment and development of fen scrub (carr). *J. Ecol.* 24: 82–116.
GODWIN, H. – 1956 – The history of the British flora. Cambridge, London.
GODWIN, H. – 1959 – Studies of the postglacial history of British vegetation. XIV. Late glacial deposits at Moss Lake, Liverpool. *Phil. Trans. R. Soc. B* 242: 127–149.
GODWIN, H. & TURNER, J. S. – 1933 – Soil acidity in relation to vegetational succession in Calthorpe Broad, Norfolk. *J. Ecol.* 21: 235–262.
GODWIN, H. & WILLIS, E. H. – 1959 – Radiocarbon dating of prehistoric wooden trackways. *Nature* 184: 490–491.
GODWIN, H. & WILLIS, E. H. – 1962 – Cambridge University natural radiocarbon measurements. *Radiocarbon* 4: 57–70.
GODWIN, H. & WILLIS, E. H. – 1965 – Cambridge University natural radiocarbon measurements. *Radiocarbon* 7: 205–212.
GOLF, G. – 1973 – Experimentelle Untersuchungen über Pflanzengesellschaften von anthropogenen Rasen in Süd-Chile. Diss. Univ. Giessen.
GOLLEY, F. B. – 1960 – Energy dynamics of a food chain of an old-field community. *Ecol. Monogr.* 30: 187–206.
GOOD, N. F. – 1968 – A study of natural replacement of chestnut in six stands in the highlands of New Jersey. *Bull. Torr. Bot. Club* 95: 240–253.
GOODALL, D. W. – 1967 – Computer simulation of changes in vegetation subject to grazing *J. Ind. Bot. Soc.* 46: 356.
GOODALL, D. W. – 1969 – Simulating the grazing situation. *Biomathematics* (*New York*) 1.
GOODMAN, G. T. & PERKINS, D. F. – 1968 – The role of mineral nutrients in *Eriophorum* communities. III, IV. *J. Ecol.* 66: 667–696.
GORDEYEVA, T. K. & LARIN, I. V. – 1965 – The natural vegetation of the Caspian semidesert, as a forage reserve for animal breeding. "Nauka", Moscow, Leningrad.
GORDIYENKO, I. I. – 1969 – The Oleshian sands and the biogeocoenotic relations in the course of their colonization. 242 pp. Izd. "Naukova Dumka", Kiew.
GORDON, R. W., BEYERS, R. J., ODUM, E. P. & EAGON, R. G. – 1969 – Studies of a simple laboratory microecosystem: bacterial activities in a heterotrophic succession. *Ecology* 50: 86–100.
GORDYAGIN, A. Y. – 1901 – Some data on studying soils and vegetation of the Western Siberia. *Trudy Ob-va Estestv. pri Kazansk. Univ.* 35(2).
GORE, A. J. P. & OLSON, J. S. – 1967 – Preliminary models for accumulation of organic matter in an *Eriophorum-Calluna ecosystem. Aquilo Ser. Bot.* 6: 297–313.
GORHAM, E. – 1957 – Development of peatlands. *Quart. Rev. Biol.* 32: 145–166.
GORODKOV, B. N. – 1926 – Reindeer pastures on the northern part of the Ural district. *Ural* 8.
GORODKOV, B. N. – 1938 – Vegetation of the arctic and of the mountain tundras in the U.S.S.R. *Rastit. SSSR* 1.
GORODKOV, B. N. – 1939 – The botanico-geographical sketch of the Chukotsk shore. *Uch. Zapiski Ped. Inst. Im. Gerzena* 21.
GORODKOV, B. N. – 1950 – The frost cracks of soils in the North. *Izv. Vsesoyuzn. Geogr. Ob-va* 82(5).
GORODKOV, B. N. – 1956 – Vegetation and soils of the Kotyelny Island (New

Siberian Islands). *Rastit. Krain. Sev. SSSR i Ee Osv.* 2.
GORODKOV, B. N. – 1958 – Vegetation and soils of Wrangel Island. *Rastit. Krain. Sev. SSSR i Ee Osv.* 3.
GÖRS, S. – 1968 – Der Wandel der Vegetation im Naturschutzgebiet Schwenninger Moos unter dem Einfluß des Menschen in zwei Jahrhunderten. *Natur-u. Landschaftsschutz – geb. Baden-Württ.* 5: 190–284.
GORTINSKY, G. B. – 1964 – On the factors determining the germination of seeds and the growth of seedlings of spruce (*Picea excelsa* Link.) in the forest of the Southern Taiga subzone. *Bot. Zh.* 49(1).
GORTINSKY, G. B. – 1969 – Essay of the analysis of spruce stands yearly productivity dynamics in the Southern Taiga biogeocoenoses. In: Experimental investigation of biogeocoenoses of Taiga. "Nauka", Moscow, Leningrad.
GORTINSKY, G. B. & TARASOV, A. I. – 1969 – On the geographical conjugation of the annual increment of spruce stands in the Southern Taiga subzone. – In: The mechanism of the interaction of plants in the taiga biogeocoenoses. "Nauka", Leningrad.
GORYSHINA, T. K. – 1971 – Seasonal dynamics of the photosynthesis and productivity in certain summer-green herbage plants of an oakwood in the Forest-Steppe zone. *Bot. Zh.* 56: 62–75.
GOVORUKHIN, V. S. – 1936 – The tundra with spot-medallions in the Northern Ural mountains. *Zemlevedeniye* 38(2).
GOVORUKHIN, V. S. – 1950 – The tundras with spot-medallions and the plicative soils of the North. *Zemlevedeniye* (5).
GRABHERR, W. – 1936 – Die Dynamik der Brandflächenvegetation auf Kalk- und Dolomitböden des Karwendels. *Beih. Bot. Cbl.* 55B: 1–94.
GRAČANIN, Z. – 1962 – Verbreitung und Wirkung der Bodenerosion in Kroatien. *Giessener Abh. Agrar-u. Wirtschaftsf.* 21: 1–400.
GRATKOWSKI, H. – 1969 – Seral shrub communities of western Oregon. *Abstr. Intern. Bot. Congress* 11: 76.
GREENLAND, D. J. & KOWAL, J. M. L. – 1960 – Nutrient content of the moist tropical forest of Ghana. *Plant and Soil* 12: 154–174.
GREGORY, D. & WALKER, D. 1966 – Bibliographia phytosociologica: Tropical hydroseres. *Exc. Bot. B* 7: 109–115.
GREGORY, R. A. – 1960 – The development of forest soil organic layers in relation to time in southeast Alaska. *Alaska Forest. Res. Center, Techn. Note* 47: 1–3.
GREIG-SMITH, P. – 1952 – Ecological observations on degraded and secondary forest in Trinidad, British West Indies. II. Structure of the communities. *J. Ecol.* 40: 316–330.
GRIBOVA, S. A. & ISASHENKO, T. I. – 1972 – The mapping of vegetation in the scale adopted for surveying. *Field Geobotany* 4: 137–330.
GRIBOVA, S. A. & KARPENKO, A. S. – 1964 – Map of the Amur bassin vegetation, scale 1 : 3 000 000 and the methods of its composition. *Geobot. Kartograf.*
GRIBOVA, S. A. & SAMARINA, G. D. – 1963 – Detailed large-scale mapping of the dynamic of vegetational cover. *Geobot. Kartograf.*
GRIGGS, R. F. – 1933 – The colonization of the Katmai ash, a new and inorganic "soil". *Amer. J. Bot.* 20: 92–113.
GRIGORYEV, A. A. – 1925 – Types of the tundra microrelief in the subarctic Eurasia, their geographical distribution and genesis. *Zemlevedeniye* 27(1–2).
GRIGORYEV, A. A. – 1946 – Subarctic. Moscow, Leningrad.
GRIME, J. P. – 1969 – Control of vegetation succession by the techniques of seed exclusion and root strangulation. *J. Appl. Ecol.* 6: 25p.
GRIME, J. P., LOACH, K. & PECKHAM, D. – 1971 – Control of vegetation succession by means of soil fabrics. *J. Appl. Ecol.* 8: 257–263.
GRODZINSKY, A. M. (ed.) – 1967 – Tezisij dokladov vtorogo vsesojusnogo simpo-

zìuma po fiziologo-biochim. osnovam formir. rastitel. soob. (fitocenosov). 96 pp. Naukova Dumka, Kiev.
GRODZINSKY, A. M. (Ed.) – 1970 – Accumulations of colines in some plant communities of virgin steppe as manifestations of chemical interactions of plants. *Akad. Nauk Ukrain. RSR Bot.* 27: 348–352.
GROSSE-BRAUCKMANN, G. – 1962 – Torfe und torfbildende Pflanzengesellschaften. *Z. Kulturtechnik* 3.
GROSSE-BRAUCKMANN, G. – 1967 – Über die Artenzusammensetzung einiger nordwestdeutscher Torfe. In: TÜXEN, R. (Ed.): Pflanzensoziologie und Palynologie. Ber. Int. Sympos. Stolzenau/W. 1962: 160–180.
GROSSE-BRAUCKMANN, G. – 1968 – Einige Ergebnisse einer vegetationskundlichen Auswertung botanischer Torfuntersuchungen, besonders im Hinblick auf Sukzessionsfragen. *Acta Bot. Neerl.* 17: 59–69.
GROSSE-BRAUCKMANN, G. – 1969 – Zur Zonierung und Sukzession im Randgebiet eines Hochmoores (nach Torfuntersuchungen im Teufelsmoor bei Bremen). *Vegetatio* 17: 33–49.
GROSPIETSCH, T. – 1967 – Die Rhizopodenanalyse der Moore und ihre Anwendungsmöglichkeit. In: TÜXEN, R. (Ed.): Pflanzensoziologie u. Palynologie. Ber. Int. Sympos. Stolzenau/W. 1962: 181–192.
GROVE, R. H. & SPECHT, R. L. – 1965 – Growth of heath vegetation. I. Annual growth curves of two heath ecosystems in Australia. *Austral. J. Bot.* 13: 261–270.
GRÜMMER, G. – 1955 – Die gegenseitige Beeinflussung höherer Pflanzen, Allelopathie. 162 pp. Jena.
GUINOCHET, M. – 1955 – Logique et dynamique du peuplement végétal. *Evol. Sci.* 7: 1–143.
GUPTA, R. K. – 1966 – Studies on the succession of the oak-conifer forests of the Garhwal Himalayas. *Trop. Ecol.* 7: 67–83.
GURICHEVA, N. P., KARAMYSHEVA, Z. V. & RACHKOVSKAYA, E. I. – 1967 – On the compilation of a legend to a large-scale vegetation map in the desert steppe subzone of the Kazakhstan. *Geobot. Kartograf.*
GUYOT, L. & MASSENOT, M. – 1950 – Sur la persistance prolongée de semences dormantes dans le sol. . . *C. R. Acad. Sci. (Paris)* 230: 1894–1896.
HAASE, E. F. – 1970 – Environmental fluctuations on southfacing slopes in the Santa Catalina Mountains of Arizona. *Ecology* 51: 959–974.
HABECK, J. R. – 1968 – Forest succession in the Glacier Park cedar-hemlock forests. *Ecology* 49: 872–880.
HABER, W. – 1959 – Vergleichende Untersuchungen der Bakterienzahlen und der Bodenatmung in verschiedenen Pflanzenbeständen. *Flora* 147: 1–34.
HAFSTEN, U. – 1965 – The Norwegian *Cladium mariscus* communities and their post-glacial history. *Arb. Univ. Bergen Mat. Nat. Ser.* 4: 1–55.
HALL, E. A. A., SPECHT, R. L. & EARDLEY, C. M. – 1964 – Regeneration of the vegetation on Koonamore Vegetation Reserve, 1926–1962. *Austral. J. Bot.* 12: 205–264.
HALL, J. B. & POPLE, W. – 1968 – Recent vegetational changes in the lower Volta River. *Ghana J. Sci.* 8: 24–29.
HANAWALT, R. B. – 1969 – Chemical inhibition of annual plants by *Arctostaphylos*. *Abstr. Intern. Bot. Congress Seattle* 11: 84.
HANELT, P. & METTIN, D. – 1970 – *Feddes Rep.* 81: 147–161.
HANES, T. L. – 1969 – Comparisons of postfire succession in coastal and desert chaparral of Southern California. *Abstr. Intern. Bot. Congress* 11: 84.
HANNAN, H. H. & DORRIS, T. C. – 1970 – Succession of a macrophyte community in a constant temperature river. *Limnol. and Oceanogr.* 15: 442–453.
HANSON, H. C. & WHITMAN, W. – 1937 – Plant succession on solonetz soils in

western North Dakota. *Ecology* 18: 516–522.
HANTKE, R. – 1954 – Die fossile Flora der obermiozänen Oehninger Fundstelle Schrotzburg. *Denkschr. Schweiz. Naturf. Ges.* 80: 27–118.
HARPER, J. L. – 1961 – Approaches to the study of plant competition. *Symp. Soc. Exp. Biol.* 15: 1–39.
HARTMANN, F. K. & JAHN, G. – 1967 – Waldgesellschaften des mitteleuropäischen Gebirgsraums nördlich der Alpen. 635 pp. Stuttgart.
HASLAM, S. M. – 1970 – Variation of population type in *Phragmites communis* Trin. *Ann. Bot. N. Ser.* 34(134).
HASTINGS, J. R. & TURNER, R. M. – 1965 – The changing mile. 317 pp. Univ. Arizona Press. Tucson.
HAUSBURG, H. – 1967 – Die Ausbreitung der Fichte im Hornisgrinde-Kniebis-Murg-Gebiet des Nordschwarzwaldes bis etwa 1800. *Mitt. Ver. Forstl. Standortskd. u. Forstpfl.-z.* 17: 3–22.
HAUSSER, K. – 1971 – Düngungsversuche... *Allg. Forst- u. Jagdztg.* 142: 1–11, 69–85, 225–233.
HAYASHI, I. – 1969 – Studies on plant succession at Sugadaira, Central Japan. *Jap. J. Ecol.* 19: 75–79.
HAYASHI, I. & NUMATA, M. – 1964 – Ecological studies on the buried-seed population in the soil from the viewpoint of plant succession. III. A mature stand of *Pinus thunbergii*. *Physiology and Ecology* 12: 185–190.
HAYASHI, I. & NUMATA, M. – 1966 – Ecological studies on the buried-seed population in the soil as related to plant succession, IV. *Ecol. Stud. Biotic Comm. in National Park* 1: 62–71.
HAYASHI, I. & NUMATA, M. – 1968 – Ecology of pioneer species of early stages in secondary succession. II. The seed production. *Bot. Magaz.* (*Tokyo*) 81: 55–66.
HÄYRÉN, E. – 1914 – Landvegetation und Flora der Meeresfelsen von Tvärminne. *Acta Soc. Fauna Flora Fenn.* 39.
HÄYRÉN, E. – 1931 – Aus den Schären Südfinnlands. *Verh. Intern. Ver. Theor. u. Angew. Limnologie* 5.
HEAL, O. W. (Ed.) – 1971 – Working meeting on analyses of ecosystems, Kevo, Finland. IBP Central Office, London.
HEIKKINEN, L. – 1969 – Die *Alchemilla*-Flora der Provinz Kainuu (Ost-Finnland) unter besonderer Berücksichtigung der polemochoren Fernverbreitung der Arten. *Mem. Soc. Fauna et Flora Fenn.* 45: 52–62.
HEILMAN, P. E. – 1966 – Change in distribution and availability of nitrogen with forest succession on north slopes in interior Alaska. *Ecology* 47: 825–831.
HEILMAN, P. E. – 1968 – Relationship of availability of phosphorus and cations to forest succession and bog formation in interior Alaska. *Ecology* 49: 331–336.
HEIM, J. – 1962 – Recherches sur les relations entre la végétation actuelle et le spectre pollinique récent dans les Ardennes belges. *Bull. Soc. Bot. Belg.* 96: 5–92.
HEIM, J. – 1966 – Analyse pollinique d'un podzol à Hergenrath. *Bull. Ass. Fr. Et. Quarternaire* 3: 208–216.
HEISER, C. B. – 1965 – Sunflowers, weeds and cultivated plants. In: BAKER, H. G. & STEBBINS, G. L. (Ed.): The genetics of colonizing species, pp. 391–398. Academic Press, New York and London.
HEJNÝ, S. – 1960 – Ökologische Charakteristik der Wasser- und Sumpfpflanzen in der Slowakischen Tiefebene. 492 pp. Naklad SAV, Bratislava.
HELBAEK, H. – 1954 – Prehistoric food plants and weeds in Denmark. *Danmarks Geol. Unders.* II, 80: 210–229.
HELLER, H. – 1969 – Lebensbedingungen und Abfolge der Flußauenvegetation in der Schweiz. *Mitt. Schweiz. Anst. Forstl. Versuchswesen* 45(1): 1–124.
HERINGER, E. P. & BARROSO, G. M. – 1968 – Sucessão da espécies do cerrado

em função do fogo, do cupim, do cultivo e da subsolagem. *Anais Congr. Nac. Soc. Bot. Brasil* 21: 132–139.

HESMER, H. – 1931 – Untersuchungen zur Waldentwicklung in Pommern unter besonderer Berücksichtigung des natürlichen Fichtenvorkommens. *Z. Forst -u. Jagdw.* 63: 553–574.

HESMER, H. – 1933a – Alter und Entstehung der Humusauflagen in der Oberförsterei Erdmannshausen. *Forstarchiv*: 323–339.

HESMER, H. – 1933b – Die natürliche Bestockung und die Waldentwicklung auf verschiedenen Märkischen Standorten. *Z. Forst- u. Jagdw.* 65: 505–651.

HESMER, H. – 1958 – Wald und Forstwirtschaft in Nordrhein-Westfalen. 540 pp. Hannover.

HESMER, H. – 1960 – Die forstliche Rekultivierung im Rheinischen Braunkohlengebiet. *Hilfe durch Grün* pp. 23–26.

HEUSSER, C. J. – 1956 – Postglacial environments in the Canadian Rocky Mountains. *Ecol. Monogr.* 26: 263–302.

HEYWARD, F. – 1939 – The relation of fire to stand composition of longleaf pine forests. *Ecology* 20: 287–304.

HILD, H. J. – 1956 – Untersuchungen über die Vegetation im Naturschutzgebiet der Krieckenbecker Seen. *Geobot. Mitteilungen* 3 : 1 –112.

HILD, H. J. – 1968 – Die Naturschutzgebiete im nördlichen Rheinland. 106 pp. Recklinghausen.

HOCKENJOS, F. – 1956 – Tannen–Buchen–Fichten–Wirtschaft im Staatswald St. Märgen *D. Forstm. Bad.-Württembg.* (*Karlsruhe*) 1956.

HOCQUETTE, M. – 1927 – Etude sur la végétation et la flore du littoral de la mer du Nord de Nieuport à Sangatte. *Arch. Bot.* 1(4).

HOFMANN, G. – 1958 – Die eibenreichen Waldgesellschaften Mitteldeutschlands. *Arch. Forstwes.* 7: 502–558.

HOFMEISTER, H. – 1970 – Pflanzengesellschaften der Weserniederung oberhalb Bremens. *Dissert. Bot.* 10: 1–116.

HOHENESTER, A. – 1960 – Grasheiden und Föhrenwälder auf Diluvial- und Dolomitsanden im nördlichen Bayern. *Ber. Bayer. Bot. Ges.* 33: 30–83.

HOLT, B. R., – 1972 – Effect of arrival time on recruitment, mortality, and reproduction in successional plant populations. *Ecology* 53: 668–673.

HORN, H. S. – 1971 – The adaptive geometry of trees. *Monogr. Population Biol.* (Princeton) 3: 1–144.

HORNSTEIN, F. v. – 1958 – Wald und Mensch. 2. Aufl. 282 pp. Ravensburg.

HORVÁT, A. O. – 1972 – Die Vegetation des Mecsekgebirges und seiner Umgebung. 376 pp. Akád. Kiadó, Budapest.

HORVAT, I. – 1959 – Die Pflanzenwelt Südosteuropas als Ausdruck der erd- und vegetationsgeschichtlichen Vorgänge. *Acta Soc. Bot. Pol.* 28: 381–408.

HORVAT, I. – 1962 – La végétation des montagnes de la Croatie d'ouest. *Acta Biol.* (*Zagreb.*) 1962.

HOSNER, J. F. & GRANEY, D. L. – 1970 – The relative growth of three forest tree species on soils associated with different successional stages in Virginia. *Amer. Midland Natural.* 84: 418–427.

HUBER, B. – 1954 – Die Jahresringe der Bäume und die Messung der Zeit-Methoden und Ergebnisse der modernen Jahrringchronologie. *Universitas* 9(10).

HUDSON, H. J., – 1962 – Succession of microfungi on ageing leaves of *Saccharum officinarum*. *Trans. Br. Mycol. Soc.* 45: 395–423.

HUDSON, H. J. & WEBSTER, J. – 1958 – Succession of fungi on decaying stems of *Agropyron repens*. *Trans. Br. Mycol. Soc.* 41: 165–177.

HUECK, K. – 1929 – Die Vegetation und die Entwicklung des Hochmoores am Plötzendiebel (Uckermark). *Beitr. Naturdenkmalpfl.* 13: 1–229.

HULT, R. – 1881 – Försök till analytik behandling af växtformationerna. *Medd. Soc. Fauna et Flora Fenn.* 8: 1–155.
HULT, R. – 1885 – Blekinges vegetation: Ett bidrag till växtformationernas utvecklinghistorie. *Medd. Soc. Fauna et Flora Fenn.* 12: 161–251.
HULT, R. – 1887 – Die alpinen Pflanzenformationen des nördlichsten Finnlands. *Medd. Soc. Fauna et Flora Fenn.* 14: 153–228.
HUNDT, R. – 1965 – Ein vegetationskundliches Verfahren zur Ermittlung des Ertragspotentials im Grünland. *Z. Landeskultur* 6: 61–85.
HÜRLIMANN, H. – 1951 – Zur Lebensgeschichte des Schilfs an den Ufern der Schweizer Seen. *Beitr. Geobot. Landesaufn. Schweiz* 30: 1–232.
HUSOVÁ, M. – 1969 – Sukzession der Waldphytozönosen im Zusammenhang mit der Bodenentwicklung auf Amphiboliten. In: TÜXEN, R. (Ed.): Experimentelle Pflanzensoziologie. *Ber. Int. Sympos. Rinteln* 1965: 198–205.
IGNATENKO, I. V. & NORIN, B. N. – 1969 – Dynamics of the tundras with spot-medallions in the easteuropean North. *Problemy Bot.* 11.
IGOSHINA, K. N. – 1934 – Botanical and economical characteristics of reindeer pastures in the district of the Obdorsk zonal station. *Trudy Inst. Polarn. Zemled. Ser. Olenevodstvo* 1.
IGOSHINA, K. N. – 1937 – Grazing forage and forage seasons in the reindeer breeding of the cis-Ural district.
IGOSHINA, K. N. – 1939 – Growth of the fodder lichens in the cis-Ural North. *Trudy Inst. Polarn. Zemled. Ser. Olenevodstvo* 4.
IGOSHINA, K. N. & FLOROVSKAYA, E. F. – 1939 – Utilization of pastures and the reindeer grazing in the cis-Ural district. *Trudy Inst. Polarn. Zemled. Ser. Olenevodstvo* 8.
ILIJANIĆ, L. – 1962 – Beitrag zur Kenntnis der Ökologie einiger Niederungswiesentypen Kroatiens. *Acta Bot. Croat.* 20/21: 95–167.
ILYINA, I. S. – 1968 – A dynamical principle of the compilation a large-scale geobotanical map. In: Geobotanical Mapping. "Nauka", Leningrad.
ILYINSKY, A. P. – 1939 – On the problem of the northern limit of coniferous-broadleafed forest. In: The collection papers. Acad. Sci. USSR Press, Moscow, Leningrad.
IPATOV, V. S. & TARKHOVA, T. N. – 1969 – A study of the variation of the soil cover from year to year in a taiga-type boreal forest. *Bot. Zh.* 54: 1939–1951.
ISACHENKO, T. I. – 1964 – Some methodological aspects of middlescale geobotanical mapping. *Geobot. Kartograf.*
ISACHENKO, T. I. – 1965 – An essay on mapping the dynamics of steppe vegetation (a large-scale mapping of a typical area in Onon-Argun steppe). In: Geobotanical mapping. "Nauka", Leningrad.
ISACHENKO, T. I. – 1967 – On the cartography of serial rows and microbelts in valley and lake-basin. *Geobot. Kartograf.*
IVANOV, V. V. – 1958 – The steppes of Western Kazakhstan in relation to the dynamics of their vegetational cover. Acad. Sci. USSR Press, Moscow, Leningrad.
IVERSEN, J. – 1941 – Landnam i Danmarks Stenalder. *Danmarks Geol. Unders.* II, 66: 1–68.
IVERSEN, J. – 1947 – Plantevaekst, dyreliv og klima i det senglaciale Danmark. *Geol. Fören. Stockholm Förhandl.* 69(1): 67–78.
IVERSEN, J. - 1958 - Pollenanalytischer Nachweis des Reliktencharakters eines jütischen Linden-Mischwaldes. - *Veröff. Geobot. Inst. Rübel* 33: 137-144.
IVERSEN, J. – 1960 – Problems of the early post-glacial forest development in Denmark. *Danm. Geol. Unders.* IV 4 (3): 1–32.
IVERSEN, J. – 1964 – Retrogressive vegetational succession in post-glacial. *J. Ecol.* 52 (*Suppl.*): 59–70.

IVERSEN, J. – 1969 – Retrogressive development of a forest ecosystem demonstrated by pollen diagrams from fossil mor. *Oikos, Suppl.* 12.

IVES, R. L. – 1941 – Forest replacement rates in the Colorado headwaters area. *Bull. Torr. Bot. Club* 68: 407–408.

IWATA, E. & ISHIZUKA, K. – 1967 – Plant succession in Hachirogata polder. Ecological studies on common reed (*Phragmites communis*) I. *Ecol. Rev.* 17: 37–46.

JACKSON, M. T. & PETTY, R. O. – 1969 – Pattern of mesic forest succession along the beech (*Fagus grandifolia* Ehrh.) border of the Indiana-Illinois forest-prairie transition. *Abstr. Intern. Bot. Congress* 11: 99.

JÄGER, H. – 1970 – Der Einfluss des Bergbaus. . . auf die Waldentwicklung in Thüringen. *Wiss. Veröff. Geogr. Inst. Dtsch. Akad. Wiss. N. F.* 27/28: 264–284..

JAHNS, W. – 1969 – Torfmoos-Gesellschaften der Esterweger Dose. *Schr.-R Veget.-kunde* 4: 49–74.

JAKUCS, P. – 1969 – Die Sproßkolonien und ihre Bedeutung in der dynamischen Vegetationsentwicklung (Polycormonsukzession). *Acta Bot. Croat.* 28: 161–170.

JAKUCS, P. – 1972 – Dynamische Verbindung der Wälder und Rasen. 228 pp. Akadémiai Kiadó, Budapest.

JÄNICHEN, H. – 1956 – Die Holzarten des Schwäbisch-Fränkischen Waldes zwischen 1650 und 1800. *Mitt. Ver. Forstl. Standortkd. u Forstpfl.-z.* 1956: 10–31.

JANKOVSKÁ, V. – 1970 – Ergebnisse der Pollen- und Großrestanalyse des Moors "Velanská cesta" in Südböhmen. *Fol. Geobot. Phytotax.* 5: 43–60.

JANSSEN, C. R. – 1960 – On the lateglacial and postglacial vegetation of South Limburg (Netherlands). *Wentia* 4: 1–112.

JANSSEN, C. R. – 1966 – Recent pollen spectra from the deciduous and coniferous-deciduous forests of north-eastern Minnesota: a study in pollen dispersal. *Ecology* 47: 804–825.

JANSSEN, C. R. – 1967 – A comparison between the recent regional pollen rain and the subrecent vegetation in four major vegetation types in Minnesota (U.S.A.). *Rev. Palaeobot. Palynol.* 2: 331–342.

JANSSEN, C. R. – 1970 – Problems in the recognition of plant communities in pollen diagrams. *Vegetatio* 20: 187–198.

JEFFERS, J. N. R. (Ed.) – 1972 – Discussion of the application of mathematical models. *Sympos. Brit. Ecol. Soc.* 12: 345–354.

JENNY, H. – 1941 – Factors of soil formation. McGraw-Hill, New York.

JENNY, H. – 1946 – Arrangement of soil series and types according to functions of the soil forming factors. *Soil Sci.* 61: 375–391.

JENNY, H. – 1958 – Role of the plant factor in the pedogenic functions. *Ecology* 39: 5–16.

JENNY, H. – 1961 – Derivation of state factor equations of soil and ecosystems. *Proc. Soil Sci. Soc. Amer.* 25: 385–388.

JENNY, H. – 1965 – Bodenstickstoff und seine Abhängigkeit von Zustandsfaktoren. *Z. Pflanzenernähr., Düng., Bodenk.* 109: 97–112.

JENNY, H., ARKLEY, R. J. & SCHULTZ, A. M. – 1969 – The pygmy forest-podsol ecosystem and its dune associates of the Mendocino Coast. *Madroño* 20: 60–74.

JENNY, H., GESSEL, S. P. & BINGHAM, F. T. – 1949 – Comparative study of decomposition rates of organic matter in temperate and tropical regions. *Soil Sci.* 68: 419–432.

JENNY, H., SALEM, A. E. & WALLIS, J. R. – 1968 – Interplay of soil organic matter and soil fertility with state factors and soil properties. *Pontificiae Acad. Sci. Scripta Varia* 32: 5–44.

JENNY-LIPS, H. – 1930 – Vegetationsbedingungen und Pflanzengesellschaften auf Felsschutt. *Beih. Bot. Cbl.* 46: 119–296.

JOCHIMSEN, M. – 1963 – Vegetationsentwicklung im hochalpinen Neuland.

(Beobachtungen an Dauerflächen im Gletschervorfeld 1958–1962). *Ber. Naturwiss. Med. Ver. Innsbruck* 53: 109–123.
JOCHIMSEN, M. – 1970 – Die Vegetationsentwicklung auf Moränenböden in Abhängigkeit von einigen Umweltfaktoren. *Veröff. Univ. Innsbruck* 46- 1–22.
JOHNSTON, R. D. & HALL, N. – 1970 – Australia (Afforestation). *Monogr. Biol.* Junk, The Hague 20: 385–414.
JONAS, F. – 1933 – Grenzhorizont und Vorlaufstorf. *Repert. Spec. Nov. Regni Veg. Beih.* 71: 194–215.
JONAS, F. – 1934a – Die paläobotanische Untersuchung brauner Flugsande und deren Stellung im Alluvium. *Repert. Spec. Nov. Regni Veg. Beih.* 76: 153–163.
JONAS, F. – 1934b – Die Entwicklung der Hochmoore am Nordhümmling. *Repert. Spec. Nov. Regni Veg. Beih.* 78: 1–88.
JONAS, F. – 1935 – Die Vegetation der Hochmoore am Nordhümmling, I. *Repert. Spec. Nov. Regni Veg. Beih.* 78(1): 1–143.
JONAS, F. – 1954 – Zur Heidefrage in Westfalen. *Natur u. Heimat* 14: 15–17.
JONASSEN, H. – 1950 – Recent pollen sedimentation and Jutland heath diagrams. *Dansk. Bot. Ark.* 13(7): 1–168.
JONES, A. S. & PATTON, E. G. – 1966 – Forest, "prairie" and soils in the Black Belt of Sumter County, Alabama, in 1832. *Ecology* 47: 75–80.
JONES, E. W. – 1945 – The structure and reproduction of the virgin forests of the north temperate zone. *New Phytol.* 44: 130–148.
JONES, E. W. – 1955 – Ecological studies on the rain forest of southern Nigeria, IV. *J. Ecol.* 43: 564–594.
JONES, J. G. W. – 1970 – The use of models in agricultural and biological research. Ed. Grassland Research Institute, Hurley.
JONES, J. M. – 1969 – The effects of saxicolous plant succession on the weathering of granitic rock. *Abstr. Intern. Bot. Congress* 11: 104.
JONES, K. – 1964 – Chromosomes and the nature and origin of *Anthoxanthum odoratum* L. *Chromosoma (Berlin)* 15: 248–274.
JØRGENSEN, S. – 1954 – A pollen analytical dating of Moglemose find from the bog Aaamosen, Zealand. *Danmarks Geol. Unders.* II, 80: 210–229.
JURASZEK, H. – 1928 – Pflanzensoziologische Studien über die Dünen bei Warschau. *Bull. Acad. Pol. Sci. Lett. Mat. Nat.* B1927: 565–610.
JURKO, A. & MÁJOWSKÝ, J. – 1956 – Die Auenwälder in den Westkarpaten I. *Acta Fac. R. Nat. Univ. Comen. Bot.* I (8/9) 2: 363–385.
KALASHNIKOV, L. N. – 1936 – The main features of the development of the vegetation of the system of ravines in the southeast of the European part of the USSR. *Priroda* 7.
KALELA, A. – 1961 – Waldvegetationszonen Finnlands und ihre klimatischen Paralleltypen. *Arch. Soc. Vanamo* 16(*Suppl.*): 65–83.
KARAMYSHEVA, Z. V. – 1960 – Formation of the steppe vegetation in the stony habitats of the Central Kazakhstan "melkosopochnik" (peneplane). *Bot. Zh.* 45(8).
KARAMYSHEVA, Z. V. – 1963 – The primary successions in the stony habitats of the Central Kazakhstan "melkosopochnik". *Acta Inst. Bot. Acad. Sci. USSR, Ser. III, Geobot.* 15.
KARAMYSHEVA, Z. V. & RACHKOVSKAJA, E. I. – 1962 – An essay on large-scale geobotanical mapping. Acad. Sci. USSR Press, Moscow, Leningrad.
KÁRPATI, I. & KÁRPATI, V. – 1958 – Elm-ash-oak grove forests turning into white poplar dominated stands. *Acta Agron.* 8: 267–283.
KÁRPATI, I. & KÁRPATI, V. – 1965 – Periodische Dynamik der zu den Agropyro-Rumicion crispi gehörenden Gesellschaften des Donau-Überschwemmungsgebietes zwischen Vác und Budapest. *Acta Bot.* 11: 165–196.
KÁRPATI, I. & KÁRPATI, V. – 1969 – Die zönologischen Verhältnisse der Donau-

auenwälder Ungarns. *Verh. Zool.-Bot. Ges. Wien* 108/109: 165–179.
KÁRPATI, I. & KÁRPATI, V. – 1970 – Methodische Probleme in der Erforschung des periodischen Jahresrhytnmus der Pflanzengesellschaften. In: TÜXEN, R. (Ed.): Gesellschaftsmorphologie. Ber. Int. Sympos. Rinteln 1966: 122–133.
KÁRPÁTI, I., PÉCSI, M. & VARGA, G. – 1962 – Zusammenhänge der Vegetation mit der Entwicklung der Inundationsschichten in der Donaubiege. *Bot. Közl.* 49: 299–308.
KÁRPÁTI, I. & SZAKÁLY, J. – 1966 – The connection of the annual periodical rhythm of vegetation with the changes of the blocking, construction, of the horizon in the plant associations of several layers of the Danube Island at Göd. *Vegetatio* 13: 215–232.
KARPENKO, A. S. – 1965 – *Geobot. Kartograph.* 1965.
KARPOV, V. G. – 1961 – The reversion of succession features and their significance for some problems of dynamics of the ground vegetation in taiga forest. *Doklady Akad. Nauk SSSR* 139(5).
KARPOV, V. G. – 1969 – Experimental phytocoenology of the dark coniferous taiga. 335 pp. Izd. Nauk, Leningrad.
KATYSHEVTZEVA, V. G. – 1955 – The vegetation of the Caspian sea coast between the Volga and Ural. *Abstract of Dissertation.*
KATZ, N. J. – 1926 – *Sphagnum* bogs of central Russia: phytosociology, ecology and succession. *J. Ecol.* 14: 177–202.
KAUL, R. N. (Ed.) – 1970 – Afforestation in arid zones. XII 435 pp. Junk, The Hague.
KEAY, R. W. J. – 1959 – Lowland vegetation on the 1922 lava flow, Cameroons Mountain. *J. Ecol.* 47: 25–29.
KEEVER, C. – 1950 – Causes of succession on old fields of the Piedmont, North Carolina. *Ecol. Monogr.* 20: 229–250.
KELLER, P. – 1932 – Der postglaziale Eichenmischwald in der Schweiz und den Nachbargebieten. *Beih. Bot. Cbl.* 49 II: 176–204.
KELLMAN, M. C. – 1970 – Plant species interrelationship in a secondary succession in coastal British Columbia. *Syesis* 2: 201–212.
KELLMAN, M. C. – 1971 – Secondary plant succession in tropical montane Mindanao. 174 pp. Austral. Nat. Univ. Press. Canberra.
KELTIKANGAS, M. & TIILILÄ, P. – 1968 – The economic sequence of common birch (*Betula verrucosa*) and Norway spruce (*Picea abies*) when planting *Oxalis-Myrtillus* type forest land. *Acta Forest. Fenn.* 82: 1–63.
KENFIELD, W. G. – 1966 – The wild gardener in the wild landscape. XI + 232 pp. Hafner, New York.
KENOYER, L. A. – 1929 – Ecological notes on Kalamazoo County, Michigan, based on the original land survey. *Papers Mich. Acad. Sci.* 11: 211–217.
KENOYER, L. A. – 1933 – Forest distribution in southwest Michigan as interpreted from the original survey. *Papers Mich. Acad. Sci.* 19: 107–111.
KENOYER, L. A. – 1939 – Plant associations in Berry, Calhoun, and Branch Counties, Michigan, as interpreted from the original land survey. *Papers Mich. Acad. Sci.* 25: 75–77.
KENOYER, L. A. – 1942 – Forest associations of Ottawa County, Michigan, at the time of the original survey. *Papers Mich. Acad. Sci.* 28: 47–49.
KERNER, A. – 1863 – Das Pflanzenleben der Donauländer. Innsbruck.
KERSHAW, K. A. – 1964 – Quantitative and dynamic ecology. E. Arnold Ltd., London.
KHANTIMER, I. S. – 1951 – Perspective species of the perennial herbs for the grassland crop rotation in the Komi A.S.S.R. In: Agriculture in the Komi A.S.S.R. Moscow.
KHANTIMER, I. S. – 1964 – On the study of the forage problem in the Far North.

Problemy Severa 8.
KICKUTH, R. – 1969 – Höhere Wasserpflanzen und Gewässerreinhaltung. *Schr. R. Ver. Dtsch. Gewässerschutz* 19: 1–14.
KIMMEL, U. – 1964 – Beobachtungen zur vegetativen Entwicklung der Pflanzenarten und zur Regeneration auf Schlagflächen. *Ber. Oberhess. Ges. Naturw. Abt.* 33: 359–371.
KIMURA, M. – 1963 – Dynamics of vegetation in relation to soil development in northern Yatsugatake Mountains. *Jap. J. Bot.* 18: 255–287.
KIRSTE, A. & WALTHER, K. – 1955 – Bestandsverschiebung auf Wiese und Weide unter dem Einfluß von Düngung und Nutzung. *Mitt. Flor. Soz. Arb. gem. N. F.* 5: 104–109.
KISELEVA, K. V., RAZUMOVSKIJ, S. M. & RABITSIN, A. P. – 1969 – Boundaries between plant communities and dynamics of vegetation. *Zh. Ob. Biol.* 30: 123–131.
KITTREDGE, J. – 1934 – Evidence of the rate of forest succession on Star Island, Minnesota. *Ecology* 15: 24–35.
KLAPP, E. – 1958 – Beobachtung von Dauerquadraten in einer Bergheide. *Abh. Naturwiss. Ver. Bremen* 35: 280–295.
KLAPP, E. – 1962 – Ertragsfähigheit und Düngungsreaktion von Wiesenpflanzen-Gesellschaften. *Z. Acker- u. Pflanzenbau* 115: 81–98.
KLAUSING, O. – 1956 – Klimatisch-bodenkundliche Gliederung der natürlichen Eichen- und Buchenwälder in den deutschen Mittelgebirgen: Ein ökologischer Beitrag zum Klimaxproblem. *Ber. Dtsch. Bot. Ges.* 69: 3–20.
KLAUSING, O. – 1959 – Untersuchungen über Vegetation und Wasserhaushalt am Volcan de San Salvador. *Ber. Dtsch. Bot. Ges.* 71: 439–452.
KLIKA, J. – 1931 – Die Pflanzengesellschaften und ihre Sukzessionen auf den entblößten Sandböden in dem mittleren Elbe-Tale. *Sborn. Českosl. Akad. Zem.* 6A: 277–302.
KLIKA, J. – 1936 – Sukzessionen der Pflanzengesellschaften auf den Fluss-Alluvionen der Westkarapathen. *Ber. Schweiz. Bot. Ges.* 46: 248–265.
KNAPP, G. & KNAPP, R. – 1955 – Über Möglichkeiten der Durchsetzung und Ausbreitung von Pflanzenindividuen auf Grund verschiedener Wuchsformen. *Ber. Dtsch. Bot. Ges.* 67: 411–420.
KNAPP, R. – 1953a – Pflanzengesellschaften. Hemmende und fördernde Einflüsse unter Pflanzen. *Umschau* 53: 624–627.
KNAPP, R. – 1953b – Über Zusammenhänge zwischen Polyploidie, Verbreitung, systematischer und soziologischer Stellung von Pflanzenarten in Mitteleuropa. *Zeitschr. Indukt. Abst.-u. Vererbungslehre* 85: 163–179.
KNAPP, R. – 1953c – Über die natürliche Entwicklung von *Arnica montana* L. und ihre Entwicklungsmöglichkeit auf verschiedenen Böden. *Ber. Dtsch. Bot. Ges.* 66: 167–178.
KNAPP, R. – 1955a – Mutual influences between species of grassland plant communities. *Bull. Ecol. Soc. America* 36: 79.
KNAPP, R. – 1955b – Über die Beständigkeit der Artenzusammensetzung von Pflanzengesellschaften. *Feddes Repert.* 58: 220–231.
KNAPP, R. – 1956a – Über den Einfluß der Temperatur auf die Entwicklung der nächsten Generation bei *Agrostemma githago* L. *Naturwissenschaften* 43: 115.
KNAPP, R. – 1957a – Über Wechselwirkungen zwischen den Einflüssen von Temperatur, Licht und Nährstoffkonzentration auf die Pflanzenentwicklung. *Ber. Dtsch. Bot. Ges.* 70: 173–190.
KNAPP, R. – 1957b – Über die Gliederung der Vegetation von Nord-Amerika. *Geobot. Mitteilungen* 4: 1–63.
KNAPP, R. – 1957c – Vegetative Entwicklung und Fertilität einiger annueller Pflanzenarten unter dem Einfluß verschiedener konstanter Temperaturen.

Experientia 13: 405–406.
Knapp, R. – 1958 – Untersuchungen über den Einfluß verschiedener Baumarten auf die unter ihnen wachsenden Pflanzen. *Ber. Dtsch. Bot. Ges.* 71: 411–421.
Knapp, R. – 1959a – Über die gegenseitige Beeinflussung von Pflanzenarten in Trockenrasen und Laubwäldern. *Ber. Dtsch. Bot. Ges.* 72: 368–382.
Knapp, R. – 1959b – Untersuchungen über den Einfluß von Tieren auf die Vegetation, I. *Angewandte Bot.* 33: 177–189.
Knapp, R. – 1960a – Die gegenseitige Beeinflussung von Pflanzen in natürlicher Vergesellschaftung. *Angewandte Bot.* 34: 179–191.
Knapp, R. – 1960b – Gegenseitige Beeinflussung von Pflanzen und ursprüngliche Struktur von Nadelwäldern am Beispiel nord-amerikanischer Bestände. *Ber. Oberhess. Ges. N. F. Naturwiss. Abt.* 30: 74–99.
Knapp, R. – 1961a – Kennzeichnung der sozialen Beziehungen, der gegenseitigen Beeinflussung und der Konkurrenzkraft der Pflanzen bei Vegetations-Analysen. *Ber. Dtsch. Bot. Ges.* 73: 418–428.
Knapp, R. – 1961b – Einfluß der Laubstreu von verschiedenen Pflanzenarten, von einigen Moosen und von Fichten- und Buchen-Asche auf die Keimung. *Geobot. Mitteilungen* 11: 8–10.
Knapp, R. – 1961c – Wirkungen von Behandlungen mit Gibberellinen auf die Entwicklung von Pflanzen. *Angewandte Bot.* 35: 221–258.
Knapp, R. – 1962a – (Ed.) Eigenschaften und Wirkungen der Gibberelline. VIII 275 pp. Springer, Berlin-Göttingen-Heidelberg.
Knapp, R. – 1962b – (Ed.) Untersuchung der Pflanzenentwicklung unter klimatisch kontrollierten Bedingungen. 46 pp. Ulmer, Stuttgart.
Knapp, R. – 1962c – Vegetation des Kleinen Walsertales, Vorarlberg, Nordalpen. *Geobot. Mitteilungen* 12: 1–53.
Knapp, R. – 1964a – Experimental evidence and field observations on factors influencing competition and the importance of competition in the natural distribution of plants. 10 *th Intern. Bot. Congr. Abstr.* pp. 253–254.
Knapp, R. – 1964b – Vegetationsuntersuchungen im mittleren Griechenland und über Probleme der Entstehung anthropogener Pflanzengesellschaften. *Ber. Oberhess. Ges. N. F. Naturwiss. Abt.* 33: 373–393.
Knapp, R. – 1965a – (Ed.) Weide-Wirtschaft in Trockengebieten. 170 pp. G. Fischer, Stuttgart.
Knapp, R. – 1965b – Die Vegetation von Nord- und Mittelamerika und der Hawaii-Inseln. 40+373 pp. G. Fischer, Stuttgart.
Knapp, R. – 1965c – Die Vegetation von Kephallinia, Griechenland. 206 pp. Koeltz, Koenigstein.
Knapp, R. – 1965d – Pflanzengesellschaften und höhere Vegetationseinheiten von Ceylon und Teilen von Ost- und Central-Afrika. *Geobot. Mitteilungen* 33: 1–31.
Knapp, R. – 1966a – Einfluß der Laubstreu einiger afrikanischer Gräser und Trockengehölz-Arten auf die Keimung. *Ber. Dtsch. Bot. Ges.* 79: 329–335.
Knapp, R. – 1966b – Höhere Vegetationseinheiten von West-Afrika. . . *Geobot. Mitteilungen* 34: 1–16.
Knapp, R. – 1967a – Experimentelle Soziologie und gegenseitige Beeinflussung der Pflanzen (incl. Allelopathie). 2. Aufl. 266 pp. Ulmer, Stuttgart. (1. Aufl. 1954).
Knapp, R. – 1967b – Wild und Vegetation in den Tropen, Subtropen und anderen Gebieten. *Ber. Oberhess. Ges. N. F. Naturwiss. Abt.* 35: 157–175.
Knapp, R. – 1967c – Die Vegetation des Landes Hessen. *Ber. Oberhess. Ges. N. F. Naturwiss. Abt.* 35: 93–148.
Knapp, R. – 1967d – Gegenseitige Beeinflussung und Temperaturwirkung bei

tropischen und subtropischen Pflanzen. 64 pp. Cramer, Lehre.
KNAPP, R. – 1968a – Änderung der Artenzusammensetzung von Rasenflächen im Vogelsberg im Verlauf von 19 und 21 Jahren. *Hess. Flor. Br.* 17: 47–52.
KNAPP, R. – 1968–1973 – Vegetationskunde (Soziologische Geobotanik). *Fortschritte Bot.* 30: 340–351, 31: 320–336, 33: 321–335, 34: 419–429, 35.
KNAPP, R. – 1969a – Induktive und experimentelle Untersuchungen über Probleme der Syndynamik von Pflanzengesellschaften. *Vegetatio* 18: 82–90.
KNAPP, R. – 1969b – Änderungen der Vegetation Hessischer Gebirge in den letzten Jahrzehnten. *Mitt. Flor.-soz. Arb.-gem. N. F.* 14: 274–286.
KNAPP, R. – 1970 – Wald und offene Flächen in der Naturlandschaft, Flußauen und Trockenrasen in Alaska und NW-Kanada. *Ber. Oberhess. Ges. N. F. Naturwiss. Abt.* 37: 87–118.
KNAPP, R. – 1971a – Einführung in die Pflanzensoziologie. 3. Aufl. 388 pp. Ulmer, Stuttgart. (1. Aufl. 1948/49, 2. Aufl. 1958).
KNAPP, R. – 1971b – Wirkungen von Rauhreif und Rauhfrost auf die Vegetation. *Oberhess. Naturwiss. Z.* 38: 101–110.
KNAPP, R. – 1971c – Influence of indigenous animals on the dynamics of vegetation in conservation areas. *Symp. Brit. Ecol. Soc.* 11: 387–390.
KNAPP, R. – 1971d – Pflanzenverteilung und Pflanzungsmuster in ihrer Wirkung auf gegenseitige Beeinflussung und Bestandesstruktur. *Angewandte Bot.* 45: 47–63.
KNAPP, R. – 1971e – Die Pflanzenwelt der Rhön. 127 pp. Cramer, Lehre.
KNAPP, R. – 1973 – Die Vegetation von Afrika. 669(43+626)pp. G. Fischer, Stuttgart.
KNAPP, R. & FURTHMANN, S. – 1954 – Experimentelle Untersuchungen über die Bedeutung von Hemmstoffen für das Wachstum und die Vergesellschaftung höherer Pflanzen. *Ber. Dtsch. Bot. Ges.* 67: 252–269.
KNAPP, R. & LIETH, H. – 1952 – Über Ursachen des verstärkten Auftretens von erdbewohnenden Cyanophyceen. *Arch. Mikrobiol.* 17: 292–299.
KNAPP, R. & LINSKENS, H. F. – 1954 – Über Aminosäuren aus der Blattstreu einiger Pflanzenarten von Wäldern auf verschiedenen Böden. *Naturwissenschaften* 41: 480–481.
KNAPP, R. & STOFFERS, A. L. – 1962 – Über die Vegetation von Gewässern und Ufern im mittleren Hessen. *Ber. Oberhess. Ges. N. F. Naturwiss. Abt.* 32: 90–141.
KNAUER, N. – 1969 – Veränderung der Artenzusammensetzung verschiedener Grünland-Pflanzengesellschaften durch Düngung mit Phosphat, Kali oder Kalk. In: TÜXEN, R. (Ed.): Experimentelle Pflanzensoziologie. Ber. Int. Sympos. Rinteln 1965: 63–72.
KNOBLOCH, E. – 1969 – Tertiäre Floren von Mähren. Moravske Musejni Spol. (*Brno*).
KNÖRZER, K. - H. – 1962 – Ein Beispiel für die Anwendung phytosoziologischer Kenntnisse bei der Grabungsforschung. *Bonner H. D. Rhein. Landesmus. u. Ver. Altertumsfreunden Rheinl.* 162: 260–265.
KNÖRZER,K. - H. – 1964 – Über die Bedeutung von Untersuchungen subfossiler pflanzlicher Großreste. *Bonner Jb.* 164: 202–214.
KNÖRZER, K. - H. – 1968a – Ein Teilergebnis der Untersuchung pflanzlicher Großreste bei der Ausgrabung an der Niederungsburg bei Haus Meer, Gemeinde Büderich. *Rheinische Ausgrabungen* 1. *Beih. Bonner* Jb. 28: 97–100.
KNÖRZER, K. - H. – 1968b – Untersuchung der Pflanzenreste. *Rheinische Ausgrabungen* 1. *Beih. Bonner* Jb. 28: 137–169.
KNÖRZER, K. - H. – 1971 – Urgeschichtliche Unkräuter im Rheinland. Ein Beitrag zur Entstehungsgeschichte der Segetalgesellschaften. *Vegetatio* 23: 89-111.
KOCH, H. – 1930 – Stratigraphische und pollenfloristische Studien an drei nordwestdeutschen Mooren. *Planta* 11: 509–527.

Koch, H. – 1934a – Mooruntersuchungen im Emsland und im Hümmling. *Int. Rev. Hydrobiol.* 31: 109–156.
Koch, H. – 1934b – Ein Profil aus dem Bourtanger Moor als Beispiel zur Moor- und Waldgeschichte an der Mittelems. *Ber. Dtsch. Bot. Ges.* 52: 101–109.
Koch, W. – 1926 – Die Vegetationseinheiten der Linthebene. *Jb. Naturwiss. Ges. St. Gallen* 61, 144 pp.
Koenekamp, A. & König, F. – 1929 – Über den Einfluß wirtschaftlicher Maßnahmen auf den Pflanzenbestand des Grünlandes. *Landw. Jb.* 70: 61–88.
Kohler, A. – 1969,1970 – Geobotanische Untersuchungen an Küstendünen Chiles zwischen 27. und 42. Grad S. Breite. 186 pp. Habil.-schrift Weihenstephan. (polykop.). *Bot. Jahrb.* 90: 55–200.
Kohler, A. & Sukopp, H. – 1964 – Über die Gehölzentwicklung auf Berliner Trümmerstandorten. *Ber. Dtsch. Bot. Ges.* 76: 389–406.
Kokin, K. A., Koljtzova, T. I. & Khlebovich, T. V. – 1970 – The composition and the dynamics of the phytoplankton along the Karelian coast of the White Sea. *Bot. Zh.* 55: 499–509.
Kolumbe, E. & Beyle, M. – 1938 – Die Bihlwege im Wittmoor (Holstein) und ihre Stellung im Pollendiagramm. *Aus Hansischem Raum. Sonderhefte der Hansischen Gilde*, 33 pp.
Körber-Grohne, U. – 1966 – Geobotanische Untersuchungen an einer prähistorischen Ausgrabung in der Marsch. In: Tüxen, R.: Ber. Int. Sympos. Stolzenau/W. 1961: 181–182.
Körber-Grohne, U. – 1967 – Geobotanische Untersuchungen auf der Feddersen Wierde. 357 pp. Wiesbaden.
Korchagin, A. A. – 1937 – Vegetation of the marine alluvia of Mesen bay and Cheshskaya bay. *Trudy Bot. Inst. Akad. Nauk SSSR, Ser. 3, Geobot.* 2.
Korchagin, A. A. – 1960a – Determination of the age of trees in the temperate zone. *Field Geobotany (Acad. Sci. SSSR, Leningrad)* 2.
Korchagin, A. A. – 1960b – Determination of the age and longevity of shrubs. *Field Geobotany (Acad. Sci. USSR, Leningrad)* 2.
Korchagin, A. A. – 1960c – Estimation of the age and longevity of mosses and liverworthes. *Field Geobotany (Acad. Sci. USSR, Leningrad)* 2.
Korchagin, A. A. – 1960d – Methods of determination of the seed productivity of forest trees and forest communities. *Field Geobotany (Acad. Sci. USSR, Leningrad)* 2.
Korchagin, A. A., Gorchakovsky, P. L. & Matveyeva, E. P. – 1968 – Symposium on the dynamics of vegetation in the FRG (20–23 March 1967). *Bot. Zh.* 53: 1669–1676.
Korchagin, A. A. & Neistadt, M. I. – 1966 – Vegetation. In: North of the European part of the USSR. "Nauka", Moscow.
Košir, Ž. – 1970 – Beitrag zur Erforschung der Urwaldstruktur reiner Buchenwälder. In: Tüxen, R. (Ed.): Gesellschaftsmorphologie. Ber. Int. Sympos. Rinteln 1966: 306–314.
Kovakina, V. A. – 1958 – Biological peculiarities of some wintergreen plants in the Far North. *Bot. Zh.* 43(9).
Kozhevnikov, A. V. – 1937 – Certain principles of governing the seasonal development of plant associations. *Sci. Papers Moscow Univ.* 11.
Krajina, V. J. – 1969 – Ecology of forest trees in British Columbia. *Ecol. W. North America* 2(1): 1–146.
Krasnov, A. N. – 1888 – An essay of history of development of flora of the southern part of Eastern Tyan-Shan. *Zapiski Russk. Geograf. Ob-va* 19.
Krasnov, A. N. – 1894 – Grass-steppe of the northern hemisphere. *Izv. Ob-va Lyubit. Estestv. Anthrop. Ethnogr.* 83.
Krausch, H. - D. – 1964, 1967, 1968, 1970 – Die Pflanzengesellschaften des

Stechlinsee-Gebietes, I-V. *Limnologica (Berlin)* 2: 145–203, 423–482; 5: 331–366; 6: 321–380: 7: 397–454.
KRAUSE, W. – 1952 – Das Mosaik der Pflanzengesellschaften und seine Bedeutung für die Vegetationskunde. *Planta* 41: 240–289.
KRIPPEL, E. – 1954 – Die Pflanzengesellschaften auf Flugsandböden des slowakischen Teiles des Marchfeldes. *Angew. Pfl. soz. (Wien) Festschr.* 1: 635–645.
KUBITZKI, K. – 1960 – Moorkundliche und pollenanalytische Untersuchungen am Hochmoor "Esterweger Dose". *Schr. Naturw. Ver. Schleswig-Holst.* 30: 12–28.
KÜCHLER, A. W. – 1969 – The vegetation of Kansas on maps. *Tr. Kansas Acad. Sci.* 72: 141–166.
KÜCHLER, A. W. – 1972 – The oscillations of the mixed prairie in Kansas. *Erdkunde* 26: 120–129.
KÜHNHOLTZ-LORDAT, G. – 1923 – Les dunes du Golfe de Lion. Paris 1923,
KUJALA, V. – 1926 – Untersuchungen über die Waldvegetation in Süd- und Mittelfinnland. *Comm. Inst. Forest. Fenn.* 10.
KUJALA, V. – 1945 – Waldvegetationsuntersuchungen in Kanada. *Ann. Acad. Sci. Fenn., Ser. A4, Biol.* 7: 1–434.
KÜNZLI, W. – 1967 – Über die Wirkung von Hof- und Handelsdüngern auf Pflanzenbestand, Ertrag und Futterqualität der Fromentalwiese. *Schweiz. Landw. Forsch.* 6: 34–130.
KURIMO, U. – 1970 – Effect of pollution on the aquatic macroflora of the Varkaus area, Finnish Lake District. *Ann. Bot. Fenn.* 7: 213–254.
KUROCHKINA, L. J. – 1966 – The vegetation of the sandy deserts of Kazakhstan. In: The vegetational cover of Kazakhstan. Vol.1. "Nauka", Alma-Ata.
KWASNITSCHKA, K. – 1955 – Die Entwicklungsdynamik der Mischbestände auf dem Buntsandstein des Ostschwarzwaldes. *Forstwiss. Cbl.* 74: 65–128.
LA MARCHE, V. C. – 1969 – Environment in relation to age of bristlecone pines. *Ecology* 50: 53–59.
LAMBINON, J. – 1956 – Aperçu sur les groupements végétaux du district maritime belge entre la Panne et Coxyde. *Bull. R. Soc. Bot. Belg.* 88: 107–127.
LANG, G. – 1955 – Über spätquartäre Funde von *Isoetes* und *Najas flexilis* im Schwarzwald. *Ber. Dtsch. Bot. Ges.* 68: 24–27.
LANG, G. – 1962 – Vegetationsgeschichtliche Untersuchungen der Magdalénienstation an der Schussenquelle. *Veröff. Geobot. Inst. ETH. St. Rübel* 37: 129–154.
LANG, G. – 1967 – Über die Geschichte von Pflanzengesellschaften auf Grund quartärbotanischer Untersuchungen. In: TÜXEN, R. (Ed.): Pflanzensoziologie und Palynologie. Ber. Int. Sympos. Stolzenau/W. 1962: 24–37.
LANG, G. – 1967 – Die Ufervegetation des westlichen Bodensees. *Arch. Hydrobiol. Suppl.* 32: 437–574.
LANGE, E. & HEINRICH, W. – 1970 – Floristische und vegetationskundliche Beobachtungen auf dem MTB Frankenberg/Sa. *Hercynia N. F.* 7: 53–86.
LANGE, E. & SCHLÜTER, H. – 1972 – Zur Entwicklung eines montanen Quellmoores im Thüringer Wald und des Vegetationsmosaiks seiner Umgebung. *Flora* 161: 562–585.
LANGENHEIM, J. H. – 1956 – Plant succession on a subalpine earthflow in Colorado. *Ecology* 37: 301–317.
LANGFORD, A. N. & BUELL, M. F. – 1969 – Integration and stability in the plant association *Adv. Ecol. Res.* 6.
LANZ, W. – 1969 – Forstdüngung. *Forstarchiv Beih.* 2, 1–36.
LARIN, I. V. – 1952 – Investigation of the effect of grazing on the vegetation. In: A brief manual of geobotanical investigations associated with shelter belt planting and the establishment of a stable forage reserve in the south of the European part of the USSR. Acad. Sci. USSR Press, Moscow, Leningrad.
LARSEN, J. A. – 1929 – Fire and forest succession in the Bitterroot Mountains of

northern Idaho. *Ecology* 10: 67–76.
Launchbaugh, J. L. – 1955 – Vegetational changes in the San Antonio prairie. *Ecol. Monogr.* 25: 39–57.
Lavrenko, E. M. – 1940 – The steppes of the USSR. In: The vegetation of the USSR. Vol. 2. Acad. Sci. USSR Press, Moscow, Leningrad.
Lavrenko, E. M. – 1947 – Principles and units of the geobotanical regionalization. Russ. In: Geobot. Rayonirovaniye SSSR. Moscow, Leningrad.
Lavrenko, E. M. – 1950 – Some observations of the effect of fire on the vegetation of the northern steppe. *Bot. Zh.* 35.
Lavrenko, E. M. – 1952 – *Trudy Bot. Inst. Akad. Nauk SSSR Ser. III Geobot.* 8.
Lavrenko, E. M. – 1957 – Vegetation of the Gobi deserts in the M.N.R. and its relation with the recent geological processes. *Bot. Zh.* 42(9).
Lavrenko, E. M. – 1959 – Main principles governing plant communities and the ways of their investigation. *Field Geobotany.* (*Moscow, Leningrad*) 1.
Lavrenko, E. M. & Yunatov, A. A. – 1952 – The fallow regime in steppes as the result of the effect of the *Microtus brandtii* on the steppe grass stand and on the soil. *Bot. Zh.* 37(2).
Lawrence, D. B. – 1958 – Glaciers and vegetation in southeastern Alaska. *Amer. Sci.* 46: 89–122.
Laws, R. M. – 1970 – Elephants as agents of habitat and landscape change in East Africa. *Oikos* 21: 1–15.
Lawton, R. M. – 1971 – Destruction or utilization of a wildlife habitat? *Sympos. Brit. Ecol. Soc.* 11: 333–336.
Leak, W. B. – 1970 – Successional change in northern hardwoods predicted by birth and death simulation. *Ecology* 51: 794–801.
Lebrun, J. – 1960 – Etudes sur la flore et la végétation des champs de lave au nord du lac Kivu. *Expl. Parc Nat. Albert* (*Bruxelles*) 2.
Leeuwen, C. G. van – 1966 – A relation to theoretical approach to pattern and process in vegetation. *Wentia* 15: 25–46.
Lemberg, B. – 1933 – Über die Vegetation der Flugsandgebiete an den Küsten Finnlands. I: Die Sukzession. *Acta Bot. Fenn* 12: 1–143.
Lemée, G. & Bichaut, N. – 1971 – Recherches sur les écosystems des réserves biologiques de la fôret de Fontainebleau. I. *Oecol. Plant.* 6: 133–149.
Lenoble, F. – 1926 – A propos des associations végétales. *Bull. Soc. Bot. France* 73: 873–893.
Léonard, A. – 1959 – Contribution à l'étude de la colonisation des laves du volcan Nyamuragira par les végétaux. *Vegetatio* 8.
Leontiev, V. L. – 1952 – Development of understory and overstory of black saxaul (*Haloxylon aphyllum* (Minkw.) Iljin) in the Kara-Kum desert. *Trudy Bot. Inst. Akad. Nauk SSSR, Ser. III, Geobot.* 8: 7–19.
Lewis, H. & Raven, P. H. – 1958 – Rapid evolution in *Clarkia*. *Evolution* 12: 319–336.
Lieth, H. (Ed.) – 1962 – Die Stoffproduktion der Pflanzendecke. 156 pp. G. Fischer, Stuttgart.
Lieth, H., Osswald, D. & Martens, H. – 1965 – Stoffproduktion, Spross-Wurzel-Verhältnis, Chlorophyllgehalt und Blattfläche von Jungpappeln. *Mitt. Ver. Forstl. Standortskd. u. Forstpfl.-z.* 15: 70–74.
Lindeman, R. L. – 1942 – The trophic-dynamic aspect of ecology. *Ecology* 23: 397–418.
Lindsey, A. A. – 1960 – The landscape and vegetation of the Docket 254 area in 1830. Report to Lands Division, U.S. Dept. of Justice (Mimeo).
Lindsey, A. A., Crankshaw, W. B. & Qadir, S. A. – 1965 – Soil relations and distribution map of the vegetation of presettlement Indiana. *Bot. Gaz.* 126: 155–163.

LINSKENS, H. F. & KNAPP, R. – 1955 – Über die Ausscheidung von Aminosäuren in reinen und gemischten Beständen verschiedener Pflanzenarten. *Planta* 45: 106–117.

LIVINGSTON, R. B. & ALLESSIO, M. L. – 1968 – Buried viable seed in successional field and forest stands, Harvard Forest, Massachusetts. *Bull. Torr. Bot. Club* 95: 58–69.

LLOYD-BINNS, B. – 1969 – Bioseres of Malawi. *Abstr. Intern. Bot. Congress* 11: 130.

LOEWE, F. – 1966 – Climate. In: MIRSKY, A. (Ed.): Soil development and ecological succession in a deglaciated area of Muir Inlet, southeast Alaska. *Ohio State Univ. Report* 20: 18–27.

LOHMEYER, W. – 1953 – Beitrag zur Kenntnis der Pflanzengesellschaften der Umgebung von Höxter. *Mitt. Flor.-soz. Arb.-gem. N.F.* 4: 59–76.

LOHMEYER, W. – 1960 – Bericht über die Aussaatversuche auf den offenen Quarzsandhalden im Bergwerksgelände der Mechernicher Werke. *Hilfe durch Grün H.* 9: 61–62.

LONA, F. & FOLLIER, M. – 1958 – Successione pollinica della serie superiore (Günz-Mindel) di Leffe (Bergamo). *Veröff. Geobot. Inst. Rübel* 34: 86–98.

LONDO, G. – 1971 – Pattern and process in dune slack vegetations along an excavated lake in the Kennemer dunes (the Netherlands). *Rijksinst. Natuurbeheer Verh.* (*Leersum*) 2: 1–279.

LOPATIN, V. D. – 1956 – The similarity index and its use for vegetation dynamic analysis. *Problemy Sov. Bot.* 1.

LOPATIN, V. D. – 1967 – To comparison of the methods for dynamic rates measuring in meadow vegetation. *Bot. Zh.* 62(7).

LOSERT, H. – 1953 – Pollenanalytische Untersuchungen am "Blanken Flat" bei Vesbeck. *Mitt. Flor.-soz. Arb.-gem. N.F.* 4.

LOSSAINT, P. – 1959 – Etude expérimentale de la mobilisation du fer des sols sous l'influence des litières forestières. 107 pp. *Thèse Univ. Strasbourg.*

LÖTSCHERT, W. – 1952 – Vegetation und pH-Faktor auf kleinstem Raum in Kiefern- und Buchenwäldern auf Kalksand, Löß und Granit. *Biol. Zbl.* 71: 327–348.

LÖTSCHERT, W. – 1969 – Pflanzen an Grenzstandorten. X, 167 pp. Fischer, Stuttgart.

LOUCKS, O. L. – 1970 – Evolution of diversity, efficiency, and community stability. *Amer. Zoologist* 10: 17–25.

LÖVE, À. & LÖVE, D. – 1971 – Polyploidie et géobotanique. *Nat. Canad.* 98: 469–494.

LOVELIUS, N. V. – 1970 – The effect of the volcanic activity on the vegetation of Kamchatka. *Bot. Zh.* 55: 1630–1633.

LÜDI, W. – 1919 – Die Sukzession der Pflanzenvereine. *Mitt. Naturforsch. Ges. Bern* 1919: 9–87.

LÜDI, W. – 1921 – Die Pflanzengesellschaften des Lauterbrunnentales und ihre Sukzession. *Beitr. Geobot. Landesaufn. Schweiz* 9: 1–364.

LÜDI, W. – 1923 – Die Untersuchung und Gliederung der Sukzessionsvorgänge in unserer Vegetation. *Verhandl. Naturforsch. Ges. Basel* 35: 277–302.

LÜDI, W. – 1930 – Die Methoden der Sukzessionsforschung in der Pflanzensoziologie. *Handb. d. Biol. Arb.-meth.* 11(5): 527–728.

LÜDI, W. – 1940 – Die Veränderungen von Dauerflächen in der Vegetation des Alpengartens Schinigeplatte innerhalb des Jahrzehnts von 1928/29–1938/39. *Ber. Geobot. F.-Inst. Rübel* 1939: 93–148.

LÜDI, W. – 1945 – Besiedlung und Vegetationsentwicklung auf den jungen Seitenmoränen des grossen Aletsch-Gletschers . . . *Ber. Geobot. F.-Inst. Rübel* 1944: 35–112.

LÜDI, W. – 1955 – Die Vegetationsentwicklung seit dem Rückzug der Gletscher

in den mittleren Alpen und ihrem nördlichen Vorland. *Ber. Geobot. Inst. Rübel* 1954.
LÜDI, W. – 1958 – Beobachtungen über die Besiedlung von Gletschervorfeldern in den Schweizeralpen. *Flora* 146: 386–407.
LUNDEGÅRDH-ERICSON, C. – 1972 – Changes during four years in the aquatic macro-vegetation in a flat in N Stockholm Archipelago. *Svensk Bot. Tidskr.* 66: 207–225.
LUTHER, H. – 1961 – Veränderungen in der Gefäßpflanzenflora der Meeresfelsen von Tvärminne. *Acta Bot. Fenn.* 62: 1–100.
LUTZ, H. J. – 1930 – Forest composition in northwest Pennsylvania as indicated by early land survey notes. *J. Forestry* 28: 1098–1103.
LUX, H. – 1969 – Zur Biologie des Strandhafers (*Ammophila arenaria*) und seiner technischen Anwendung in Dünenbau. In: TÜXEN, R. (Ed.): Experimentelle Pflanzensoziologie. Ber. Int. Sympos. Rinteln 1965: 138–144.
MAC FADYEN, W. A. – 1950 – Vegetation patterns in the semidesert plains of British Somaliland. *Geogr. J.* 116: 199–211.
MÄDLER, K. – 1939 – Die pliozäne Flora von Frankfurt am Main. *Abh. Senckenberg. Naturf. Ges.* 446: 1–202.
MAHN, E.-G. – 1966 – Beobachtungen über die Vegetations- und Bodenentwicklung eines durch Brand gestörten Silikattrockenrasen-Standortes. *Arch. Natursch. Landsch.-Forsch.* 6: 61–90.
MAHN, E.-G. – 1969 – Untersuchung zur Bestandsdynamik einiger charakteristischer Segetalgesellschaften unter Berücksichtigung des Einsatzes von Herbiziden. *Arch. Natursch. Landsch.-Forsch.* 9: 3–42.
MAJOR, J. – 1951 – A functional, factorial approach to plant ecology. *Ecology* 32: 392–412.
MAJOR, J. – 1953 – The relationship between factors of soil formation and vegetation, with an analysis from the west slope of Mt. Hamilton, California. *Ph. D. Thesis Univ. California, Berkeley.*
MAJOR, J. – 1958 – Plant ecology as a branch of botany. *Ecology* 39: 352–363.
MAJOR, J. – 1970 – Essay review of RODIN & BAZILEVICH: The illusive mineral equilibrium. *Ecology* 51: 160–163.
MÁLEK, J. – 1966 – Vegetationsentwicklung auf den Lokalitäten der im 15. und 16. Jahrhundert eingegangenen Ortschaften in Südwestmähren. *Acta Musei Moraviae* 51: 153–180.
MÁLEK, J. – 1970 – Entwicklung der Wälder Südwestmährens unter dem Einfluß des Menschen. *Acta Sci. Nat. Acad. Sci. Bohemosl., Brno, N. Ser.* 4(5): 1–45.
MALYSHEVA, G. S. – 1968 – Methodical text-book on the composing of phytophenological maps. "Nauka", Leningrad.
MARBUT, C. F. – 1932 – Morphology of laterites. *Proc. 2nd Int. Congr. Soil Sci.* 1930: 72–80.
MARGALEF, R. – 1963 – On certain unifying principles in ecology. *Amer. Nat.* 97: 357–374.
MARSCHNER, J. F. – 1930 – Original forests of Minnesota (Map compiled from land office survey notes on file with U.S. Forest Service, USDA, Washington, D.C.).
MARSHALL, D. R. & JAIN, S. K. – 1970 – Seed predation and dormancy in the population dynamics of *Avena fatuna* and *A. barbata. Ecology* 51: 886–891.
MARSHALL, J. K. – 1968 – Factors limiting the survival of *Corynephorus canescens* . . in Great Britain . . . *Oikos* 19: 206–216.
MARTENS, R. – 1969 – Quantitative Untersuchungen zur Gestalt, zum Gefüge und Haushalt der Naturlandschaft (Imoleser Subapennin). *Hamburger Geogr. Schr.* 21: 1–238.
MARTIN, N. D. – 1959 – An analysis of forest succession in Algonquin Park,

Ontario. *Ecol. Monogr.* 29: 187–218.
MARVET, A. V. – 1964 – Some methods of reflecting the vegetation dynamic on large-scale geobotanical maps. *Progr. i Tez. Nauchn. Konf. Posv. 100 Let. Dnya Rozhd. N.I. Kuznetzova, Tartu.*
MARVET, A. V. – 1967 – Reflecting of the vegetation dynamic on large-scale maps. Tartu.
MARVET, A. V. – 1968 – On the construction of a legend to detailed large-scale maps, reflecting the vegetation dynamic. *Geobot. Kartograf.*
MASING, V. V. – 1968 – Classification rows of the territorial units in geobotany. *Uchen. Zapisky Tart. Univ.* 211.
MATHON, C. C. – 1949 – L'autodynamisme des complexes écologiques et les groupements végétaux durables. *Feuille Nat.* 4(9/10): 89–92.
MATHON, C. C. – 1952 – Description, écologie et dynamique de quelques phytocénoses en Haute-Provence occidentale (Montagne de Lure). *Année Biol., 3e Sér.* 28: 165–174.
MAYER, H. – 1964 – Bergsturzbesiedlungen in den Alpen. *Mitt. Staatsforstverw. Bayerns* 34: 191–203.
MAYER, H. – 1966 – Waldgeschichte des Berchtesgadener Landes (Salzburger Kalkalpen). *Forstwiss. Forsch. Beih. Forstwiss. Zbl.* 22.
MAYER, H. – 1969 – Tannenreiche Wälder am Südabfall der mittleren Ostalpen. 259 pp. München, Basel, Wien.
MAYER, H. – 1970 – *Mitt. Int. Ver. Veget.-kd. Ostalpin-Dinar. Sekt.* 10(2): 55–63.
MAYER, H., SCHLESINGER, B. & THIELE, K. – 1967 – Dynamik der Waldentstehung und Waldzerstörung auf den Dolomitschuttflächen im Wimbachgries (Berchtesgadener Alpen). *Jahrb. Ver. z. Schutze Alpenpfl. u. Tiere* 32.
MCANDREWS, J. H. – 1966 – Post-glacial history of prairies, savanna and forest in northwestern Minnesota. *Mem. Torr. Bot. Club* 22: 1–72.
MCCORMICK, J. – 1968 – Vegetation in fallowed vineyards, South Bass Island, Ohio. *Ohio J. Sci.* 68: 1–11.
MCNEILLY, T. – 1968 – Evolution in closely adjacent plant populations. III. *Agrostis tenuis* on a small copper mine. *Heredity* 23: 99–108.
MCNEILLY, T. & ANTONOVICS, J. – 1968 – Evolution in closely adjacent plant populations. IV. Barriers to gene flow. *Heredity* 23: 205–218.
MCNEILLY, T. & BRADSHAW, A. D. – 1968 – Evolutionary processes in populations of copper tolerant *Agrostis tenuis* Sibth. *Evolution* 22: 108–118.
MCPHERSON, J. K. & MULLER, C. H. – 1969 – Allelopathic effects of *Adenostoma fasciculatum*, "chamise", in the California chapparal. *Ecol. Monogr.* 39: 177–198.
MEDINA, K. & LIETH, H. – 1963 – Contenido de clorofila de algunas asociaciones vegetales de Europa Central y au relación con la productividad. *Qual. Plant. Mat. Veget.* 9: 217–229.
MEDWECKA-KORNÁS, A. – 1960 – Some problems of forest climaxes in Poland. *Silva Fenn.* 105: 68–72.
MEDWECKA-KORNÁS, A. & KORNÁS, J. – 1963 – Plant communities of the Ojcow National Park and their successions. *Bull. Acad. Pol. Sci. II* 11: 353–355.
MEHER-HOMJI, V. M. – 1963 – A contribution to the study of plant succession and climax vegetation with special reference to the dry parts of North-West and South-East India. *Trop. Ecol.* 4: 40–54.
MEISEL, K. – 1966 – Ergebnisse von Daueruntersuchungen in nordwestdeutschen Acker-Unkrautgesellschaften. In: TÜXEN, R. (Ed.): Anthropogene Vegetation. Ber. Int. Sympos. Stolzenau/W. 1961: 86–93.
MEISEL-JAHN, S. – 1955 – Die pflanzensoziologische Stellung der Hauberge des Siegerlandes. *Mitt. Flor.-soz. Arb.-gem. N.F.* 5: 145–150.
MENKE, B. – 1963 – Beiträge zur Geschichte der Erica-Heiden Nordwestdeutschlands. *Flora* 153: 521–548.

Menke, B. – 1967 – Ein Beitrag zur eemzeitlichen Vegetations- und Klimageschichte nach dem Profil von Ostrohe/Schleswig-Holstein, II. In: Gripp, K., Schütrumpf, R. & Schwabedissen, H. (Ed.): Frühe Menschheit und Umwelt. pp. 126–135. Graz.

Menke, B. – 1968 – Ein Beitrag zur pflanzensoziologischen Auswertung von Pollendiagrammen, zur Kenntnis früherer Pflanzengesellschaften in den Marschenrandgebieten der schleswig-holsteinischen Westküste und zur Anwendung auf die Frage der Küstenentwicklung. *Mitt. Flor.-soz. Arb.-gem. N.F.* 13: 195–224.

Menke, B. – 1969 – Vegetationskundliche und vegetationsgeschichtliche Untersuchungen an Strandwällen. *Mitt. Flor.-soz. Arb.-gem. N.F.* 14: 95–120.

Meusel, H. – 1940 – Die Grasheiden Mitteleuropas: Versuch einer vergleichend-pflanzengeographischen Gliederung. *Bot. Arch.* 41: 357–418, 419–519.

Meyer, B. & Willerding, U. – 1961 – Bodenprofile, Pflanzenreste und Fundmaterial von neuerschlossenen neolithischen und eisenzeitlichen Siedlungsstellen im Göttinger Stadtgebiet. *Göttinger Jahrb.* 1961.

Mihai, G. – 1970 – Considérations sur l'évolution et la succession des associations végétales du bassin de la rivière du Başeu (Roumaine, district Botoşani). *Stud. Com. Muz. Şti. Nat. Bacau* 11.

Mikkelsen, V. M. – 1954 – Studies on the sub-atlantic history of Bornholm's vegetation. *Danm. Geol. Unders.* II 80: 210–229.

Mikola, P. & Hintikka, V. – 1956 – The development of a microbial population in decomposing forest litter. *Comm. Inst. Forest. Fenn.* 46: 1–15.

Miller, G. R. & Miles, J. – 1970 – Regeneration of heather (*Calluna vulgaris* (L.) Hull) at different ages and seasons in northeast Scotland. *J. Appl. Ecol.* 7: 51–60.

Milner, C. – 1972 – The use of computer simulation in conservation management. *Sympos. Brit. Ecol. Soc.* 12: 249–275.

Mina, V. N. – 1955 – The cycle of nitrogen and ash elements in oak forests of the forest steppe. *Pochvovedenie* 1955(6): 32–44.

Mina, V. N. – 1967 – Influences of stemflow in soil. *Soviet Soil Sci.* 1967: 1321–1329.

Mirkin, B. M. – 1970 – Successions of the vegetation in river flood-plains. *Bot. Zh.* 55: 1405–1418.

Mirkin, B. M. & Popova, T. V. – 1969 – The inter-institutional conference on the problem of dynamics of the vegetational cover held in Vladimir on September 20–23, 1968. *Bot. Zh.* 54: 648–650.

Mitchell, G. F. – 1965 – Littleton Bog, Tipperary: an Irish agricultural record. *J. R. Soc. Antiqu. Ir.* 95: 121–132.

Miyai, K. – 1936 – The lava-fields and its vegetation of Sakurajima. *Bull. Biogeograph. Soc. Japan* 6: 285–307.

Molisch, H. – 1937 – Der Einfluß einer Pflanze auf die andere. Allelopathie. VIII, 106 pp. Fischer, Jena.

Möller, C. M., Müller, D. & Nielsen, J. – 1954 – Ein Diagramm der Stoffproduktion im Buchenwald. *Ber. Schweiz. Bot. Ges.* 64: 487–494.

Monk, C. D. – 1968 – Successional and environmental relationships of the forest vegetation of north central Florida. *Amer. Midland Natural.* 79: 441–457.

Moor, M. – 1958 – Pflanzengesellschaften schweizerischer Flußauen. *Mitt. Schweiz. Anst. Forstl. Vers.-w.* 34: 221–360.

Moor, M. – 1969 – Zonation und Sukzession am Ufer stehender und fließender Gewässer. *Vegetatio* 17: 26–32.

Moore, J. J. – 1967 – Zur pflanzensoziologischen Bewertung irischer nacheiszeitlicher Pollendiagramme. In: Tüxen, R. (Ed.): Pflanzensoziologie und Palynologie. Ber. Int. Sympos. Stolzenau/W. 1962: 96–105.

Moore, P. D. – 1968 – Human influence upon vegetational history in North Cardiganshire. *Nature* 217: 1006–1009.

Moore, P. D. & Chater, E. H. – 1969 – The changing vegetation of west-central Wales in the light of human history. *J. Ecol.* 57: 361–379.

Moravec, J. – 1969 – Succession of plant communities. *Folia Geobot. Phytotax.* 4: 133–164.

Moravec, J. & Rybníčková, E. – 1964 – Die *Carex davalliana*-Bestände im Böhmerwaldvorgebirge, ihre Zusammensetzung, Ökologie und Historie. *Preslia* 36.

Morel, G. & Bourlière, F. – 1962 – Relations écologiques des avifaunes sedentaire et migratrice dans une savane sahélienne du bas Sénégal. *La Terre et la Vie* 4: 371–393.

Morozov, G. F. – 1912 – Fundamentals of forest study. St.-Petersburg.

Morozov, G. F. – 1928 – The forest sciences. 4th ed. Leningrad.

Morrison, M. E. S. – 1959 – Evidence and interpretation of "landnam" in the northeast of Ireland. *Bot. Not.* 112: 185–204.

Moss, C. E. – 1907 – Succession of plant formations in Britain. *Rept. Brit. Assoc. Adv. Sci.* 106: 742–743. – 1908 – *Bot. Cbl.* 107: 255-256.

Moss, C. E. – 1910 – The fundamental units of vegetation: Historical development of the concepts of the plant association and the plant formation. *New Phytol.* 9: 18–53.

Mothes, K., Arnold, G. & Redmann, H. – 1937a – Zur Bestandesgeschichte ostpreussischer Wälder. *Jahresber. Preuß. Bot. Ver.* 1930–1936.

Mothes, K., Arnold, G. & Redmann, H. – 1937b – Zur Bestandesgeschichte ostpreussischer Wälder. *Schr. Phys. Ökon. Ges. Königsberg* (*Pr.*) 69: 267–282.

Mullenders, W. – 1962 – Les relations entre la végétation et les spectres polliniques en Forêt du Mont-Dieu (Département des Ardennes, France) .*Bull. Soc. Bot. Belg.* 94: 131–138.

Mullenders, W. & Munaut, A. V. – 1960 – Première contribution à l'étude palynologique des sols forestiers du district picardo-brabançon. *Bull. Soc. Bot. Belg.* 92: 271.

Muller, C. H. – 1940 – Plant succession in the *Larrea-Flourensia* climax. *Ecology* 21: 206–212.

Muller, C. H. – 1952 – Plant succession in arctic heath and tundra in northern Scandinavia. *Bull. Torr. Bot. Club* 79: 296–309.

Muller, C. H. – 1953 – The association of desert annuals with shrub. *Amer. J. Bot.* 40: 53–60.

Muller, C. H. – 1965 – Inhibitory terpenes volatilized from *Salvia* shrubs. *Bull. Torr. Bot. Club* 92: 38–45.

Muller, C. H. – 1966 – The role of chemical inhibition (allelopathy) in vegetational composition. *Bull. Torr. Bot. Club* 93: 332–351.

Muller, C. H. – 1969 – Allelopathy as a factor in ecological process. *Vegetatio* 18: 348–357.

Muller, C. H. & Del Moral, R. – 1966 – Soil toxicity induced by terpenes from *Salvia leucophylla*. – *Bull. Torr. Bot. Club* 93: 130–137.

Müller, H. – 1970 – Ökologische Veränderungen im Otterstedter See im Laufe der Nacheiszeit. *Ber. Naturhist. Ges. Hannover* 114: 33–47.

Müller, K. – 1924 – Das Wildseemoor bei Kaltenbronn im Schwarzwald. 161 p. Karlsruhe.

Müller, T. & Görs, S. – 1960 – Pflanzengesellschaft stehender Gewässer in Baden-Württemberg. *Beitr. Natkd. Forsch. SW-Dtschld.* 19: 60–100.

Müller-Stoll, W. R. – 1936 – Untersuchungen urgeschichtlicher Holzreste nebst Anleitung zu ihrer Bestimmung. *Prähist. Zeitschr.* 27: 3–57.

Mulligan, G. A. & Findlay, J. N. – 1970 – Reproductive systems and colon-

ization in Canadian weeds. *Canad. J. Bot.* 48: 859–860.
MUNAUT, A. V. – 1959 – Première contribution à l'étude palynologique des sols forestiers du district picardo-brabançon. *Bull. Soc. Forest. Belg.* 66: 361–379.
MUNAUT, A. V., DURIN, L. & EVRARD, J. C. – 1968 – Recherches palaeoécologiques et pedologiques en Forêt d'Audigny (Aisne, France). *Bull. Soc. Bot. Nord Fr.* 21: 105–133.
MUTCH, R. W. – 1970 – Wildland fires and ecosystems: a hypothesis. *Ecology* 51: 1046–1051.
NAUMOV, N. P., DMITRIEV, P. P. & LOBACHEV, V. S. – 1970 – Changes in biocoenoses of Aral Karakumy following the extermination of great gerbils. *Zool. Zh.* 49: 1758–1766.
NAVEH, Z. – 1961 – Toxic effects of *Adenostoma fasciculatum* (chamise) in the California chaparral. *Bull. Res. Council D Bot.* 9(4): 190.
NECHAYEVA, N. T. – 1958 – The dynamics of the pasture vegetation of the Kara-Kumy in response to the meteorological conditions. Acad. Sci. Turkmen. SSR Press, Ashkhabad.
NECHAYEVA, N. T. – 1967 – *Artemisia-Salsola* pastures of the Northwestern Turkmenistan. *Transactions Institute Animal Breeding Acad. Sci. Turkmen. SSR* 1.
NEILL, R. L. & RICE, E. L. – 1971 – Possible role of *Ambrosia psilostachya* on pattern and succession in old-fields. *Amer. Midland Natural.* 86: 344–357.
NEKRASOVA, T. P. – 1957 – The importance of precipitation for the seed yield of pine in the pine forest of the arid zones of West Sibiria. *Transactions of Silviculture of West Sibiria* 3.
NEUSCHWANDER, H. E. – 1957 – The vegetation of Dodge County, Wisconsin 1833–1837. *Trans. Wisc. Acad. Sci., Arts, Lett.* 46: 233–254.
NEUHÄUSL, R. – 1965 – Vegetation der Röhrichte und Sublitoral-Magnicariceten im Wittingauer Becken. *Vegetace CSSR A* 1: 11–117.
NEUWEILER, E. – 1925 – Über Hölzer in prähistorischen Fundstellen. *Veröff. Geobot. Inst. Rübel* 3: 509–519.
NICHOLS, G. E. – 1923 – A working basis for the ecological classification of plant communities. *Ecology* 4: 11–23, 154–179.
NICHOLS, H. – 1967 – Vegetational change, shoreline displacement and the human factor in the Late Quaternary history of south-west Scotland. *Trans. Roy. Soc. Edinb.* 67: 145–187.
NIEMI, A. – 1969 – Influence of the Soviet tenancy on the flora for the Porkala area. *Acta Bot. Fenn.* 84: 1–52.
NIERING, W. A. & EGLER, F. E. – 1955 – A shrub community of *Viburnum lentago* stable for twenty-five years. *Ecology* 36: 356–360.
NIKLEWSKI, J. & ZEIST, W. v. – 1970 – A late quaternary pollen diagram from northwestern Syria. *Acta Bot. Neerlandica* 19(5).
NILSSON, T. – 1967 – Pollenanalytische Datierung mesolithischer Siedlungen im Randgebiet des Ageröds mosse im mittleren Schonen. *Acta Univ. Lund* II 16: 1–80.
NOIRFALISE, A. & SOUGNEZ, N. – 1961 – Les forêts riveraines de Belgique. *Bull. Jard. Bot. Et. Bruxelles* 30: 199–288.
NUMATA, M. – 1969 – Progressive and retrogressive gradient of grassland vegetation measured by degree of succession. *Vegetatio* 19: 96–127.
NUMATA, M., HAYASHI, I., KOMURA, T. & OKI, K. – 1964 – Ecological studies on the buried-seed population in the soil as related to plant succession, I. *Jap. J. Ecol.* 14: 207–215.
NUMATA, M. & MUSHIAKI, Y. – 1967 – Experimental studies on early stages of secondary succession IV. *J. Coll. Arts and Sci. Chiba Univ. Nat. Sci. Ser.* 5: 143–157.
OBERDORFER, E. – 1936 – Erläuterung zur vegetationskundlichen Karte des

Oberrheingebietes bei Bruchsal. *Beitr. Naturdenkmalpfl.* 16: 41–125.
OBERDORFER, E. – 1951 – Die Schafweide im Hochgebirge. *Forstwiss. Cbl.* 70: 117–124.
ODUM, E. P. – 1962 – Relationships between structure and function in ecosystems. *Jap. J. Ecol.* 12: 108–118.
ODUM, E. P. – 1969 – The strategy of ecosystem development. *Science* 164: 262–270.
ODUM, E. P. – 1971 (1970) – Fundamentals of ecology. 3rd ed. XIV+574 pp. Saunders, Philadelphia, London, Toronto. (2nd ed. 1959.)
OHBA, T., MIYAWAKI, A. & TÜXEN, R. – 1973 – Pflanzengesellschaften der Japanischen Dünen-Küsten. *Vegetatio* 26: 3–143.
OLBERG, A. – 1943 – Bestandesgeschichte des Preuss. Hochschulforstamtes Chorin. *Schriftr. Akad. Dtsch. Forstwiss.* 6(1): 1–342.
OLDFIELD, F. – 1970 – The ecological history of Blelham Bog National Nature Reserve. In: WALKER, D. & WEST, R. G. (Ed.): Studies in the vegetational history of the British Isles. pp. 141–157. Cambridge, London.
OLMSTEAD, C. E. & RICE, E. L. – 1970 – Relative effects of known plant inhibitors on species from first two stages of old-field succession. *Southw. Natural.* 15: 165–173.
OLSCHOWY, G. – 1960 – Grundsätze der Landschaftspflege für den Abbau und die Rekultivierung von Tagebaugebieten. *Hilfe durch Grün* H. 9: 7–9.
OLSON, J. S. – 1958 – Rates of succession and soil changes on southern Lake Michigan sand dunes. *Bot. Gaz.* 119: 125–170.
OLSON, J. S. – 1964 – Gross and net production of terrestrial vegetation. *J. Ecol.* 52 (*Suppl.*): 99–118.
O'NEILL, R. V. – 1971 – A stochastic model of energy flow in predator compartments of an ecosystem. *Statistical Ecology* 3: 107–121.
OOSTING, H. J. – 1956 – The study of plant communities. 2nd ed. VIII, 440 pp. San Francisco. (1st ed. 1948.)
OPRAVIL, E. – 1961 – Die Wälder der Gegend von Znojmo im Hallstatt. *Acta Musei Morav.*: 81–100.
OREKHOVSKY, A. R. – 1969 – The characteristic features of the growth of common reed (*Phragmites communis* Trin.). *Bot. Zh.* 54: 185–195.
ORLOV, A. J. – 1966 – The effect of excessive moisture and other soil factors on the root systems and on the productivity of spruce forests of the southern taiga zone. In: The effects of excessive soil moisture on the productivity of forests. Moscow.
ORSHAN, G. & DISKIN, S. – 1970 – Seasonal changes in productivity under desert conditions. 10 pp. Sep. o. O.
OSVALD, H. – 1923 – Die Vegetation des Hochmoores Komosse. *Svenska Växtsoc. Sällsk. Handl.* 1.
OVERBECK, F. – 1950 – Die Moore. Geologie und Lagerstätten Niedersachsens 3. *Nieders. Amt Landesplanung u. Statistik. Veröff. Reihe* A 3, 4 Abt. 112 pp.
OVERBECK, F. & SCHMITZ, H. – 1931 – Zur Geschichte der Moore, Marschen und Wälder Nordwestdeutschlands. I. Das Gebiet von der Niederweser bis zur unteren Ems. *Mitt. Prov. Naturdenkmalspflege Hannover* 3: 1–179.
OVINGTON, J. D. – 1957 – Dry-matter production by *Pinus sylvestris* L. *Ann. Bot.* 21(82): 287–314.
OVINGTON, J. D. – 1959 – The circulation of minerals in plantations of *Pinus sylvestris* L. *Ann. Bot.* 23: 229–239.
OVINGTON, J. D. & MADGWICK, H. A. I. – 1959 – The growth and composition of natural stands of birch. *Plant and Soil* 10: 271–283, 389–399.
OWNBEY, M. – 1950 – *Amer. J. Bot.* 37: 487–499.
PACHOSKY, (= PACZOSKI) I. K. – 1891 – Stages of flora development. *Vestnik Est.* 8.

Pachosky, I. K. – 1908 – Steppe in the Black Sea region. Botanical-geographical Sketch. Odessa.
Pachosky, I. K. – 1910 – Principal features of flora development in the south-western Russia. Kherzon.
Pachosky, I. K. – 1917 – The description of the vegetation of the Kherson Government. II. Steppes. Kherson.
Pachosky, I. K. – 1921 – Basic phytosociology. Izd. Stud. Kom. s.-kh. Tekhnikuma, Kherson.
Pallmann, H. – 1934 – Über Bodenbildung und Bodenserien in der Schweiz. *Ernährung d. Pflanze (Berlin)* 30: 225–234.
Pallmann, H. – 1942 – Grundzüge der Bodenbildung. *Schweiz. Landwirtsch. Monatshefte* 1942(6/7): 142–166.
Pallmann, H. & Haffter, P. – 1933 – Pflanzensoziologische und bodenkundliche Untersuchungen im Oberengadin mit besonderer Berücksichtigung der Zwergstrauchgesellschaften der Ordnung Rhodoreto-Vaccinietalia. *Ber. Schweiz. Bot. Ges.* 42: 357–466.
Parenti, R. L. – 1969 – Inhibitional effects of *Digitaria sanguinalis* and its possible role in plant succession. *Abstr. Intern. Bot. Congress* 11: 166.
Parenti, R. L. & Rice, E. L. – 1969 – Inhibitional effects of *Digitaria sanguinalis* and its possible role in old-field succession. *Bull. Torr. Bot. Club* 96: 70–78.
Parkinson, D. & Thomas, A. – 1969 – Studies of fungi in the root region. VIII. Qualitative studies on fungi in the rhizosphere of dwarf bean plants. *Plant and Soil* 31: 299–310.
Parks, J. M. – 1969 – Effects of certain plants of old-field succession on the growth of blue-green algae. *Abstr. Intern. Bot. Congress* 11: 166.
Parks, J. M. & Rice, E. L. – 1969 – Effects of certain plants of old-field succession on the growth of blue-green algae. *Bull. Torr. Bot. Club* 96: 345–360.
Passarge, H. – 1957 – Über Kahlschlaggesellschaften im baltischen Buchenwald von Dargun (Ost-Mecklenburg). *Phyton* 7: 142–151.
Passarge, H. – 1960 – Zur soziologischen Gliederung binnenländischer Corynephorus-Rasen im Nordostdeutschen Flachland. *Verh. Bot. Ver. Prov. Brandenburg* 98/100: 113–124.
Patil, G. P., Pielou, E. C. & Waters, W. E. – 1971 – Statistical ecology. 3: Many species populations, ecosystems, and systems analysis. 462 pp. Pennsylv. State Univ. Press, University Park and London.
Patten, B. C. (Ed.) – 1971, 1972 – Systems analysis and simulation in ecology. Vol. 1, 592 pp., Vol. 2, 610 pp. Academic Press, New York, London.
Patten, D. T. – 1969 – Succession from sagebrush to mixed conifer forest in the northern Rocky Mountains. *Amer. Midland Natural.* 82: 229–240.
Paul, K. H. – 1944, 1953 – Morphologie und Vegetation der Kurischen Nehrung, I, II. *Nova Acta Leopoldina N.F.* 13: 215–378, 16: 261–378.
Pavillard, J. – 1919 – Remarques sur la nomenclature phytogéographique. Montpellier.
Pavillard, J. – 1927 (1928) – Les tendances actuelles de la phytosociologie. *Arch. Bot.* 6.
Pawłowski, B. – 1966 – Dynamics of plant communities. In: Szafer, W. (Ed.): The vegetation of Poland. Warszawa.
Pegau, R. E. – 1970 – Succession in two exclosures near Unalakleet, Alaska. *Canad. Field Nat.* 84: 175–177.
Pennington, W. – 1970 – Vegetation history in the north-west of England: a regional synthesis. In: Walker, D.: Studies in the vegetational history of the British Isles. p. 41–79. London.
Penzig, C. – 1902 – Die Fortschritte der Flora des Krakatau. *Ann. Jard. Bot. Buitenzorg* 13.

Perring, F. H. – 1967 – The Irish problem. In: Tüxen, R. (Ed.): Pflanzensoziologie und Palynologie. Ber. Int. Sympos. Stolzenau/W. 1962: 257–268.
Persson, Å. – 1964 – The vegetation at the margin of the receding glacier Skaftafellsjökull, SE Island. *Bot. Not.* 117: 323—354.
Perttula, U. – 1941 – Untersuchungen über die generative und vegetative Vermehrung der Blütenpflanzen in der Wald-, Hainwiesen- und Hainfelsenvegetation. *Ann. Acad. Sci. Fenn.* 58(1).
Pettersson, B. – 1958 – Dynamik och konstans i Gotlands flora och vegetation. *Acta Phytogeogr. Suec.* 40: 1–288.
Pettersson, B. – 1965 – Recent changes in flora and vegetation. *Acta Phytogeogr. Suec.* 50: 288–294.
Pfaffenberg, K. – 1934 – Das Interglacial von Tidofeld (Jeverland) in Oldenburg. *Abh. Nat. Ver. Bremen* 29: 122–128.
Pfaffenberg, K. – 1939a – Das Interglacial von Haren (Emsland). *Abh. Nat. Ver. Bremen* 31: 360–376.
Pfaffenberg, K. – 1939b – Entwicklung und Aufbau des Lengener Moores. *Abh. Nat. Ver. Bremen* 31: 114–151.
Pfaffenberg, K. – 1942 – Die geologische Lagerung und pollenanalytische Altersbestimmung der Moorleiche von Bockhornesfeld. *Abh. Nat. Ver. Bremen* 32: 77–90.
Pfaffenberg, K. – 1947 – Getreide- und Samenfunde aus der Kulturschicht des Steinzeitdorfes am Dümmer. *Jber. Naturhist. Ges. Hannover* 94/98: 69–82.
Pfaffenberg, K. – 1952 – Pollenanalytische Untersuchungen an nordwestdeutschen Kleinstmooren. Ein Beitrag zur Waldgeschichte des Syker Flottsandgebietes. *Mitt. flor.-soz. Arbeitsgem. N.F.* 3: 27–43.
Pfaffenberg, K. – 1954 – Das Wurzacher Ried. Eine stratigraphische und paläobotanische Untersuchung. *Geolog. Jb.* 68: 479–500.
Pfaffenberg, K. & Dienemann, W. – 1964-Das Dümmerbecken. *Forschungen z. Landes- u. Volkskunde* I. 78: 121 pp.
Pfaffenberg, K. & Hassenkamp, W. – 1934 – Über die Versumpfungsgefahr des Waldbodens im Syker Flottsandgebiet. *Abh. Nat. Ver. Bremen* 29: 89–121.
Pfeiffer, H. – 1928 – Von der Besiedlung und Flora von Maulwurfshügeln. *Feddes Rep. Beih.* 51: 34–38.
Pfeiffer, H. – 1943 – Syngenetische Beobachtungen an der Bitterkrautgesellschaft. *Ber. Dtsch. Bot. Ges.* 61: 101–110.
Pfeiffer, H. – 1963 – Vom gesetzlichen Verhalten der Pioniere bei Neuland-Besiedlung. *Mitt. Flor.-soz. Arb.-gem. N.F.* 10: 87–91.
Pflüger, R. – 1966 – Einfluß von unter der Wirkung von Erle angereicherten Stickstoffverbindungen und von Ammoniumnitratgaben auf andere Pflanzenarten. *Ber. Oberhess. Ges. N.F. Naturwiss. Abt.* 34: 277–294.
Philippi, G. – 1969 – Besiedlung alter Ziegeleigruben in der Rheinniederung zwischen Speyer und Mannheim. *Mitt. Flor.-soz. Arb.-gem. N.F.* 14: 238–254.
Phillips, J. – 1934, 1935 – Succession, development and climax, and the complex organism: An analysis of concepts. I-III. *J. Ecol.* 22: 554–571; 23: 210–246, 488–508.
Phillips, W. S. – 1963 – Photographic documentation of vegetational changes in the Northern Great Plains. *Univ. Arizona Agric. Exp. Sta. Report (Tucson)* 214.
Pichi-Sermolli, R. E. – 1948a – Flora e vegetazione delle serpentine e delle altre ofioliti dell'Alta Valle del Tevere (Toscana). *Webbia* 6: 1–378.
Pichi-Sermolli, R. E. – 1948b – An index for establishing the degree of maturity in plant communities. *J. Ecol.* 36: 85–90.
Pielou, E. C. – 1966 – Species-diversity and pattern-diversity in the study of ecological succession. *J. Theoret. Biol.* 10: 370–383.
Pielou, E. C. – 1969 – An introduction to mathematical ecology. Wiley, New York.

PIEMEISEL, R. L. – 1951 – Causes affecting change and rate of change in a vegetation of annuals in Idaho. *Ecology* 32: 53–72.
PIETSCH, W. – 1965 – Die Erstbesiedlungs-Vegetation eines Tagebau-Sees. Synökologische Untersuchungen im Lausitzer Braunkohlenrevier. *Limnologica (Berlin)* 3: 177–222.
PIGNATTI, S. – 1952 – Sulla vegetazione psammofilia litoranea del N.-Africa franc. *N. Giorn. Bot. Ital. N.* 59: 167–168.
PIGNATTI, S. – 1954 – Introduzione allo studio fitosociologico della pianura Veneta orientale con particolare riguardo alla vegetazione litoranea. *Arch. Bot.* 28/29.
PIGOTT, C. D. – 1970 – Soil formation and development on the carboniferous limestone of Derbyshire, II. *J. Ecol.* 58: 529–541.
PITELKA, F. A. – 1964 – The nutrient recovery hypothesis for arctic microtine cycles. I. Introduction. *Brit. Ecol. Soc. Sympos.* 4: 55–56.
PLAISSANCE, G. – 1961 – Guide des forêts de France. 411 pp. Paris.
PLOCHMANN, R. – 1956 – Bestockungsaufbau und Baumartenwandel nordischer Urwälder. *Forstwiss. Forsch. Beih. Forstw. Cbl.* 6: 1–112.
POHL, F. – 1937 – Die Pollenerzeugung der Windblütler. *Beih. Bot. Cbl.* 56 A: 365–470.
POLUNIN, N. – 1936 – Plant succession in Norwegian Lapland. *J. Ecol.* 24: 372–391.
POMELOVA, G. I. – 1971 – Seasonal dynamics of soil algae in the fields under crop rotation. *Bot. Zh.* 56: 126–133.
PONOMARIOVA, V. V. – 1970 – The forest as an eluvium-resistant type of vegetation. *Bot. Zh.* 55: 1585–1595.
POP, E., BOŞCAIU, N., RATIU, F. & DIAÇONEASA, B. – 1965 – Correlation between the recent pollen spectra and the vegetation of the National Park of "Retezat". *Rev. Roum. Biol.* 10: 187–197.
POPOV, T. I. – 1914 – The origin and development of the aspen scrub within the Voronezh Region. *Transactions of the Dokuchayev Comittee,* 2. Petrograd.
POPOVA, T. V. – 1970 – Changes in the herb layer of flood plain forests on the R. Bolshoy Inser in the course of successions. *Bjull. Mosk. Ob. Isp. Prir., O. Biol.* 75(4): 88–94.
POPOVICI, R. – 1932 – Beiträge zur Waldgeschichte Nord-Rumäniens. *Bul. Fac. Şti. Cernăuti* 6.
POPOVICI, R. – 1934 – Ein weiterer Beitrag zur Waldgeschichte unseres Landes. *Bul. Fac. Şti. Cernăuti.*
POST, L. VON & SERNANDER, R. – 1910 – Pflanzenphysiognomische Studien auf Torfmooren in Närke. *Livretguide Exc. Suède XIe Congr. Géol. Int.* 14.
POTZGER, J. E., POTZGER, M. E. & MCCORMICK, J. – 1956 – The forest primeval of Indiana as recorded in the original U.S. land surveys and an evaluation of previous interpretations of Indiana vegetation. *Bull. Butler Univ. Bot. Studies* 13: 95–111.
PREISING, E. – 1942 – Die Begrünung offener Sandböden im ostdeutschen Flachland. *Die Strasse* 1942: 1–6.
PROCHAL, P. – 1970 – Influence of ground water fluctuations on plant communities of Dziewińskie meadows. *Roczn. Nauk Roln.* 77: 331–355.
PROCTOR, M. C. F. & LAMBERT, C. – 1961 – Pollen spectra from recent *Helianthemum* communities. *New Phytol.* 60: 21–26.
PRODAN, M. – 1968 – Forest Biometrics. XI+447 pp. Pergamon Press, New York.
PROZOROVSKY, A. V. – 1938 – The methods of geobotanical investigation of semideserts and deserts. In: The methods of field geobotanical investigations. Acad. Sci. USSR Press, Moscow, Leningrad.

Puri, G. S. – 1950 – Surface geology, vegetation and plant succession. *Indian For.* 70: 199–209, 254–262.
Putwain, P. D. & Harper, J. L. – 1970 – Studies in the dynamics of plant populations. III. The influence of associated species on populations of *Rumex acetosa* L. and *R. acetosella* L. in grassland. *J. Ecol.* 58: 251–264.
Pyatin, A. M. – 1970 – On the content of viable seeds in the soils of meadow and woodland pastures used without system. *Bjull. Mosk. Ob. Isp. Prir. O. Biol.* 75(3): 85–95.
Pyavchenko, N. I. – 1955 – Palsa peatbogs. Moscow.
Pyavchenko, N. I. – 1960 – Biological circulation of nitrogen and ash elements in boggy soils. *Pochvovedenie* 1960(6): 21–32; *Soviet Soil Sci.* 1960: 593–602.
Quantin, A. – 1935 – L'évolution de la végétation à l'étage de la chênaie dans le Jura méridional. *Comm. S.I.G.M.A.* 37: 1–382.
Quézel, P. – 1965 – La végétation du Sahara. XI+333 pp. Stuttgart.
Quimby, J. R., Kramer, N. W., Stephens, J. C., Lahr, K. A. & Karper, R. E. – 1958 – *Bull. Texas Agr. Exp. Sta.* 912: 1–35.
Quimby, J. R., Reitz, L. P. & Laude H. H. – 1962 – Effect of source of seed on productivity of hard red winter wheat. *Crop Sci.* 2: 201–209.
Raabe, E.-W. – 1954 – Sukzessionsstudien am Sandkatener Moor. *Arch. Hydrobiol.* 49: 349–375.
Raabe, E.-W. – 1960 – Über die Regeneration überschwemmter Grünländereien in der Treene-Niederung. *Schr. Naturw. Ver. Schlesw.-Holst.* 31: 25–55.
Raabe, E.-W. – 1965 – Sukzessionsstudien an Salzrasen. *Heimat (Neumünster)* 72: 312–316.
Rabotnov, T. A. – 1936 – Ecological observations on lichens in Yakutia. *Sov. Bot.* 6.
Rabotnov, T. A. – 1950a – The life cycle of perennial herbaceous plants in meadow coenoses. *Trudy Bot. Inst. Akad. Nauk USSR, Ser. III, Geobot.* 6: 7–204.
Rabotnov, T. A. – 1950b – The problems of investigating the composition of population for the purposes of phytocoenology. In: The problems of botany. Vol. 1. Acad. Sci. USSR Press, Moscow, Leningrad.
Rabotnov, T. A. – 1955 – Fluctuations of meadows. *Bjull. Mosk. O. I. Prir. O. Biol.* 60(3).
Rabotnov, T. A. – 1958 – Changes of the flood meadow because of invasion of *Heracleum sibiricum* L. — *Dokl. Akad. Nauk. S.S.S.R.* 121(4).
Rabotnov, T. A. – 1960 – Some data on the experimental study of syngenesis in meadows. *Bjull. Mosk. O. I. Prir. O. Biol.* 65(3).
Rabotnov, T. A. – 1965 – On the dynamics of structure of the polydominant meadow coenoses. *Bot. Zh.* 50(10).
Rabotnov, T. A. – 1972 – The study of fluctuations (yearly variations) of plant communities. *Field Geobotany (Leningrad)* 4: 95–136.
Rademacher, B. – 1957 – Die Bedeutung allelopathischer Erscheinungen in der Pflanzenpathologie. *Z. Pflanzenkrankh.* 64: 427–439.
Radke, G. J. – 1972 – Genese der Waldmoore des nördlichen Schwarzwaldes. *Ber. Deutsch. Bot. Ges.* 85: 157–164.
Radvanyi, A. – 1970 – Small mammals and regeneration of white spruce forests in western Alberta. *Ecology* 51: 1102–1105.
Ramensky, L. G. – 1925 – The main principles governing the vegetational cover. Voronezh.
Ramensky, L. G. – 1938 – Introduction to the complex pedology-geobotanical study of lands. Moscow.
Ranwell, D. – 1958 – Movement of vegetated sand dunes at Newborough Warren, Anglesey. *J. Ecol.* 46: 83–100.
Ranwell, D. – 1959, 1960 – Newborough Warren, Anglesey. *J. Ecol.* 47: 571–

601, 48: 117–141, 385–395.
RANWELL, D. – 1961 – Spartina salt marshes in southern England, I. *J. Ecol.* 49: 325–340.
RATTER, J. – 1969 – *Notes R. Bot. Garden Edinburgh* 29: 213–232.
RAUP, H. M. – 1941 – Botanical problems in boreal America. *Bot. Rev.* 7: 147–248
RAUP, H. M. – 1951 – Vegetation and cryoplanation. *Ohio J. Sci.* 51: 105–116.
RAZUMOVA, L. M. – 1970 – On the vegetation successions on sands in Baltic Soviet Republics. *Bot. Zh.* 55: 1333–1335.
REHDER, H. – 1962 – Der Girstel, ein natürlicher Föhrenwaldkomplex am Albis bei Zürich. *Ber. Geobot. Inst. ETH St. Rübel* 33: 17–64.
REICHLE, D. (Ed.) – 1970 – Proceedings of the Gatlinburg symposium on forest ecosystems. *Studies in Ecology* 1.
REINHOLD, F. – 1942 – Die Bestockung der Kursächsichen Wälder im 16.Jahrhundert. 178 pp. Dresden.
REINHOLD, F. – 1949 – Zusammensetzung und Aufbau eines natürlichen Eichen-Buchenwaldes auf der Baar bei Donaueschingen. *Forstwiss. Cbl.* 68: 691–698.
REISS, J. – 1970 – Orientierende Untersuchungen über die Erstbesiedlung von Torfstichen durch Pflanzen. *Ber. Bayer. Bot. Ges.* 42: 139–140.
REMEZOV, N. P., BYKOVA, L. N. & SMIRNOVA, K. M. – 1959 – Consumption and circulation of nitrogen and ash elements in the forests of the European part of the USSR. Izd. Mosk. Univ., Moscow.
REMEZOV, N. P., SAMOYLOVA, Y. M., SUIRIDOVA, I. K. & BOGASHOVA L. G. – 1964 – Dynamics of the interaction of oak forest and soil. *Pochvovedenie* 1964(3): 1-14; *Soviet Soil Sci.* 1964: 222-232.
REMMERT, H. – 1972 – *Umschau* 72: 41–44.
REMPE, H. – 1937 – Untersuchungen über die Verbreitung des Blütenstaubes durch die Luftströmungen. *Planta* 27: 93–147.
RICE, E. L. – 1965 – Inhibition of nitrogen-fixing and nitrifying bacteria by seed plants. II. Characterization and identification of inhibitors. *Physiol. Plant.* 18: 255–268.
RICE, E. L. – 1967 – Chemical warfare between plants. *Bios* (*Mt. Vernon*) 38: 67–74.
RICE, E. L. – 1972 – Allelopathic effects of *Andropogon virginicus* and its persistence in old-fields. *Amer. J. Bot.* 59: 752–755.
RICE, E. L. & PARENTI, R. L. – 1967 – Inhibition of nitrogen-fixing and nitrifying bacteria by seed plants. V. Inhibitors produced by *Bromus japonicus* Thunb. *Southwestern Naturalist* 12: 97–103.
RICE, E. L., PENFOUND, W. T. & ROHRBAUCH, L. M. – 1960 – Seed dispersal and mineral nutrition in succession in abandoned fields in central Oklahoma. *Ecology* 41: 224–228.
RICHARD, J.-L. – 1961 – Les forêts acidophiles du Jura: Étude phytosociologique et écologique. *Beitr. Geobot. Landesaufn. Schweiz* 38: 1–164.
RICHARDS, P. W. – 1952 – The tropical rain forest: An ecological study. 450 pp. Cambridge Univ. Press, London.
RICHARDSON, J. A. – 1970 – Die Entwicklung von Orchideen-Populationen in Tongruben in der Grafschaft Durham (England). *Orchidee* 21: 173–178.
RICHTER, W. – 1966 – Die natürliche Begrünung der erzgebirgischen Bergwerkshalden. *Hercynia N.F.* 3: 114–164.
RINTANEN, T. – 1970 – On the vegetation and ecology of frost ground sites in eastern Finnish Lapland. *Ann. Bot. Fenn.* 7: 1–24.
RIVALS, P. – 1952 – Études sur la végétation naturelle de l'île de la Réunion. *Trav. Lab. Forest. Toulouse* V 3(1): 1–274.
ROBYNS, A. – 1963 – Essai de monographie du genre *Bombax* s.l. (*Bombacaceae*). *Bull. Jard. Bot. Bruxelles* 33: 1–316.

Robyns, W. – 1932 – La colonisation végétale des laves récentes du volkan Rumoka. *Mém. Inst. R. C. Belge Nat. Méd.* 1(1): 1–32.
Rödel, H. – 1970 – Waldgesellschaften der Sieben Berge bei Alfeld und ihre Ersatzgesellschaften. *Dissert. Bot.* (*Lehre*) 7: 1–144.
Rodin, L. E. – 1946 – Brulage de la végétation comme moyen d'amélioration des pâturages à Graminées et *Artemisia*. – *Sov. Botanica* 3.
Rodin, L. E. – 1948 – The materials for the investigation of the vegetation of the northern and Trans-Unguzian Kara-Kumy. *Acta Inst. Bot. Acad. Sci. USSR, Ser. III, Geobot.* 5.
Rodin, L. E. – 1961 – The dynamics of the desert vegetation. Acad. Sci. USSR Press, Moscow, Leningrad.
Rodin, L. E. – 1963 – The vegetation of the deserts of Western Turkmenistan. Acad. Sci. USSR Press, Moscow, Leningrad.
Rodin, L. E. & Bazilevich, N. I. – 1965 – Production and mineral cycling in terrestrial vegetation. Izd. Nauka, Moscow.
Rodin, L. E. & Bazilevich, N. I. – 1966 – The biological productivity of the main vegetation types in the northern hemisphere of the world. *Forestry Abstr.* 27: 369–372.
Rodin, L. E. & Bazilevich, N. I. – 1967 – Dynamics of the organic matter and biological cycling of ash elements and nitrogen in the main types of the world's vegetation. (Transl. by E. G. Fogg.) Oliver & Boyd, London.
Rohlmann, C. – 1958 – Entstehungsgeschichte des Seeangers bei Ebergötzen im Rahmen der spät- und nacheiszeitlichen Waldgeschichte des Eichsfeldes um Seeburg. *Staatsexamensarbeit Univ. Göttingen.*
Rollet, B. – 1971 – La regeneración natural en bosque denso siempreverde de llanura de la Guayana Venezolana. *Inst. Forest. Lat.-Am. Merida Bol.* 35: 39–73.
Ross, B. A., Bray, J. R. & Marshall, W. H. – 1970 – Effects of longterm deer exclusion on a *Pinus resinosa* forest in northcentral Minnesota. *Ecology* 51: 1088–1093.
Runge, A. – 1969 – Pilzsukzessionen auf Eichenstümpfen. *Abh. Landesmus. Natk. Münster Westf.* 31(2): 3–10.
Runge, F. – 1959 – Pflanzengeographische Probleme in Westfalen. *Abh. Landesmus. Naturkd. Münster Westf.* 21: 3–51.
Runge, F. – 1962 – Vegetationsänderungen in den Bockholter Bergen bei Münster. *Natur u. Heimat* 22: 60–64.
Runge, F. – 1963a, 1967 – Die Artmächtigkeitsschwankungen in einem nordwestdeutschen Enzian-Fiederzwenken-Rasen, I, II. *Vegetatio* 11: 237–240, 15: 124–128.
Runge, F. – 1963b – Die Vegetationsentwicklnug auf einer Brandstelle in einer Bergheide. *Arch. Naturschutz* 3: 173–177.
Runge, F. – 1967a – Vegetationsschwankungen im Rhynchosporetum. *Mitt. Flor.-soz. Arb.-gem. N.F.* 11/12: 49–53.
Runge, F. – 1967b – Vegetationsschwankungen in Hochheiden des Sauerlandes. *Decheniana* 118: 145–151.
Runge, F. – 1968 – Schwankungen der Vegetation sauerländischer Talsperren. *Arch. Hydrobiol.* 65: 223–239.
Runge, F. – 1969a – Über die Wirkung des "Abflämmens" von Wegrainen (Dauerquadrat-Beobachtungen). In: Tüxen, R. (Ed.): Experimentelle Pflanzensoziologie. Ber. Int. Sympos. Rinteln 1965: 213–219.
Runge, F. – 1969b – Vegetationsschwankungen in einem Melico-Fagetum. *Vegetatio* 17: 151–156.
Runge, F. – 1969c – Vegetationsschwankungen in einer nassen Heide. *Natur u. Heimat* 29: 28–30.
Runge, F. – 1969d – Vegetationsänderungen in einer aufgelassenen Wiese.

Mitt. Flor.-soz. Arb.-gem. N.F. 14: 287–290.
RUNGE, F. – 1969e – Die Wirkung der Graureiherkolonie auf die Vegetation. *Natur u. Heimat* 29: 130–131.
RUNGE, F. – 1970 – Die pflanzliche Besiedlung eines Straßenbanketts. *Natur u. Heimat* 30: 54–56.
RUNGE, F. – 1972 – Dauerquadratbeobachtungen bei Salzwiesen-Assoziationen. In: TÜXEN, R. (Ed.): Grundfragen u. Methoden in der Pflanzensoziologie. Ber. Int. Sympos. Rinteln 1970: 419–425.
RUTTNER, F. – 1940 – Grundriß der Limnologie. Berlin. (2.Aufl. 1952.)
RYBNIČEK, K. & RYBNIČKOVÁ, E. – 1968 – The history of flora and vegetation on the Bláto mire in southeastern Bohemia, Czechoslovakia. *Folia Geobot. Phytotax.* 3: 117–142.
RYBNÍČKOVÁ, E. – 1966 – Pollen-analytical reconstruction of vegetation in the upper regions of the Orlické hory mountains, Czechoslovakia. *Folia Geobot. Phytotax.* 1: 289–310.
SAITO, K. & TACHIBANA, H. – 1969 – Changes in forest vegetation and soil due to human impact in the Sendai area, Miyagi prefecture, Northeast Japan. *Ecol. Rev. (Sendai)* 17: 131–152.
SALASKIN, A. S. – 1937 – Rapidity of the fodder lichens growth. *Sov. Olenevodstvo* 11.
SALISBURY, E. J. – 1925 – Note on the edaphic succession in some dune soils with special reference to the time factor. *J. Ecol.* 13: 322–328.
SALISBURY, E. J. – 1952 – Downs and dunes. 328 pp. G. Bell & Sons, London.
SALISBURY, E. J. – 1970 – The pioneer vegetation of exposed muds and its biological features. *Phil. Trans. R. Soc.* 259: 207–255.
SAMBUK, F. V. – 1933 – Pastures of the first Nenetz reindeer collective farm. *Olenyi Pastb. Sev. Kraya* 2.
SAMPSON, A. W. – 1919 – Plant succession in relation to range management. *U.S. Dept. Agr. Bull.* 791.
SAN JOSÉ, J. J. & FARINAS, M. R. – 1971 – Estudio sobre los cambios de la vegetación protegida de la quema y el pastoreo en la Estación Biológica de los Llanos. *Bol. Soc. Venez. Ci. Nat.* 29(119–120): 136–146.
SARVIS, J. T. – 1941 – Grazing investigations on the northern Great Plains. *N.D. Agr. Exp. Sta. Bull.* 308.
SATYANARAYAN, Y. & GAUR, Y. D. – 1969 – Phytosociological changes of monsoon vegetation in semi-rocky habitats. *J. Indian Bot. Soc.* 47: 371–381.
SAUBERER, A. – 1942 – Die Vegetationsverhältnisse der Unteren Lobau. 55 pp. Kühne, Wien u. Leipzig.
SAVCHENKO, I. V. – 1972 – Changes of the *Stipa* pastures of Trans-Baikal regions as result of grazing. *Bot. Zh.* 57: 1133–1138.
SAVKINA, Z. P. – 1951 – Improvement of the flood meadows plant cover by means of protective shrub stripes. *Kormovaya Basa* 10.
SAVKINA, Z. P. – 1953 – Forage reserve of the cattle breeding in the Far North and the principal ways for its consolidation. *Doklady VI Sess. Uchen. Sov. Inst. Selsk. Khos. Krayn. Severa* 2.
SAVKINA, Z. P. – 1960 – Clearing of the shrub flood lands of the Far North rivers for hay meadows with the protective shrubs stripes. In: LARIN, I. V.: Voprosy senokosno. Moscow.
SCAMONI, A. – 1966 – Diskussionsbemerkung. In: TÜXEN, R. (Ed.): Anthropogene Vegetation. Ber. Int. Sympos. Stolzenau/W. 1961: 262.
SCHIECHTL, H. M. – 1958 – Grundlagen der Grünverbauung. *Mitt. Forstl. Vers.-Anst. Mariabrunn* 55: 1–273.
SCHLÜTER, H. – 1966 – Untersuchungen über die Auswirkungen von Bodenkalkungen auf die Bodenvegetation in Fichtenforsten. *Kulturpflanze* 14: 47–60.

Schmid, E. – 1942 – Über einige Grundbegriffe der Biocoenologie. *Ber. Geobot. Forsch.-inst. Rübel* 1941: 12–26.
Schmid, E. – 1950 – Zur Vegetationsanalyse numidischer Eichenwälder. *Ber. Geobot. Forsch.-inst. Rübel* 1949: 23–39.
Schmid, E. – 1955 – Der Ganzheitsbegriff in der Biocoenologie und in der Landschaftskunde. *Geogr. Helvetica* 10: 153–162.
Schmidt, A. – 1964 – Zytotaxonomische Beiträge zu einer Neugliederung der Sektion *Melanium* der Gattung *Viola*. – *Ber. Dtsch. Bot. Ges.* 77: 94–99.
Schmidt, G. – 1964 – Pionierstadien und Sukzessionen am Steilufer. *Wiss. Z. Univ. Rostock* 13: 209–217.
Schmidt, H. – 1968 – Grundlegung für pflanzensoziologische Untersuchungen im Ruhroberkarbon. *Beih. Ber. Naturh. Ges. Hannover* 5: 329–335.
Schmithüsen, J. – 1948 – Fliesengefüge der Landschaft und Ökotop. *Ber. Dtsch. Landeskd.* 5.
Schmithüsen, J. – 1950 – Das Klimaxproblem vom Standpunkt der Landschaftsforschung aus betrachtet. *Mitt. Flor.-soz. Arb.-gem. N.F.* 2: 176–182.
Scholz, H. – 1960 – Die Veränderungen in der Ruderalflora Berlins. *Willdenowia* 2: 379–397.
Schreier, K. – 1955 – Die Vegetation auf Trümmerschutt zerstörter Stadtteile in Darmstadt. *Schr.-r. d. Naturschutzst. Darmstadt* 3(1): 1–49.
Schretzenmayr, M. – 1950 – Sukzessionsverhältnisse der Isarauen südlich Lenggries. *Ber. Bayer. Bot. Ges.* 28: 19–63.
Schreuder, G. F. – 1968 – Optimal forest investment decisions through dynamic programming. *Yale Univ. School of Forest. Bull.* 72: 70 pp.
Schröder, D. – 1934 – Eine *Calluna*-Heide unter der Zuidersee. Pollenanalytische Untersuchungen an einem Profil vom Wieringermeerpolder. *Abh. Nat. Ver. Bremen* 29: 83–88.
Schubert, E. – 1933 – Zur Geschichte der Moore, Marschen und Wälder Nordwestdeutschlands II. *Mitt. Prov.-st. Naturdenkmalspflege Hannover* 4: 3–148.
Schubert, R. – 1960 – Die zwergstrauchreichen azidophilen Pflanzengesellschaften Mitteldeutschlands. *Pfl.-soz. (Jena)* 11: 1–235.
Schultz, A. M. – 1964 – The nutrient-recovery hypothesis for arctic microtine cycles. II. Ecosystem variables in relation to arctic microtine cycles. *Brit. Ecol. Soc. Sympos.* 4: 57–68.
Schultz, V. – 1971 – A bibliography of selected publications on population dynamics, mathematics and statistics in ecology. *Statistical Ecology* (*Pennsylv. State Univ. Press*) 3: 417–425.
Schulz, J. P. – 1960 – Ecological studies on forests in northern Suriname. *Verh. K. Nederl. Akad. Wet. A. Natkd.* 53: 1–267.
Schwaar, J. – 1969 – Die Gerolsteiner Moß, Eifel, in moor- und vegetationskundlicher Sicht. *Ber. Dtsch. Bot. Ges.* 82: 249–264.
Schwaar, J. – 1970 – Nachwärmezeitliche Vegetationsgeschichte des Salmwaldes/Eifel. *Ber. Dtsch. Bot. Ges.* 83: 89–107.
Schwabe, A. – 1972 – Vegetationsuntersuchungen in den Salzwiesen der Nordseeinsel Trischen. *Abhandl. Landesmus. Naturkd. Münster Westf.* 34: 9–22.
Schwerdtfeger, F. – 1968 – Ökologie der Tiere, II: Demökologie: Struktur und Dynamik tierischer Populationen. 460 pp. Parey, Hamburg, Berlin.
Schwickerath, M. – 1952 – Untersuchungen über Erstberasungen von Talsperren . . . (Eifel). *Arch. Hydrobiol.* 46: 103–124.
Schwickerath, M. – 1954 – Die Landschaft und ihre Wandlung im Bereich des Meßtischblatts Stolberg. 118 pp. Aachen.
Scifres, C. J., Brock, J. H. & Hahn, R. R. – 1971 – Influence of secondary succession on honey mesquite (*Prosopis glandulosa*) invasion in North Texas. *J. Range Manag.* 24: 206–210.

Scott, B. R. M. – 1955 – Amount and chemical composition of the organic matter contributed by oveistory and understory vegetation to forest soil. *Bull. Yale Univ. School For.* 62.
Sears, P. B. – 1925 – The natural vegetation of Ohio. I. A map of the virgin forest. *Ohio J. Sci.* 25: 139–149.
Šeda, Z. – 1970 – Einige Probleme der Vegetationsforschung an den Talsperrenufern. *Folia Fac. Sci. Nat. Univ. Brunensis* 11(3): 33–41.
Seddon, B. – 1967 – Prehistoric climate and agriculture: a review of recent palaeoecological investigations. In: Taylor, J. A.: Weather and Agriculture. Oxford.
Segal, S. – 1965 – Een vegetatieonderzoek van hogere waterplanten in Nederland. *Wetensch. Med. K. Nederl. Nat.-hist. Ver.* 57: 1–80.
Segal, S. – 1968 – On structure, zonation and succession in vegetation of higher aquatics. *Abstr. Ecol. and Control of Aquatic Veget. UNESCO-Meeting (Paris)*:12 pp.
Segal, S. – 1969 – Ecological notes on wall vegetation. 325 pp. Junk, The Hague.
Segal, S. – 1970 – Strukturen und Wasserpflanzen. In: Tüxen, R. (Ed.): Gesellschaftsmorphologie. Ber. Int. Sympos. Rinteln 1966: 157–169.
Segal, S. & Groenhart, M. C. – 1967 – Het Zuideindigerwiede, een uniek verlandingsgebied. *Gorteria* 3: 165–181.
Seibert, P. – 1955 – Die Niederwaldgesellschaften des südwestfälischen Berglandes. *Allgem. Forst- u. Jagdztg.* 126: 1–11.
Seibert, P. – 1958 – Die Pflanzengesellschaften im Naturschutzgebiet Pupplinger Au. *Landschaftspfl. u. Veget.-kd. (München)* 1: 1–79.
Seibert, P. – 1962 – Die Auenvegetation an der Isar nördlich von München und ihre Beeinflussung durch Menschen. *Landschaftspflege u. Veget.-kde (München)* 3: 1–124.
Seibert, P. – 1969 – Die Auswirkungen des Donau-Hochwassers 1965 auf Ackerunkrautgesellschaften. *Mitt. Flor.-soz. Arb.-gem. N.F.* 14: 121–135.
Selle, W. – 1936 – Die nacheiszeitliche Wald- und Moorentwicklung im südöstlichen Randgebiet der Lüneburger Heide. *Jb. Preuß. Geol. Landesanst.* 56: 371–422.
Selle, W. – 1939 – Ergänzung zur nacheiszeitlichen Wald- und Moorentwicklung im südöstlichen Randgebiet der Lüneburger Heide. Pollenanalyse eines kleinen Moores bei Grussendorf. *Jb. Preuß. Geol. Landesanst.* 59: 272–288.
Selle, W. – 1940 – Die Pollenanalyse von Ortstein-Bleichsandschichten. *Beih. Bot. Centralbl.* 60: 525–549.
Selleck, G. W. – 1960 – The climax concept. *Bot. Rev.* 26: 534–545.
Selm, H. R. De & Shanks, R. E. – 1961 – Accumulation and cycling of organic matter and chemical constituents during early vegetational succession on a radioactive waste disposal area. *Contrib. Bot. Lab. Univ. Tennessee N. Ser.* 228: 83–96.
Semenova Tian-Shanskaya, A. M. – 1954 – The re-establishement of vegetation on steppe fallow lands as related to the problem of interspecific conversion. *Bot Zh.* 38(6).
Serve, L. & Baudière, A. – 1972 – Végétation et dynamique périglaciaire dans les milieux de haute montagne (Pyrénées orientales et Sierre Nevada). *C.R. Acad. Sci. (Paris) D*, 274: 2585–2586.
Shanks, R. E. – 1953 – Forest composition and species association in the beech-maple forest region of western Ohio. *Ecology* 34: 455–466.
Shantz, H. L. – 1917 – Plant succession and abandoned roads in eastern Colorado. *J. Ecol.* 5: 19–43.
Shantz, H. L. & Turner, B. L. – 1958 – Photographic documentation and vegetational changes in Africa over a third of a century. *Univ. Arizona Coll. Agric. Report (Tucson)* 169.

SHEA, K. R. – 1960 – Fungus succession and the significance of environment in the deterioration of logs. *For. Res. Note Weyerhaeuser* Co. 30: 7 pp.
SHEKHOV, A. G. – 1970 – The effect of time of irrigation on the overgrowing of ponds in the Don delta. *Bot. Zh.* 55: 1152–1157.
SHENNIKOV, A. P. – 1930 – The Volga-meadows in the middle course of the Volga. Uljyanovsk.
SHENNIKOV, A. P. – 1941 – Grassland science. Univ. Press, Leningrad.
SHENNIKOV, A. P. – 1964 – Introduction to geobotany. Leningrad.
SHIGO, A. L. – 1962 – Observations on the succession of fungi on hardwood pulpwood bolts. *Plant Dis. Reptr.* 46: 379–380.
SHIGO, A. L. – 1967 – Successions of organisms in discoloration and decay of wood. *Intern. Rev. For. Res.* 2: 237–299.
SHIMWELL, D. W. – 1971 – Description and classification of vegetation. 322 pp. Sidgwick & Jackson, London.
SHREVE, F. – 1942 – The desert vegetation of North America. *Bot. Rev.* 8: 195–246.
SHREVE, F. – 1951 – Vegetation and flora of the Sonoran Desert. I. Vegetation of the Sonoran Desert. *Publ. Carnegie Inst. Washington* 591(1): 1–192.
SHURE, D. J. – 1971 – Insecticide effects on early succession in an old-field ecosystem. *Ecology* 52: 271–279.
SIEGRIST, R. – 1913 – Die Auenwälder der Aare mit besonderer Berücksichtigung ihres genetischen Zusammenhanges mit anderen flussbedingten Pflanzengesellschaften. *Mitt. Aargau. Naturforsch. Ges.* 13: 183 pp.
SIEGRIST, R. & GESSNER, H. – 1925 – Über die Auen des Tessin-Flusses. *Veröff. Geobot. Inst. Rübel* 3: 127–169.
SIIRA, J. – 1970 – Studies in the ecology of the sea-shore meadows of the Bothnian Bay with special reference to the Liminka area. *Aquilo Ser. Bot.* 9: 1–109.
SIMONCSICS, P. – 1960 – Palynologische Untersuchungen an der miozänen Braunkohle des Salgotarjaner Kohlenreviers. II. Sukzession der Pflanzenvereine des Miozänmoores von Katalinbanya. *Acta Univ. Szeged Acta Biol. N. Ser.* 6: 99–106.
SJÖRS, H. – 1955 – Remarks on ecosystems. *Svensk Bot. Tidskr.* 49: 155–169.
SJÖRS, H. – 1963 – Bogs and fens on Attawapiskat River, northern Ontario. *Nat. Mus. Canada Bull.* 186: 45–133.
SKORNIAKOVA, A. G. & BOGACHIOV, B. K. – 1969 – On the effect of herbicides on some agro-phytocoenoses. *Bot. Zh.* 54: 222–231.
SKOTTSBERG, C. – 1941 – Plant succession on recent lava flows in the island of Hawaii. *Göteborgs Kgl. Vet. och Vitterh.-Samh. Handl. Sj. f. Ser. B* 1(8): 1–32.
SLAVIKOVÁ, J. – 1960 – Rekonstruktion des Eiben-Buchenwaldes (Taxeto-Fagetum Etter 1947) an der mittleren Moldau (Vltava). *Preslia* 32: 389–397.
SMITH, A. G. – 1964 – Problems in the study of the earliest agriculture to post-glacial habitat changes. *Rep. VIth Intern. Congr. Quat. Warsaw* 1961, 2: 461–471.
SMITH, A. G. – 1970 – The influence of Mesolithic and Neolithic man on British vegetation: a discussion. In: WALKER, D., WEST, R. G.: Studies in the vegetational history of the British Isles. pp. 81–96. Cambridge, London.
SMITH, J. – 1944 – The grass-*Acacia* cycle in Gedaref and notes on fire protection. *Rep. Soil Conserv. Comm. Sudan Governmt.* 87.
SMOLIAK, S. – 1956 – Influence of climatic conditions on forage production of shortgrass rangeland. *J. Range Manag.* 9: 89–91.
SOCHAVA, V. B. – 1930 – On the tundras with spot-medallions in the Anadyr district. *Trudy Polarn. Kom. Akad. Nauk SSSR* 2.
SOCHAVA, V. B. – 1933 – The lichen foods of reindeer. *Sov. Olenevodstov* 2.
SOCHAVA, V. B. – 1944 – An essay of phylogenetical systematic of plant association. *Sov. Bot.* 1.
SOCHAVA, V. B. – 1953 – The vegetation of the forest zone. In: The fauna of the USSR. Vol. 4. Acad. Sci. USSR Press, Leningrad.

Sochava, V. B. – 1956 – Dark coniferous forests. In: The vegetational cover of the USSR. Acad. Sci. USSR Press, Moscow, Leningrad.
Sochava, V. B. – 1962 – Mapping problems in geobotany. In: Prinzipy i metody geobot. kartogr. Moscow, Leningrad.
Sochava, V. B. – 1963 – Prospects in geobotanical mapping. *Geobot. Kartograf.*
Sochava, V. B. – 1968 – Plant community and dynamics of natural systems. *Doklady Inst. Geogr. Sibiri i Daln. Vost.* 20.
Sochava, V. B., Lipatova, V. & Gorshkova, A. – 1962 – A tentative study of the complete productivity of the superterranean part of the herbage cover. *Bot. Zh.* 47(4).
Soó, R. – 1940 – Vergangenheit und Gegenwart der pannonischen Flora und Vegetation. *Nova Acta Leopold. N.F.* 9: 1–50.
Soó, R. – 1964, 1966, 1968 – Synopsis systematico-geobotanica florae vegetationisque Hungariae, I, II, III. Budapest.
Specht, R. L. & Rayson, P. – 1957 – Dark Island heath (Ninetymile Plain, South Australia). I. Definition of the ecosystem. *Austral. J. Bot.* 5: 52–85.
Specht, R. L., Rayson, P. & Jackman, M. E. – 1958 – Dark Island heath (Ninety-mile Plain, South Australia). VI. Pyric succession. *Austral. J. Bot.* 6: 59–88.
Speidel, B. – 1966 – Änderungen des Pflanzenbestandes von Dauerwiesen bei langjähriger Düngung. *Bayer. Landw. Jahrb.* 43: 214–222.
Spurr, S. H. – 1952 – Origin of the concept of forest succession. *Ecology* 33: 426–427.
Spurr, S. H. – 1952 – Forest inventory. Ronald, New York.
Spurr, S. H. – 1954 – The forests of Itasca in the nineteenth century as related to fire. *Ecology* 35: 21–25.
Stählin, A. – 1959 – Die Entwicklung von Weideansaaten unter dem Einfluß von Standortsfaktoren, Düngung und Bewirtschaftung. *Z. Acker- u. Pflanzenbau* 109: 127–140.
Stählin, A. – 1969 – Maßnahmen zur Bekämpfung von Grünlandunkräutern. *Das Wirtschaftseigene Futter (Frankfurt/M.)* 15: 249–344.
Stalter, R. & Batson, W. T. – 1969 – Transplantation of salt marsh vegetation, Georgetown, South Carolina. *Ecology* 50: 1087–1089.
Stark, N. – 1968 – The environmental tolerance of the seedling stage oi *Sequoiadendron giganteum. Amer. Midland Natural.* 80: 84–95.
Stark, P., Firbas, F. & Overbeck, F. – 1932 – Die Vegetationsentwicklung des Interglacials von Rinnersdorf in der östlichen Mark Brandenburg. *Abh. Nat. Ver. Bremen* 28: 105–130 (*Sonderheft*).
Stearns, F. – 1949 – Ninety years change in a northern hardwood forest in Wisconsin. *Ecology* 30: 358–360.
Stearns, F. – 1960 – Effect of seed environment during maturation on seedling growth. *Ecology* 41: 221–222.
Stebbins, G. L. – 1971 – Chromosomal evolution in higher plants. Arnold, London.
Stefanović, V. – 1970 – Über relikte Waldgesellschaften aus dem Postglazial in Bosnien. *Mitt. Ostalp.-dinar. Pflanzensoz. Arbeitsgem.* 10(2): 79–84.
Steinberg, K. – 1944 – Zur spät- und nacheiszeitlichen Vegetationsgeschichte des Untereichsfeldes. *Hercynia* 3: 529–587.
Stephan, B. & Stephan, S. – 1971 – Die Vegetationsentwicklung im Naturschutzgebiet Stolzenburg und ihre Bedeutung für die Schutzmaßnahmen. *Decheniana* 123: 281–305.
Stern, C. – 1959 – Variation and hereditary transmission. *Proc. Amer. Phil. Soc.* 103: 183–189.
Steubing, L. – 1949 – Beiträge zur Ökologie der Wurzelsysteme des flachen

Sandstrandes. *Z. Naturforsch.* 4b: 114–123.
STEVENS, M. E. – 1963 – Podsol development on a moraine near Juneau, Alaska. *Soil Sci. Soc. Amer. Proc.* 27: 357–358.
STEWART, L. O. – 1935 – Public land surveys; history, instructions, methods. 202 pp. Iowa Collegiate Press.
STILLGER, E. – 1970 – Hangwälder im Gebiet der Lahn zwischen Diez und Nassau. 260 pp. Diss. Giessen.
STORK, A. – 1963 – Plant immigration in front of retreating glaciers with examples from the Kebnekajse area, northern Sweden. *Geogr. Annaler* 45(1): 1–22.
STRACKE, W. – 1969 – Ergebnisse eines Dauerversuches zur Frage der reduzierten Bodenbearbeitung, 1. *Thaer-Arch.* 13: 303–309.
STRAKA, H. – 1963 – Über die Veränderungen der Vegetation im nördlichen Teil der Insel Sylt in den letzten Jahrzehnten. *Schr. Naturwiss. Ver. Schlesw.-Holst.* 34: 19–43.
STRAKA, H. – 1970 – Bearbeitung der 2. Auflage von: WALTER, H.: Arealkunde, floristisch-historische Geobotanik. 478 pp. Ulmer, Stuttgart.
STRAUS, A. – 1935 – Vorläufige Mitteilung über den Wald des Oberpliozäns von Willershausen (Westharz). *Mitt. Dtsch. Dendrol. Ges.* 47: 182–186.
STRAUS, A. – 1954 – Beobachtungen an der Pliozänflora von Willershausen. *Geologie* 3: 526–535.
STUCKEY, R. L. & WENTZ, W. A. – 1969 – Effect of industrial pollution on the aquatic and shore angiosperm flora in the Ottawa River, Allen and Putnam Counties, Ohio. *Ohio J. Sci.* 69: 226–242.
STÜSSI, B. – 1970 – Vegetationsdynamik in Dauerbeobachtung. *Ergebn. Wiss. Unters. Schweiz. Nationalpark* 13: 1–384.
SUKACHEV, V. N. – 1911 – On the influence of the permafrost on the soil. *Izv. Akad. Nauk SSSR, Ser.* 6, 5(1).
SUKACHEV, V. N. – 1915 – Introduction to phytosociology. Petrograd.
SUKACHEV, V. N. – 1928 – Plant communities. Introduction to phytosociology. 2nd ed. Izd. Kniga, Moscow.
SUKACHEV, V. N. – 1931 – A manual of investigation of the forests type. 3rd. ed. Moscow, Leningrad.
SUKACHEV, V. N. – 1934 – Über einige Grundbegriffe der Phytocoenologie. *Bull. Acad. Sci. USSR. Cl. Math. Nat.* 7.
SUKACHEV, V. N. – 1938 – Principle concepts in the plant cover study. *Rastit. SSSR* 1.
SUKACHEV, V. N. – 1942 – Idea of development in phytocoenology. *Sov. Bot.* 1–2.
SUKACHEV, V. N. – 1950 – On some principal problems in phytocoenology. *Problemy Bot.* 1.
SUKACHEV, V. N. – 1954 – Some general theoretical problems in phytocoenology. *Voprosy Bot.* 1.
SUKACHEV, V. N. & DYLIS, N. – 1964 – Fundamentals of forest biogeocoenology. Nauka, Moscow.
SUKOPP, H. – 1959 – Vergleichende Untersuchungen der Vegetation Berliner Moore unter besonderer Berücksichtigung der anthropogenen Veränderungen. *Bot. Jahrb.* 79: 36–126.
SUMMERFIELD, R. J. – 1972 – Biological inertia: an example. *J. Ecol.* 60: 793–798.
SUNDING, P. – 1972 – Vegetation changes on Kalvøya Island, Baerum, SE Norway 1961–1971. *Blyttia* 30: 15–30.
SUOMINEN, J. – 1970 – On *Elymus arenarius* (*Gramineae*) and its spread in Finnish inland areas. *Ann. Bot. Fenn.* 7: 143–156.
SWAN, F. R. – 1970 – Post-fire response of four plant communities in south-central New York State. *Ecology* 51: 1074–1082.
SWAN, J. M. A. & GILL, A. M. – 1970 – The origins, spread, and consolidation

of a floating bog in Harvard Pond, Petersham, Massachusetts. *Ecology* 51: 829–840.

Switzer, G. I., Nelson, I. E. & Smith, W. H. – 1966 – The characterization of dry matter and nitrogen accumulation by loblolly pine (*Pinus taeda* L.). *Proc. Amer. Soc. Soil Sci.* 30: 114–119.

Switzer, G. I., Nelson, I. E. & Smith, W. H. – 1968 – The mineral cycle in forest stands. In: Forest fertilization, theory and practice (TVA Symposium). pp. 1–9.

Syers, J. K., Adams, J. A. & Walker, T. W. – 1970 – Accumulation of organic matter in a chronosequence of soils developed on wind-blown sand in New Zealand. *J. Soil Sci.* 21: 146–153.

Szafer, W. – 1925 – Über den Charakter der Flora und des Klimas der letzten Interglazialzeit bei Grodno in Polen. *Bull. Acad. Pol. Sci. Ser. B. Bot.*

Szafer, W. – 1930 – *Dulichium spathaceum* Pers. w polskim interglacjale. *Acta Soc. Bot. Pol.* 7(4).

Szafer, W. – 1931 – The oldest interglacial in Poland. *Bull. Acad. Pol. Sci. Lett. Cl. Sci. Math. Nat. B, Sci. Nat.* (I): 1–50.

Szafer, W. – 1947 – The pliocene flora of Kroscienko in Poland. I–II 162 u. 213 pp. Polsk. Akad. Umiej. 72.B1, Krakow.

Szafer, W. – 1949 – The pliocene flora of Kroscienko in Poland. *Vegetatio* 1(2–3).

Tachibana, H. – 1968 – Weed invasion upon the mountain areas in Mt. Hakkôda. *Ecol. Rev.* 17: 95–108.

Tagawa, H. – 1964 – A study of the volcanic vegetation in Sakurajima, southwest Japan. I. Dynamics of vegetation. *Mem. Fac. Sci. Kyushu Univ. Ser. E Biol.* 3(3/4): 165–228.

Takenouchi, M. – 1923 – On the change of vegetation of the Tarumai volcanic range. *Bot. Magaz.* (*Tokyo*) 37: 161–181.

Talbot, L. M. – 1965 – A survey of past and present wildlife research in East Africa. *East Afr. Wildlife J.* 3: 61–85.

Taliev, V. I. – 1897 – On the relic vegetation in glacial period. *Trudy Ob-va Ispyt. Prir. Kharkovsk. Univ.* 29.

Taliev, V. I. – 1901 – The boundary of forest and steppe in Valkovsky district of Kharkovsk region. *Trudy Ob-va Ispyt. Prir. Kharkovsk. Univ.* 36.

Taliev, V. I. – 1904 – An unsettled problem of Russian botanical geography (forest and steppe). *Lesn. Zh.* 3–4.

Tallis, J. H. & McGuire, J. – 1972 – Central Rossendale: the evolution of an upland vegetation. I. The clearance of woodland. *J. Ecol.* 60: 721–737.

Tamm, C. O. – 1950 – Growth and plant nutrient concentration in *Hylocomium proliferum* Lindl. in relation to tree canopy. *Oikos* 2: 60–64.

Tanfiliev, G. I. – 1894 – Forest boundaries in the south of Russia. St.-Petersburg.

Tansley, A. G. – 1920 – The classification of vegetation and the concept of development. *J. Ecol.* 8: 118–149.

Tansley, A. G. – 1929 – Succession: the concept and its values. *Proc. Intern. Plant Sci.* 1926: 677–686.

Tansley, A. G. – 1935 – The use and abuse of vegetational concepts and terms. *Ecology* 16: 284–307.

Tansley, A. G. – 1939 – The British Islands and their vegetation. 930 pp. Univ. Press, Cambridge.

Tansley, A. G. – 1941 – Note on the status of salt-marsh vegetation and the concept of 'formation'. *J. Ecol.* 29: 212–214.

Tansley, A. G. & Chipp, T. F. (Ed.) – 1926 – Aims and methods in the study of vegetation. London.

Tarrant, R. F. & Miller, R. E. – 1963 – Accumulation of organic matter and soil nitrogen beneath a plantation of red alder and Douglas-fir. *Soil Sci. Soc.*

Amer. Proc. 27: 231–234.
Tauber, H. – 1965 – Differential pollen dispersion and the interpretation of pollen diagrams. *Danmarks Geol. Unders.* II 89: 1–69.
Tchou Yen-Tcheng – 1948/1949 – Etudes écologiques et phytosociologiques sur les forêts riveraines du Bas-Languedoc. *Vegetatio* 1: 2–28, 93–128, 217–257, 347–384.
Teal, J. & Teal, M. – 1969 – Life and death of the salt marsh. X, 278 pp. Little Brown & Co., Boston, Toronto.
Teichmüller, M. – 1958 – Rekonstruktion verschiedener Mooı typen des Hauptflözes der niederrheinischen Braunkohle. *Fortschr. Geol. Rheinl.-Westf.* 2: 599–612.
Tešič, Z., Ristanović, B. & Ritter-Studnička, H. – 1967 – Contribution to the study of microflora on dolomite and serpentine habitat undeı different stages of vegetative succession in Bosna and Herzegovina. *Mikrobiologija (Beograd)* 4: 1–20.
Tezuka, Y. – 1961 – Development of vegetation in relation to soil formation in the volcanic island of Oshima, Izu, Japan. *Jap. J. Bot.* 17: 371–402.
Thienemann, A. – 1925 – Die Binnengewässer Mitteleuropas. 225 pp. Stuttgart.
Thill, A. – 1970 – La frêne. 86 pp. Duculot, Gembloux.
Thompson, J. W. – 1940 – Relic prairie areas in central Wisconsin. *Ecol. Monogr.* 10: 685–717.
Thor, E. & Summers, D. D. – 1971 – Changes in forest composition on Big Ridge Natural Study Area, Union County, Tennessee. *Castanea* 36: 114–122.
Thornburgh, D. A. – 1969 – Causes of successional patterns in *Abies-Tsuga* forests of the Washington Cascade range. *Abstr. Intern. Bot. Congress* 11: 218.
Thurston, M. J. M. – 1969 – The effect of liming and fertilisers on the botanical composition of permanent grassland. In: Tüxen, R. (Ed.): Experimentelle Pflanzensoziologie. Ber. Int. Sympos. Rinteln 1965: 58–61.
Tikhomirov, B. A. – 1938 – Über die Waldphase der postglazialen Vegetationsgeschichte Nordsibiriens und ihre Relikte in der rezenten Tundra. *Sov. Bot.* 1938(2).
Tikhomirov, B. A. – 1957 – On the structure of the plant communities in the Arctic. In: Akad. V. N. Sukachevu k 75 let. so dnya rozhd. Moscow, Leningrad.
Tikhomirov, B. A. – 1958 – On dynamics of the surface forms in the Arctic in connection with the hillock-baydzharakhs genesis. Voprosy Fisich. Geogr. k 75 let. so dnya rozhd. Akad. A. A. Grigoryeva. Moscow.
Tikhomirov, B. A. – 1959 – Interrelations between the animals and the plant cover in tundra. Leningrad.
Tikhomirov, B. A. – 1960 – Influence of *Citellus undulatus* Pall. on the flora and vegetation of the Chukot tundra. *Trudy MOIP*3.
Timofeyev, V. P. – 1936 – The regeneration of spruce in a spruce-broadleafed forest. *Sov. Bot.* 5.
Tischler, W. – 1947 – Über die Grundbegriffe synökologischer Forschung. *Biol. Zbl.* 66: 49–56.
Tisdale, E. W., Fosberg, M. A. & Poulton, C. E. – 1966 – Vegetation and soil development on a recently glaciated area near Mount Robson, British Columbia. *Ecology* 47: 517–523.
Tiurin, A. V. – 1970 – Seasonal phenomena in the forests of the Russian Plain. *Bot Zh.* 55: 1797–1811.
Tolonen, K. – 1968 – Zur Entwicklung der Binnenfinnland-Hochmoore. *Ann. Bot. Fenn.* 5: 17–33.
Tomaselli, R. – 1956 – Introduzione allo studio della fitosociologia. 319 pp. Milano.

Toorn, J. van der, Donougho, B. & Brandsma, M. – 1969 – Verspreiding van wegbermplanten in Oostelijk Flevoland. *Gorteria* 4: 151–160.
Trautmann, W. – 1952 – Pollenanalytische Untersuchungen über die Fichtenwälder des Bayerischen Waldes. *Planta* 41: 83–124.
Trautmann, W. – 1957a – Natürliche Waldgesellschaften und nacheiszeitliche Waldgeschichte des Eggegebirges. *Mitt. Flor.-soz. Arb.-gem. N.F.* 6/7.
Trautmann, W. – 1957b – Bibliographie zum Problem Pflanzensoziologie und Pollenanalyse (Palynologie). *Mitt. Flor.-soz. Arb.-gem. N.F.* 6/7: 362–368.
Trautmann, W. – 1962 – Natürliche Waldgesellschaften und nachwärmezeitliche Waldgeschichte am Nordwestrand der Eifel. *Veröff. Geobot. Inst. ETH St. Rübel* 37: 250–266.
Trautmann, W. – 1966 – Anthropogene Wandlungen der natürlichen Vegetation nach pollenanalytischen Untersuchungen. In: Tüxen, R. (Ed.): Anthropogene Vegetation. Ber. Int. Sympos. Stolzenau/W. 1961: 257–262.
Trautmann, W. – 1969 – Zur Geschichte des Eichen-Hainbuchenwaldes im Münsterland auf Grund pollenanalytischer Untersuchungen. *Schr.-R. Vegetationskde.* 4: 109–129.
Tregubov, S. – 1941 – Les forêts vierges montagnardes des Alpes Dinariques. *Comm. S.I.G.M.A.* 78: 1–116.
Treub, M. – 1888 – Notice sur la nouvelle flore de Krakatau. *Ann. Jard. Bot. Buitenzorg* 7.
Troels-Smith, J. – 1961 (1964) – The influence of prehistoric man on vegetation in central and northwestern Europe. *Rep. VIth Intern. Congr. Quatern. Warsaw* 2: 487–490.
Troll, C. – 1963a – Landscape ecology and land development with special reference to the tropics. *J. Trop. Geogr. (Singapore)* 17.
Troll, C. – 1963b – Über Landschaftssukzession. *Arb. Rhein. Landesk.* 19: 1–8.
Trulevich, N. V. – 1962 – A comparative characteristic of the changes of vegetative cover and the population dynamics in regard to the basic species of the dry steppe pastures in the area of the Issyk-kyl and Narynsk troughs. *Bjull. Mosk. O.I. Prir. O. Biol.* 67(4).
Trulevich, N. V. – 1966 – The proportions of vegetative and generative adult individuals in the population of *Artemisia tianschanica* Krasch. In: The materials of the biogeography of the Issyk-kyl Kettle-hole. Acad. Sci. Kirg. SSR Press, Frunse.
Tuganayev, V. V. – 1970 – The changes in the composition of the most widespread weed components of agro-phytocoenoses of the Tatar A.S.S.R. during the last 40–50 years. *Bot. Zh.* 55: 1820–1823.
Turner, J. – 1962 – The *Tilia* decline: an anthropogenic interpretation. *New Phytol.* 61: 328–341.
Turner, J. – 1964 – The anthropogenic factor in vegetational history. I. Tregaron and Whixall mosses. *New Phytol.* 63: 73–90.
Turner, J. – 1965 – A contribution to the history of forest clearance. *Proc. Roy. Soc. B* 161: 343–354.
Turner, J. – 1970 – Post-Neolothic disturbance of British vegetation. In: Walker, D., West, R. G.: Studies in the vegetational history of the British Isles, pp. 97–116. Cambridge, London.
Tutin, T. G. – 1941 – The hydrosere and current concepts of the climax. *J. Ecol.* 29: 268–279.
Tutin, T. G. – 1954 – The relationships of *Poa annua* L. – *Congr. VIII Int. Bot. Rapp. et Comm.* 9/10: 88.
Tüxen, J. – 1958 – Stufen, Standorte und Entwicklung von Hackfrucht- und Garten-Unkrautgesellschaften. *Angew. Pfl.-soz.* (Stolzenau) 16: 1–164.
Tüxen, J. – 1967 – Naturschutzgebiet “Duvenstedter Brook”. Vegetationstypen.

I, II. 118 pp. Hamburg.
TÜXEN, R. – 1928 – Über die Vegetation der nordwestdeutschen Binnen-Dünen. *Jber. Geogr. Ges. Hannover* 1928: 71.
TÜXEN, R. – 1930 – Über einige nordwestdeutsche Waldassoziationen von regionaler Verbreitung. *Jber. Geogr. Ges. Hannover* 1929: 3–64.
TÜXEN, R. – 1931 – Die Grundlagen der Urlandschaftsforschung. *Nachr. Niedersachsens Urgeschichte* 5: 59–105.
TÜXEN, R. – 1932 – Wald- und Bodenentwicklung in Nordwestdeutschland. *Ber. Wanderversammlung NW-Dtsch. Forstver.* 37: 17–37.
TÜXEN, R. – 1933 – Klimaxprobleme des nw-europäischen Festlands. *Nederl. Kruidk. Arch.* 43: 293–308.
TÜXEN, R. – 1935 – Die Bedeutung der Pollenanalyse für die Urgeschichte. *Die Kunde* 3: 102–104.
TÜXEN, R. – 1935 – Vegetationskartierung Nordwest-Deutschlands und ihre wirtschaftliche Auswertung. *Zesde Intern. Bot. Congr. Amsterdam Proc.* 2: 73–74.
TÜXEN, R. – 1937 – Die Pflanzengesellschaften Nordwestdeutschlands. *Mitt. Flor.-soz. Arb.-gem. Niedersachsens* 3: 170 pp.
TÜXEN, R. – 1938 – Die Vegetation Nordwestdeutschlands in ihren Beziehungen zu Klima, Gestein und Böden und ihre kartographische Darstellung. *Ber. Dtsch. Bot. Ges.* 56: (28)–(30).
TÜXEN, R. – 1939 – Pflanzensoziologie und Bodenkunde in ihrer Bedeutung für die Urgeschichte. In: SCHWANTES: Urgeschichtsstudien beiderseits der Niederelbe. pp. 18–37. Hildesheim.
TÜXEN, R. – 1947 – Der Pflanzensoziologische Garten in Hannover und seine bisherige Entwicklung. *Jber. Naturhist. Ges. Hannover* 94/98: 113–287.
TÜXEN, R. – 1950 – Neue Methoden der Wald- und Forstkartierung. *Mitt. Flor.-soz. Arb.-gem. N.F.* 2: 217–219.
TÜXEN, R. – 1952 – Hecken und Gebüsche. *Mitt. Geogr. Ges. Hamburg* 50: 85–117.
TÜXEN, R. – 1954 – Pflanzengesellschaften und Grundwasser-Ganglinien. *Angew. Pfl.-soz.* (*Stolzenau*) 8: 64–98.
TÜXEN, R. – 1956 – Die heutige potentielle natürliche Vegetation als Gegenstand der Vegetationskartierung. *Angew. Pfl.-soz.* (*Stolzenau*) 13: 5–55.
TÜXEN, R. – 1957a – Die heutige potentielle natürliche Vegetation als Gegenstand der Vegetationskartierung. *Ber. Dtsch. Landesk.* 19: 200–246.
TÜXEN, R. – 1957b – Die Schrift des Bodens. *Angew. Pfl.-soz.* (*Stolzenau*) 14.
TÜXEN, R. – 1958 – Pflanzengesellschaften oligotropher Heidetümpel Nordwestdeutschlands. *Veröff. Geobot. Inst. Rübel* 33: 207–231.
TÜXEN, R. – 1960 – Zur Geschichte der Sand-Trockenrasen (Festuco-Sedetalia) im nordwestdeutschen Diluvium. *Mitt. Flor.-soz. Arb.-gem. N.F.* 8.
TÜXEN, R. – 1961a – Bemerkungen zu einer Vegetationskarte Europas. In: Méthodes de la Cartographie de la Végétation (Coll. I.C.N.R.S.), pp. 61–73. Paris.
TÜXEN, R. – 1961b – Bibliographia syndynamica phytosociologica. *Exc. Bot. B* 3: 221–233.
TÜXEN, R. – 1964 – Die Schrift des Bodens. In: SCHWIND, M.: Schrift des Bodens, Sprache neuer Malerei. Hannover.
TÜXEN, R. – 1966 – Die Lüneburger Heide, Werden und Vergehen einer Landschaft. In: TÜXEN, R. (Ed.): Anthropogene Vegetation. Ber. Int. Sympos. Stolzenau/W. 1961: 379–395.
TÜXEN, R. – 1967a – Pflanzensoziologische Deutung interglazialer Bodenprofile. In: TÜXEN, R. (Ed.): Pflanzensoziologie und Palynologie. *Ber. Int. Sympos. Stolzenau/W.* 1962: 215–218.
TÜXEN, R. – 1967b – Pflanzensoziologische Beobachtungen an südwestnorwegischen Küsten-Dünengebieten. *Aquilo Ser. Bot.* 6: 241–272.

TÜXEN, R. – 1967c – Die westeuropäische Küste als Kampf- und Lebensraum. *Geogr. Rev. Jap.* 40: 167–182.
TÜXEN, R. – 1967d – Corynephoretea canescentis. *Mitt. Flor.-soz. Arb.-gem. N.F.* 11/12: 22–24.
TÜXEN, R. – 1969 – Stand und Ziele geobotanischer Forschung in Europa. *Ber. Geobot. Inst. Rübel* 39: 13–26.
TÜXEN, R. – 1970 – Anwendung des Feuers im Naturschutz. *Ber. Naturhist. Ges. Hannover* 114: 99–104.
TÜXEN, R. – 1970 – Pflanzensoziologische Beobachtungen an isländischen Dünengesellschaften. *Vegetatio* 20: 251–278.
TÜXEN, R. – 1973a – Pflanzengesellschaften Nordwestdeutschlands, Lief. 1. Lehre.
TÜXEN, R. – 1973b – Bibliographia phytosociologica synchronologica. *Excerpta Bot. B* 13.
TÜXEN, R. & BÖCKELMANN, W. – 1957 – Scharhörn. Die Vegetation einer jungen ostfriesischen Vogelinsel. *Mitt. Flor.-soz. Arb.-gem. N.F.* 6/7: 183–204.
TÜXEN, R. & DIEMONT, H. – 1937 – Klimaxgruppe und Klimaxschwarm. *Jber. Nat.-hist. Ges. Hannover* 88/89: 73–87.
TÜXEN, R. & PREISING, R. – 1942 – Grundbegriffe und Methoden zum Studium der Wasser- und Sumpfpflanzen-Gesellschaften. *Dtsch. Wasserwirtschaft* 37: 10–17, 57–69.
TYLER, G. – 1967 – On the effect of phosphorus and nitrogen supplied to baltic shore-meadow vegetation. *Bot. Notiser* 120: 433–447.
TYRTIKOV, A. P. – 1968 – The swamping of forests of West Siberia on the dynamics of permanently frozen ground. *Vestn. Mosk. Univ.* VI 6: 70–76.
UBRISZY, G. – 1956 – Die ruderalen Unkrautgesellschaften Ungarns, II. Studien über Ökologie und Sukzession. *Acta Agron. (Budapest)* 5: 393–418.
UGOLINI, F. C. – 1966 – Soils. In: MIRSKY, A.: Soil development and ecological succession in a deglaciated area of Muir Inlet, southeast Alaska. *Ohio State Univ. Inst. Polar Studies* 20: 28–72.
UGOLINI, F. C. – 1968 – Soil development and alder invasion in a recently deglaciated area of Glacier Bay, Alaska. In: TRAPPE, J. M., FRANKLIN, J. F., TARRANT, R. F., HANSEN, G. M.: Biology of alder. p. 115–140. U.S. For. Serv., Portland.
UMAMAHESWARARAO, M. & SREERAMULU, T. – 1968 – Recolonization of algae on denuded rocky surfaces of the Visakhapatnam Coast. *Botanica Marina* 11.
URANOV, A. A. & SMIRNOVA, O. V. – 1969 – Classification and basic features of the development of perennial plant populations. *Bjull. Mosk. O. I. Prir. O. Biol.* 74: 119–134.
VADAS, R. L. – 1969 – Algal succession in sublittoral kelp communities in the San Juan archipelago. *Abstr. Intern. Bot. Congress* 11: 224.
VALK, A. G. VAN DER & BLISS, L. C. – 1971 – Hydrarch succession and net primary production of oxbow lakes in central Alberta. *Canad. J. Bot.* 49: 1177–1199.
VARTIAINEN, T. – 1967 – Observations on the plant succession of the islands of Krunnit in the Gulf of Bothnia. *Aquilo Ser. Bot.* 6: 158–171.
VASILEVICH, V. I. – 1968 – On methods of quantitative analysis of vegetation dynamics. In: *Materialy po Dynam. Rast. Pokrova.* Vladimir.
VEATCH, J. O. – 1953 – Soils and land of Michigan. East Lansing.
VICHEREK, J. – 1971 – Grundriß einer Systematik der Strandgesellschaften am Schwarzen Meer. *Folia Geobot. Phytotax.* 6: 127–145.
VIERECK, L. A. – 1966 – Plant succession and soil development on gravel outwash of the Muldrow Glacier, Alaska. *Ecol. Monogr.* 36: 181–199.
VIERECK, L. A. – 1970 – Forest succession and soil development adjacent to the

Chena River in interior Alaska. *Arctic and Alpine Res.* 2(1): 1–26.
VILLAR, H. DEL – 1929 – Geobotánica. 339 pp. Editorial Labor, Barcelona, Buenos Aires.
VILLARET-VON ROCHOW, M. – 1957/58 – Die Pflanzenreste der bronzezeitlichen Pfahlbauten von Valeggio am Mincio. *Ber. Geobot. F.-Inst. Rübel* 1957.
VINK, W. – 1970 – Blumea 18: 225–354.
VISSER, A. DE – 1971 – De grote uitbreiding van *Puccinellia fasciculata* (Torr.) Bickn. in Zeeland als gevolg van inundatie en overstroming. *Gorteria* 5: 231–234.
VLIEGER, J. – 1937 – Über einige Waldassoziationen der Veluwe. *Mitt. Flor.-soz. Arb.-gem. Niedersachsen* 3: 193–203.
VODERBERG, K. – 1955 – Die Vegetation der . . Insel Bock. *Feddes Rep. Beih.* 135: 232–260.
VODERBERG, K. & FRÖDE, E. – 1957 – Die Vegetationsentwicklung auf der Insel Bock. *Feddes Rep.* 138: 214–229.
VOIGT, G. K. & STEUCEK, G. L. – 1969 – Nitrogen distribution and accretion in an alder ecosystem. *Soil Sci. Soc. Amer. Proc.* 33: 946–947.
VOLK, O. H. – 1931 – Beiträge zur Ökologie der Sandvegetation der oberrheinischen Tiefebene. *Z. Bot.* 24: 81–185.
VOLK, O. H. – 1939 – Soziologische und ökologische Untersuchungen an der Auenvegetation im Churer Rheintal und Domleschg. *Jber. Naturf. Ges. Graubünden* 76: 1–51.
VORONOV, A. G. – 1954 – The effect of rodents on the vegetational cover of pastures and hay meadows. In: The problems of improvement of the forage reserve in the steppe, semidesert and desert zones of the USSR. Moscow, Leningrad.
VOS, A. DE & JONES, T. (Ed.) – 1968 – Proceedings of the symposium on wildlife management and land use. *East Afr. Agr. Forest J.* 33 (*Spec. Issue*): 1–297.
VUORELA, I. – 1970 – The indication of farming in pollen diagrams from southern Finland. *Acta Bot. Fenn.* 87: 1–40.
VYAS, L. N. – 1962, 1963 – On succession trends in the vegetation of various habitats of Alwar. *Univ. Rajasthan Stud. Biol.*: 1–8.
VYSHIVKIN, D. D. – 1953 – The process of colonization of ravine slopes in the region of the shelter belt Kamyshin-Stalingrad. *Bjull. Mosk. O. I. Prir. O. Biol.* 58(5).
VYSOTZKY, G. N. – 1909 – On phytotopological maps, methods of their composition and its practical values. *Pochvovedenie* 2.
VYSOTZKY, G. N. – 1915 – Yergenya. *Trudy Buro po Prikl. Bot.* 8.
VYSOTZKY, G. N. – 1930 – Skizzen über die hydrologischen Grundlagen der Bodenkunde. *Pochvovedenie* 4.
WAGENITZ-HEINECKE, R. – 1958 – Zur Vegetationsentwicklung auf Brand- und Schlagflächen in märkischen Kiefernwäldern. *Wiss. Z. Pädag. Hochsch. Potsdam* 4: 55–64.
WAGNER, F. H. – 1969 – Ecosystem concepts in fish and game management. In: The ecosystem concept in natural resource management. Academic Press, New York.
WAGNER, H. – 1950 – Die Vegetationsverhältnisse der Donauniederung des Machlandes. *Mitt. Bundesvers.-inst. Kulturtechn.* 5: 1–32.
WALKER, D. – 1966 – The late-quaternary history of the Cumberland Lowland. *Phil. Trans. R. Soc. B* 251: 1–210.
WALKER, D. – 1970 – Direction and rate in some British postglacial hydroseres. In: WALKER, D., WEST, R. G.: Studies in the vegetational history of the British Isles. pp. 117–139. Cambridge, London.
WALTER, H. – 1937 – Pflanzensoziologie und Sukzessionslehre. *Z. Bot.* 31: 545.

WALTER, H. – 1942, 1943 – Die Vegetation Osteuropas unter Berücksichtigung von Klima, Boden und wirtschaftlicher Nutzung. 180 pp. Berlin.
WALTER, H. – 1954 – Klimax und zonale Vegetation. *Angew. Pfl.-soz.* (*Wien*) *Festschr.* 1: 144–150.
WALTER, H. – 1962, 1968 – Die Vegetation der Erde in ökologischer Betrachtung. Vol. 1: 538 pp. Vol. 2: 1001 pp. Fischer, Jena.
WARD, R. T. – 1956 – The beech forests of Wisconsin; changes in forest composition and the nature of the beech border. *Ecology* 37: 407–419.
WARMING, E. – 1891 – De psammophile formationer i Danmark. *Vidensk. Medd. Naturh. For. Copenhagen* 1891: 153–202.
WARMING, E. – 1896 – Lehrbuch der ökologischen Pflanzengeographie: Eine Einführung in die Kenntnis der Pflanzenvereine. 412 pp. Borntraeger, Berlin.
WATERBOLK, H. T. – 1954 – De praehistorische mens en zijn milieu. 153 pp. *Proefschr. Groningen. Assen.*
WATT, A. S. – 1925 – The development and structure of beech forests in the Sussex Downs. *J. Ecol.* 13: 27–73.
WATT, A. S. – 1947 – Pattern and process in the plant community. *J. Ecol.* 35: 1–22.
WATT, A. S. – 1955 – Bracken versus heather, a study of plant sociology. *J. Ecol.* 43: 490–506.
WATT, A. S. – 1960 – Population changes in acidophilous grass heath in Breckland 1936–1957. *J. Ecol.* 48: 605–629.
WATT, A. S. – 1971 – Factors controlling the floristic composition of some plant communities in Breckland. *Sympos. Brit. Ecol. Soc.* 11: 137–152.
WATT, K. E. F. (Ed.) – 1966 – Systems analysis in ecology. 276 pp. Academic Press, New York, London.
WAUTIER, J. – 1951 – Á propos de la dynamique des biocoenoses limniques: La notion de climax en biocoenotique dulçaquicole. *Verh. Intern. Ver. Theor. Angew. Limnol.* 11: 446–448.
WEAVER, J. E. – 1917 (1918) – A study of the vegetation of southeastern Washington and adjacent Idaho. *Univ. Nebraska Stud.* 17: 1–133.
WEAVER, J. E. – 1943 – Replacement of true prairie by mixed prairie in eastern Nebraska and Kansas. *Ecology* 24: 421–434.
WEAVER, J. E. – 1950 – Stabilization of midwestern grasslands. *Ecol. Monogr.* 20: 251–270.
WEAVER, J. E. – 1954a – A seventeen-year study of plant succession in prairie. *Amer. J. Bot.* 41: 31–38.
WEAVER, J. E. – 1954b – North American prairie. 348 pp. Johnson Publ. Co., Lincoln, Nebr.
WEAVER, J. E. & ALBERTSON, F. W. – 1936 – Effects of the great drought on the prairies. *Ecology* 17: 567–639.
WEAVER, J. E. & ALBERTSON, F. W. – 1939 – Major changes in grassland as a result of continued drought. *Bot. Gaz* .100: 576–591.
WEAVER, J. E. & ALBERTSON, F. W. – 1940 – Deterioration of grassland from stability to denudation with decrease in soil moisture. *Bot. Gaz.* 101: 598–624.
WEAVER, J. E. & ALBERTSON, F. W. – 1940a – Deterioration of midwestern ranges. *Ecology* 21: 216–236.
WEAVER, J. E. & ALBERTSON, F. W. – 1943 – Resurvey of grasses, forbs, and underground plant parts at the end of the great drought. *Ecol. Monogr.* 13: 63–117.
WEAVER, J. E. & ALBERTSON, F. W. – 1944 – Nature and degree of recovery of grassland from the great drought of 1933–1940. *Ecol. Monogr.* 14: 393–479.
WEAVER, J. E. & ALBERTSON, F. W. – 1956 – Grasslands of the Great Plains: their nature and use. 395 pp. Johnson Publ. Co., Lincoln, Nebr.

WEAVER, J. E. & BRUNER, W. E. – 1945 – A seven year quantitative study of succession in grassland. *Ecol. Monogr.* 15: 297–319.
WEAVER, J. E. & CLEMENTS, F. E. – 1938 – Plant ecology. 2nd Ed. 610 pp. McGraw-Hill, New York. (1st Ed. 1929.)
WEAVER, J. E., ROBERTSON, J. H. & FOWLER, R. L. – 1940 – Changes in true prairie vegetation during drought as determined by list quadrats. *Ecology* 21: 357–362.
WEAVER, J. E., STODDART, L. A. & NOLL, W. – 1935 – Response of the prairie to the great drought of 1934. *Ecology* 16: 612–629.
WEBB, L. J., TRACEY, J. G. & WILLIAMS, W. T. – 1972 – Regeneration and pattern in the subtropical rain forest. *J. Ecol.* 60: 675–695.
WEBSTER, J. – 1956, 1957 – Succession of fungi on decaying cocksfoot culms, I, II. *J. Ecol.* 44: 517–544, 45: 1–30.
WEE, Y. C. – 1971 – Weed succession observations on arable peat land. *Malayan Forest.* 33: 63–69.
WEINBERGER, P. – 1965 – Über den Einfluß klimatischer Faktoren auf die Saatgutqualität und den Nachbauwert zweier Hafersorten. *Angew. Bot.* 39: 16–56.
WELLS, T. C. E. – 1971 – A comparison of the effects of sheep grazing and mechanical cutting on the structure and botanical composition of chalk grassland. *Sympos. Brit. Ecol. Soc.* 11: 497–515.
WELTEN, M. – 1955 – Pollenanalytische Untersuchungen über die neolothischen Siedlungsverhältnisse am Burgäschisee. In: GUYANU, W. U.: Das Pfahlbauproblem. Basel.
WELTEN, M. – 1958 – Pollenanalytische Untersuchung alpiner Bodenprofile: historische Entwicklung des Bodens und säkulare Sukzession der örtlichen Pflanzengesellschaften. *Veröff. Geobot. Inst. Rübel* 33: 253–274.
WELTEN, M. – 1962 – Bodenpollen als Dokumente der Standorts- und Bestandesgeschichte. *Veröff. Geobot. Inst. ETH St. Rübel* 37: 330–345.
WENDELBERGER, G. – 1950 – Zur Soziologie der kontinentalen Halophyten-Vegetation Mitteleuropas unter besonderer Berücksichtigung der Salzpflanzen-Gesellschaften am Neusiedler See. *Denkschr. Öst. Akad. Wiss. Math.-Nat.* 108(5): 1–180.
WENDELBERGER, G. – 1953 – Über einige hochalpine Pioniergesellschaften aus der Glockner- und Muntanitz-Gruppe in den Hohen Tauern. *Verh. Zool.-Bot. Ges. Wien* 93: 100–109.
WENDELBERGER, G. & HARTL, H. – 1969 – Untersuchungen im Brandgebiet Aletschwald. I. Untersuchung über den Samenanflug. *Schweiz. Z. Forstwes.* 120: 453–475.
WENDELBERGER-ZELINKA, E. – 1952 – Die Vegetation der Donau-Auen bei Wallsee. 196 pp. Landesverlag, Wels.
WENT, F. W. – 1957 – The experimental control of plant growth. XVII, 343 pp. Waltham (Mass.).
WENT, F. W. – 1969 – A long term test of seed longevity. II. *Aliso* 7(1): 1–12.
WEST, N. E. – 1968 – Rodent-influenced establishment of ponderosa pine and bitter-brush seedlings in central Oregon. *Ecology* 49: 1009–1011.
WEST, R.-G. – 1964 – The interglacial vegetation of Britain and continental Europe compared. *Rep. VIth Intern. Congr. Quatern.* 2: 503–507.
WESTHOFF, V. – 1961 – Dünenbepflanzung in den Niederlanden. *Angew. Pfl.-soz. (Stolzenau)* 17: 14–21.
WESTHOFF, V. – 1966 – Diskussionsbemerkung. In: TÜXEN, R. (Ed.): Anthropogene Vegetation. Ber. Int. Sympos. 1961 Stolzenau/W.: 259–260.
WESTHOFF, V. – 1969a – Verandering en duur. 19 pp. Junk, Den Haag.
WESTHOFF, V. – 1969b – Langjährige Beobachtungen an Aussüssungs-Dauerprobeflächen beweideter und unbeweideter Vegetation an der ehemaligen

Zuiderzee. In: Tüxen, R. (Ed.): Experimentelle Pflanzensoziologie. *Ber. Int. Sympos. Rinteln* 1965: 246–253.
Westhoff, V. – 1971 – The dynamic structure of plant communities in relation to the objectives of conservation. *Sympos. Brit. Ecol. Soc.* 11: 3–14.
Westhoff, V. & Beeftink, W. G. – 1950 – De vegetatie van duinen, slikken en schorren op de Kaloot en in het Noordsloe. *De Levende Natuur* 53: 124–133, 225–233.
Westlake, D. F. – 1963 – Comparisons of plant productivity. *Biol. Rev.* 38: 385–425.
Westman, W. E. – 1968 – Invasion of fir forest by sugar maple in Itasca Park, Minnesota. *Bull. Torr. Bot. Club* 95: 172–186.
Whitemore, T. C. – 1969 – *Biol. J. Linn. Soc.* 1: 223–231.
Whitford, G. – 1949 – Distribution of woodland plants in relation to succession and clonal growth. *Ecology* 30: 199–208.
Whitman, W., Hanson, H. C. & Peterson, R. – 1943 – Relation of drought and grazing to North Dakota range lands. *N.D. Agr. Exp. Sta. Bull.* 320.
Whittaker, R. H. – 1951 – A criticism of the plant association and climatic climax concepts. *Northwest Sci.* 25: 17–31.
Whittaker, R. H. – 1953 – A consideration of the climax theory: The climax as a population and pattern. *Ecol. Monogr.* 23: 41–78.
Whittaker, R. H. – 1954a – The ecology of serpentine soils. IV. The vegetational response to serpentine soils. *Ecology* 35: 257–288.
Whittaker, R. H. – 1954b – Plant populations and the basis of plant indication. *Angew. Pfl.-soz.* (*Wien*). Festschr. 1: 183–206.
Whittaker, R. H. – 1956 – Vegetation of the Great Smoky Mountains. *Ecol. Monogr.* 26: 1–80.
Whittaker, R. H. – 1957 – Recent evolution of ecological concepts in relation to the eastern forests of North America. *Amer. J. Bot.* 44: 197–206.
Whittaker, R. H. – 1960 – Vegetation of the Siskiyou Mountains, Oregon and California. *Ecol. Monogr.* 30: 279–338.
Whittaker R. H. – 1962 – Classification of natural communities. *Bot. Rev.* 28: 1–239.
Whittaker, R. H. – 1967 – Gradient analysis of vegetation. *Biol. Rev. Cambridge Phil. Soc.* 42: 207–264.
Whittaker, R. H. – 1969 – Evolution of diversity in plant communities. *Brookhavens Sympos. Biol.* 22: 178–196.
Whittaker, R. H. – 1970a – The population structure of vegetation. In: Tüxen, R. (Ed.): Gesellschaftsmorphologie. Ber. Int. Sympos. Rinteln 1966: 39–62. Junk, The Hague.
Whittaker, R. H. – 1970b – The biochemical ecology of higher plants. In: Sondheimer, E. & Simeone, J. B.: Chemical ecology, pp. 43–70. Academic Press, New York, London.
Whittaker, R. H. & Niering, W. A. – 1965 – Vegetation of the Santa Catalina Mountains, Arizona. II. A gradient analysis of the south slope. *Ecology* 46: 429–452.
Whittaker, R. H. & Niering, W. A. – 1968 – Vegetation of the Santa Catalina Mountains, Arizona. IV. Limestone and acid soils. *J. Ecol.* 56: 523–544.
Whittaker, R. H. & Woodwell, G. M. – 1968 – Dimension and production relations of trees and shrubs in the Brookhaven forest, New York. *J. Ecol.* 56: 1–25.
Whittaker, R. H. & Woodwell, G. M. – 1969 – Structure, production and diversity of the oak-pine forest at Brookhaven, New York. *J. Ecol.* 57: 155–174.
Wiedemann, E. – 1951 – Ertragskundliche und waldbauliche Grundlagen der Forstwirtschaft. Frankfurt.

Wiegert, R. G. & Monk, C. D. – 1972 – Litter production and energy accumulation in three plantations of longleaf pine (*Pinus palustris* Mill). *Ecology* 53: 949–953.
Wijmstra, T. A., Smit, A., Hammen, T. van der & Geel, B. van – 1971 – Vegetational succession, fungal spores and short-time cycles in pollen diagrams from the Wietmarscher Moor. *Acta Bot. Neerl.* 20: 401–410.
Willerding, U. – 1960 – Beiträge zur jüngeren Geschichte der Flora und Vegetation der Flußauen (Untersuchungen aus dem Leinetal bei Göttingen). *Flora* 149: 435–476.
Willerding, U. – 1965 – Die Pflanzenreste aus der bandkeramischen Siedlung. *Neue Ausgrabungen u. Forsch. Niedersachsen* 2: 44–60.
Willerding, U. – 1966 – Pflanzenreste aus bronzezeitlichen und eisenzeitlichen Gruben. *Neue Ausgrabungen u. Forsch. Niedersachsen* 3: 49–62.
Willerding, U. – 1967 – Beiträge zur jüngeren Geschichte der Flora und Vegetation der Flußauen (Untersuchungen aus dem Leinetal bei Göttingen). In: Tüxen, R. (Ed.): Pflanzensoziologie und Palynologie. *Ber. Int. Sympos. Stolzenau/W.* 1962: 71–77.
Willerding, U. – 1969 – Pflanzenreste aus frühgeschichtlichen Siedlungen des Göttinger Gebietes. *Neue Ausgrabungen u. Forsch. Niedersachsen* 4.
Willerding, U. – 1970 – Vor- und frühgeschichtliche Kulturpflanzenfunde in Mitteleuropa. *Neue Ausgrabungen u. Forsch. Niedersachsen* 5.
Williams, W. T. – 1967a – The computer botanist. *Austral. J. Sci.* 29(3):266–271.
Williams, W. T. – 1967b – Numbers, taxonomy, and judgement. *Bot. Rev.* 33: 379–386.
Williams, W. T., Lance, G. N., Webb, L. J., Tracey, J. G. & Dale, M. B. – 1969a – Studies in the numerical analysis of complex rain-forest communities. III. The analysis of successional data. *J. Ecol.* 57: 515–535.
Williams, W. T., Lance, G. N., Webb, L. J., Tracey, J. G. & Connell, J. H. – 1969b – Studies in the numerical analysis of complex rainforest communities. IV. A method for the elucidation of small-scale forest pattern. *J. Ecol.* 57: 635–654.
Willis, A., Yemm, E. W. & Hope-Simpson, J. F. – 1959 – Braunton Burrows: The dune system and its vegetation. *J. Ecol.* 47: 1–24.
Wilmanns, O. & Bammert, J. – 1965 – Zur Besiedlung der Freiburger Trümmerflächen. *Ber. Naturf. Ges. Freiburg* 55: 399–411.
Wilson, R. E. – 1969 – Allelopathy as expressed by *Helianthus annuus* and its role in old-field succession. *Abstr. Intern. Bot. Congress* 11: 240.
Wilson, R. E. – 1970 – Succession in stands of *Populus deltoides* along the Missouri River in southeastern South Dakota. *Amer. Midland Natural.* 83: 332–342.
Winter, A. G. & Schönbeck, F. – 1953 – Untersuchungen über den Einfluß von Kaltwasserextrakten aus dem Getreidestroh und anderer Blattstreu. . . . *Naturwiss.* 40: 513–514.
Winterhoff, W. – 1968 – Die Flora und Vegetation des Bergsturzes am Schickeberg. *Nachr. Akad. Wiss. Göttingen* II 1968(7): 121–170.
Woess, E. v. – 1941 – *Z. Indukt. Abstammungs- u. Vererb.-lehre* 39: 444.
Wohlenberg, E. – 1969a – Deichbau und Deichpflege auf biologischer Grundlage. *Handb. Landespfl. u. Natursch.* 4: 185–196.
Wohlenberg, E. – 1969b – Landgewinnung an der Küste durch biologische Verlandung. *Handb. Landespfl. u. Natursch.* 4: 196–204.
Woike, S. – 1958 – Pflanzensoziologische Studien in der Hildener Heide. *Geobot. Mitteilungen* 8: 1–113.
Woodwell, G. M. – 1969 – Prediction of the effects of pollution on the structure and comparative physiology of ecosystems. *Abstr. Intern. Bot. Congress* 11: 243.
Woodwell, G. M. & Whittaker, R. H. – 1968a – Primary production in

terrestrial ecosystems. *Amer. Zoologist* 8: 19–30.
WOODWELL, G. M. & WHITTAKER, R. H. – 1968b – Effects of chronic gamma irradiation on plant communities. *Quart. Rev. Biol.* 43: 42–55.
WOOTEN, J. W. – 1970 – Experimental investigations of the *Sagittaria graminea* complex: transplant studies and genecology. *J. Ecol.* 58: 233–242.
WRABER, M. – 1968 – The part of vegetation studies for restauration of the mountain basins. *Ann. Acad. Ital. Sci. Forest.* 17: 275–289.
WRIGHT, H. E. – 1967 – The use of surface samples in Quaternary pollen analysis. *Rev. Palaeobot. and Palynol.* 2: 321–330.
WRIGHT, H. E. – 1968 – The roles of pine and spruce in the forest history of Minnesota and adjacent areas. *Ecology* 49: 937–955.
YAMAKAWA, T. & NAKAMURA, M. – 1940 – Plant communities developed on lava flows of Sakurajima. *Ecol. Rev.* 6: 103–124.
YAROSHENKO, P. D. – 1946 – On change in vegetation. *Bot. Zh.* 31: 29–40.
YAROSHENKO, P. D. – 1953 – Basic study of vegetation. 2nd ed. Gosud. Izd. Geogr. Lit., Moscow.
YAROSHENKO, P. D. – 1956 – The successions of the vegetational cover of Transcaucasia and their relationships with the changes of soil and climate and with the anthropogenous factors. Acad. Sci. USSR Press, Moscow, Leningrad.
YAROSHENKO, P. D. – 1961 – Geobotany. Izd. Akad. Nauk SSSR, Leningrad.
YAROSHENKO, P. D. & GRABAR, V. A. – 1969 – Change in the plant cover of the Transcarpathians. Izd. Nauka, Leningrad.
YLI-VAKKURI, P. – 1961 – Experimental studies on the emergence and initial development of tree seedlings in spruce and pine stands. *Acta Forest. Fenn.* 75.
YOSHII, Y. – 1932 – Revegetation of volcano Kamagatake after the great eruption in 1929. *Bot. Magaz. (Tokyo)* 46: 208–215.
YOSHII, Y. – 1939, 1940 – Plant communities on volcanoes in Japan. *Ecol. Rev.* 5: 203–217, 277–290, 6: 59–72, 146–160.
YOSHII, Y. – 1961 – Ökologische Studien über die Dünen-Vegetation von Ota. *Bot. Magaz. (Tokyo)* 30: 311–340, 359–389.
ZACH, L. W. – 1950 – A northern climax, forest or muskeg? *Ecology* 31: 304–306.
ZAGWIJN, W. H. – 1963 – Pleistocene stratigraphy in the Netherlands. *Verh. K. Nederl. Geol. Mijnb. Gen. Geol. Ser.* 21: 173–196.
ZARZYCKI, K. – 1961 – Etude sur la végétation des dunes anciennes en Petite Camargue. *Acta Soc. Bot. Pol.* 30(3/4).
ZAYKOVA, V. A. & LOPATIN, V. D. – 1970 – Changes in the ecological species composition and structure of meadow communities under the influence of anthropogenic factors. *Bot. Zh.* 55: 972–981.
ZEIDLER, H. – 1956 – Pollenanalyse und Standortskunde. *Waldhygiene* 1: 8.
ZEIST, W. VAN – 1967a – Archaeology and palynology in the Netherlands. *Rev. Palaeobot. Palynol.* 4: 45–65.
ZEIST, W. VAN – 1967b – Late quaternary vegetation history of western Iran. *Rev. Palaeobot. Palynol.* 2: 301–311.
ZEIST, W. VAN – 1970 – Betrekkingen tussen palynologie en vegetatiekunde. *Misc. Papers Landbouwhogeschool Wageningen* 5: 127–140.
ZEIST, W. VAN & WRIGHT, H. E. – 1967 – Über Probleme der Vegetation und Pollenanalyse in Minnesota und angrenzenden Gebieten. In: TÜXEN, R. (Ed.): Pflanzensoziologie und Palynologie. Ber. Int. Sympos. Stolzenau/W. 1962: 121–133.
ZEUNER, F. E. – 1958 – Dating the past. 516 pp. 4th ed. Methuen, London; Hafner, New York.
ZINKE, P. J. – 1969 – Nitrogen storage of several California forest soil-vegetation systems. In: Biology and ecology of nitrogen. pp. 40–53. National Acad. Sci. Ed. Washington, D.C.

Zoller, H. – 1960a – Pollenanalytische Untersuchungen zur Vegetationsgeschichte der insubrischen Schweiz. *Denkschr. Schweiz. Naturforsch. Ges.* 83(2): 45–156.
Zoller, H. – 1960b – Die wärmezeitliche Verbreitung von Haselstrauch, Eichenmischwald, Fichte und Weißtanne in den Alpenländern. *Bauhinia* 1: 189–207.
Zoller, H. – 1962 – Pollenanalytische Untersuchungen zur Vegetationsentwicklung tiefgelegener Weißtannenwälder im Schweizerischen Mittelland. *Veröff. Geobot. Inst. ETH St. Rübel* 37: 346–358.
Zoller, H. – 1967 – Pollenanalytische Untersuchungen zum Kastanienproblem am Alpen-Südfuß. In: Tüxen, R. (Ed.): Pflanzensoziologie und Palynologie. Ber. Int. Sympos. Stolzenau/W. 1962: 48–55.
Zoller, H. & Kleiber, H. – 1967 – Über die postglaziale Einwanderung und Ausbreitung der Rotbuche (*Fagus silvatica* L.) am südlichen Alpenrand. *Bauhinia* 3(2).
Zoller, H. & Kleiber, H. – 1971 – Vegetationsgeschichtliche Untersuchungen in der montanen und subalpinen Stufe der Tessintäler. *Verh. Naturf. Ges. Basel* 81(1): 90–154.
Zollitsch, B. – 1967/68 – Soziologische und ökologische Untersuchungen auf Kalkschiefern in hochalpinen Gebieten. I. *Ber. Bayer. Bot. Ges.* 40: 67–100.
Zollitsch, B. – 1969 – Die Vegetationsentwicklung im Pasterzenvorfeld. *Wiss. Alpenver.-h.* 21: 267–290.
Zólyomi, B. – 1951 – Les phytocénoses des montagnes de Buda et le réboisement des endroits dénudés. *Acta Biol. Acad. Sci. Hung.* 1(1–4): 9–67.
Zólyomi, B. – 1953 – Die Entwicklungsgeschichte der Vegetation Ungarns seit dem letzten Interglazial. *Acta Biol. Acad. Sci. Hung.* 4: 367–430.
Zöttl, H. – 1951 – Die Vegetationsentwicklung auf Felsschutt in der alpinen und subalpinen Stufe des Wettersteingebirges. *Jb. Ver. Schutz d. Alpenpfl.-u.-tiere* 16: 10–74.
Zöttl, H. – 1964 – Waldstandort und Düngung. *Cbl. Ges. Forstwesen* 81: 1–24.
Zotz, L. – 1930 – Der Aufbau bronzezeitlicher Grabhügel, ein Kriterium zur Altersbestimmung des Ortsteins und zur Rekonstruktion vorgeschichtlicher Vegetation in NW-Deutschland. *Mitt. Flor.-soz. Arb.-gem. Niedersachsen* 2: 90–97.
Zubkov, A. I. – 1932 – Tundras of Gusinaya Zemlya. *Trudy Bot. Muzeya Akad. Nauk SSSR* 25.

INDEX

With few exceptions, names of syntaxonomical classes only are registered in the index. The numbers of pages with descriptions and notes on particular syntaxonomical orders, alliances and associations in the text or in the figures can be found in the index behind the name of the class concerned (e.g. for Fagetalia sylvaticae behind Querceto-Fagetea sylvaticae).

E

F

G

Q

R